The Nature of Technology

The Nature of Technology

Implications for Learning and Teaching

Edited by;

Michael P. Clough
Joanne K. Olson
Dale S. Niederhauser
Iowa State University, USA

SENSE PUBLISHERS
ROTTERDAM/BOSTON/TAIPEI

A C.I.P. record for this book is available from the Library of Congress.

ISBN: 978-94-6209-267-9 (paperback)
ISBN: 978-94-6209-268-6 (hardback)
ISBN: 978-94-6209-269-3 (e-book)

Published by: Sense Publishers,
P.O. Box 21858,
3001 AW Rotterdam,
The Netherlands
https://www.sensepublishers.com/

Printed on acid-free paper

In memory of Neil Postman who said what needed to be said despite it not being popular. His writings and personal communication were the impetus for this book.

MPC and JKO

To David B. Owen, for his friendship and significant influence on my thinking that has sharpened my work and made many insights possible. *Ancora imparo.*

JKO

For Isaac, his contemporaries, and future generations who must learn to wisely use technology rather than permitting it to use them.

MPC and JKO

In memory of contributing author David H. Jonassen, a prodigious and prolific scholar. He was my friend.

DSN

TABLE OF CONTENTS

Introduction 1
Michael P. Clough, Joanne K. Olson & Dale S. Niederhauser

Section I Philosophical & Historical Issues in the Nature of Technology

Chapter 1: Informing Ourselves To Death 7
Neil Postman

Chapter 2: Know thy Cyborg-Self: Thoughts on Socrates and Technological Literacy 15
Gordon Hull

Chapter 3: The Interpersonal Divide 35
Michael Bugeja

Chapter 4: The Nature of Technoscience (NOTS) 51
Suvi Tala

Chapter 5: Computer Savvy but Technologically Illiterate: Rethinking Technology Literacy 85
Teresa J. Shume

Chapter 6: Transforming Learning with Technology: Beyond Modernism and Post-Modernism, or Whoever Controls the Technology Creates the Reality 101
David H. Jonassen

Section II Technology's Faustian Bargain for Education

Chapter 7: Visually Dominant, Dynamic and Yet Deceptive: The Nature of Simulation Technology as Displayed in Secondary School Science Teaching 113
Jocelyn Wishart

Chapter 8: Social Networking Technology and Societal Expectations for Teachers as Role Models 129
Joanne K. Olson, Michael P. Clough, & Kimberly A. Penning

Chapter 9: And Now this . . . Problem: Neil Postman, Technology and the Secondary School Mathematics Curriculum 163
Michael Todd Edwards, Suzanne R. Harper, & Robert M. Klein

Chapter 10: A Cautionary Note: Technology's Tendency to Undermine Serious Study and Teaching 189
Joanne K. Olson & Michael P. Clough

Chapter 11: Smart Boards, Money and the Pedagogy of Watching 201
Amy Noelle Parks

Chapter 12: The Purposes of Schooling and The Nature of Technology: The End of Education? 217
Joanne K. Olson

Chapter 13: Learning *From* Technology or Learning *With* Technology: Theoretical Perspectives on the Nature of Using Technology in the Classroom 249
Dale S. Niederhauser

Chapter 14: The Nature of Social Communication Technologies and Cyberbullying: Filtered Cues and Disinhibited Actions 269
Warren J. Blumenfeld

Section III Teacher Education and the Nature of Technology

Chapter 15: Convergence of Postman and Vygotsky Perspectives Regarding Contemporary Media's Impact on Learning and Teaching 293
Benjamin C. Herman

Chapter 16: Ideologies in The Conceptualization and Use of Educational Technology: Using Huxley, Orwell and Forster to Inform a Humanizing Framework for Educational Technology Practice 329
Heather Tillberg-Webb & Johannes Strobel

Chapter 17: Implications of the Nature of Technology for Teaching and Teacher Education 345
Jerrid W. Kruse

Section IV Teaching the Nature of Technology

Chapter 18: Teaching about the Nature of Technology: Issues and Pedagogical Practices 373
Michael P. Clough

Chapter 19: Promoting Middle School Students' Understanding of the Nature of Technology 391
Jerrid W. Kruse

Chapter 20: Confusion in the Classroom About the Natures of Science and Technology: Implications for Scientific and Technological Literacy 411
James Jadrich & Crystal Bruxvoort

Chapter 21: Technology Criticism in the Classroom 427
John T. Spencer

Author Biographies 441

Recommended Reading 447

Index 449

INTRODUCTION

The first portion of this book's title, *The Nature of Technology*, may appear odd to readers. Articles, books and other information media abound addressing particular technologies and how to use them. This book has a different and more important purpose. Meaningful technology education is far more than learning how to use technology. It includes an understanding of what technology is, how and why technology is developed, how individuals and society direct, react to, and are sometimes unwittingly changed by technology. In this book we place these and other questions regarding the nature of technology in the context of learning, teaching and schooling. Our intent is to introduce educators to the nature of technology, its relevance to teaching and learning, and how they can effectively teach students about the social and ethical issues that are always present with technology.

Thus, the intent of this book is akin to efforts in the science education community to promote teaching and learning about the nature of science. Both science and technology have enormous and pervasive impacts on society and culture. All science education reform documents state that promoting scientific literacy demands attention to the nature of science. A scientifically literate citizenry should understand what science is; how science works; the limitations of science; how science and technology are different, yet related; and how science impacts and is impacted by society. Much research exists regarding effective nature of science teaching and learning, but while the phrase *nature of science* is widely recognized by science teachers, accurate and effective instruction regarding the nature of science is still not widespread.

The phrase *nature of technology* has only recently entered the conversation among educators, and attention to the nature of technology among educators and education researchers is still in its infancy. The National Educational Technology Standards recommend addressing the social, ethical, and human issues inherent in technology, but are vague regarding specific issues that ought to be addressed in educating students about the nature of technology. Thus, unsurprisingly, educators rarely consider the nature of technology, use this understanding to make appropriate pedagogical decisions, or attempt to help their students understand this important matter. The nature of technology and its impact on education must become a significant object of inquiry among educators, and students must come to understand the nature of technology so that they can make informed decisions regarding how technology may influence thinking, values and action, and when and how technology should be used in their personal lives and in society. Prudent choices regarding technology cannot be made without understanding the issues that this book raises.

When educators and the general public do consider the pros and cons of technology, they usually do so only in Orwellian (Orwell, 1949) terms - the explicit and overt

Michael P. Clough, Joanne K. Olson and Dale S. Niederhauser (Eds.), The Nature of Technology: Implications for Learning and Teaching, 1–4.

ways that technology affects individuals, groups and society. The impetus behind this book draws on Postman's (1985) argument in *Amusing Ourselves to Death* to take on the more difficult task of addressing technology education in Huxlean (Huxley, 1932) and Bradburean (Bradbury, 1953) terms. That is, how does technology change individuals, social institutions, and cultures when it is embraced without critique? Asking teachers and students to critically examine technology in Orwellian terms is fairly easy, but asking them to do so in Huxlean and Bradburean terms is far more difficult. In the first case, the technology is forced upon us by outside forces such as businesses, schools, employers, governments, the marketplace, etc. In the latter case, we willingly embrace technology for a variety of reasons (because it is interesting, novel, labor saving, entertaining, eases communication, and/or reduces some burden). In Orwellian terms, the technology is easily seen as needing to be critically examined. In Huxlean and Bradburean terms, people wrongly believe they have already done so.

Neil Postman tirelessly wrote and spoke about the nature of technology, both in general terms and in terms of schooling. He clearly was not against technology, and wrote in the *End of Education* (1995) that being against technology makes no more sense than being against food. But critically thinking about food—what we eat, when we eat, what portions we consume, and knowing when to push ourselves away from the table—is crucial for individual and societal well-being. Postman repeatedly warned that unexamined adoption of technology, much like indiscriminate eating, has severe negative ramifications for how we live, and that these ramifications extend beyond individuals to impact society and culture. The first chapter of this book is a speech by Postman, and we hope it and the many references to his work throughout this book will encourage widespread reading of his and similar work.

As Postman and others have noted, most people only see technology in a positive light, and rarely step back and consider the trade-offs that result. As Rees (2011) noted in his praise of *TechNo-Fix: Why Technology Won't Save Us or the Environment*:

> This is the new age of "unreason." ...Even as the impacts of technology destroy the ecosphere, the faithful preach that technology alone can salvage civilization.

As a personal example of this unexamined faith in technology, over a decade ago, the first two editors of this book wrote a cautionary note regarding the use of technology in education (Olson & Clough, 2001). The reaction to that published work was surprising to us. Despite the analytical and measured position taken in that article, we were seen as attacking a cherished belief that technology would of course improve education. People wanted to debate us, we were asked to give radio interviews, and we even heard whispers in our own department that we had written a negative piece about technology in education. We were accused of being "Luddites" and "Fuddy-duddyism" (a personal favorite). Why, we wondered, does making technology the object of analysis result in the swift emotional response of many to immediately dismiss the authors as taking a hostile negative position?

The hostile emotional response to those who do seriously consider the pros and cons of technology demands that anyone who dare make technology an object of inquiry must apologize in some way for making the case that technology is not neutral or, in the words of Postman, technology is always a Faustian bargain—that when we uncritically embrace technology, we also unknowingly agree to its inherent consequences. This is particularly intriguing when one considers that technology optimists never apologize for their uncritical adoration of technology.

Those who question technology and/or choose not to use it are often labeled with derogatory terms like "laggard" (Rogers, 2003) or "resister" (Rossing, 2012). Technology enthusiasts often ignore legitimate issues and arguments raised about technologies, and when forced to address such issues and arguments, they brush them aside, wrongly claiming that any negative consequence is merely due to how the technology is being used. The upshot is that they impetuously dismiss reasoned arguments that technology is not always good, certainly not neutral, and requires analysis so that we can use it rather than, in the words of Postman, let it use us. And yet, we still feel the need to assure readers and explicitly state that the purpose of this book is one of analysis, rather than to promote an "anti-technology" position.

Perhaps, as Postman asserted, people do worship technology. In *The End of Education*, he wrote:

> At some point it becomes far from asinine to speak of the god of Technology—in the sense that people believe technology works, that they rely on it, that it makes promises, that they are bereft when denied access to it, that they are delighted when they are in its presence, that for most people it works in mysterious ways, that they condemn people who speak against it, that they stand in awe of it, and that, in the born-again mode, they will alter their lifestyles, their schedules, their habits, and their relationships to accommodate it. If this be not a form of religious belief, what is? (Postman, 1995, p. 38)

This revering of technology is most evident in pervasive attitudes that conceptualize solutions to most personal and societal problems in terms of technological development. In schooling, this reverence for technology is apparent in narrow efforts to *redesign* schools, teaching, curriculum and even children to achieve greater efficiency and a better product (i.e., higher test scores); and also in STEM education efforts that hijack the science curriculum with engineering objectives, promote STEM education primarily in terms of job training and future technological development, and marginalize the value of the humanities.

The nature of technology raises serious issues for schooling, teaching, learning and teacher education that are in desperate need of significant attention among educators and education researchers. This book is intended to raise such issues and stimulate thinking and action among teachers, teacher educators, and education researchers. Toward those ends, the six chapters making up section one in the book introduce philosophical and historical issues in the nature of technology. The eight chapters in section II continue this effort but with explicit attention to their implications, both pro and con, for education. Section III consists of three chapters addressing the role of teacher education for promoting attention to the nature of technology among teachers and the accurate and effective teaching about the nature of technology. The authors of chapters appearing in section IV put forward practical considerations for teaching the nature of technology to students. That section IV contains only four chapters is evidence that attention to the nature of technology in education is in its early stages, and we hope that a second edition of this book will contain far more examples of successful efforts to teach the nature of technology.

We sincerely appreciate the patience of the chapter authors throughout this book project. When we first sent out the call for book proposals nearly five years ago, few in the education community appeared to even understand what the nature of technology meant and addressed. The rejection rate for submitted book chapter proposals exceeded 80 percent. As books, articles, and other forms of popular media outlets began raising questions about how technology was changing individual, societal and cultural values, what we think, how we think, and even our relationships with others (see the recommended reading list on pages 447–448), and as these ideas began to make their way into educators' consciousness, we received many additional chapter contributions. We are cautiously optimistic that a more balanced attitude toward technology (attention to its Faustian bargain as Postman would say) will become more widespread and that this book will assist in that end.

Michael P. Clough, Joanne K. Olson & Dale S. Niederhauser

REFERENCES

Bradbury, R. (1953). *Fahrenheit 451*. Simon & Schuster, New York.

Huxley, A. (1932). *Brave New World*. HarperCollins, New York.

Olson, J. K. & Clough, M. P. (2001). Technology's Tendency to Undermine Serious Study: A Cautionary Note. *The Clearing House*, 75(1), 8–13.

Orwell, G. (1949). *Nineteen Eighty-Four*, Penguin Books, New York.

Postman, N. (1985). *Amusing Ourselves to Death: Public Discourse in the Age of Show Business*. Penguin, New York.

Postman, N. (1995). *The End of Education: Redefining the Value of School*. Vintage, New York.

Rees, W. (2011). Front page review in Huesemann, M., & Hueseman, J. (2011). *Techno-Fix: Why Technology Won't Save Us or the Environment*. New Society Publishers, Gabriola Island, BC, Canada.

Rogers, E. M. (2003). *Diffusion of Innovations* (5th Ed). New York: Free Press.

Rossing, J. P. (2012). Mobile Technology and Liberal Education. *Liberal Education, 98*(1), 68–72.

SECTION I

PHILOSOPHICAL & HISTORICAL ISSUES IN THE NATURE OF TECHNOLOGY

CHAPTER 1

NEIL POSTMAN

INFORMING OURSELVES TO DEATH[1]

The great English playwright and social philosopher George Bernard Shaw once remarked that all professions are conspiracies against the common folk. He meant that those who belong to elite trades – physicians, lawyers, teachers, and scientists – protect their special status by creating vocabularies that are incomprehensible to the general public. This process prevents outsiders from understanding what the profession is doing and why – and protects the insiders from close examination and criticism. Professions, in other words, build forbidding walls of technical gobbledegook over which the prying and alien eye cannot see.

Unlike George Bernard Shaw, I raise no complaint against this, for I consider myself a professional teacher and appreciate technical gobbledegook as much as anyone. But I do not object if occasionally someone who does not know the secrets of my trade is allowed entry to the inner halls to express an untutored point of view. Such a person may sometimes give a refreshing opinion or, even better, see something in a way that the professionals have overlooked.

I believe I have been invited to speak at this conference for just such a purpose. I do not know very much more about computer technology than the average person – which isn't very much. I have little understanding of what excites a computer programmer or scientist, and in examining the descriptions of the presentations at this conference, I found each one more mysterious than the next. So, I clearly qualify as an outsider.

But I think that what you want here is not merely an outsider but an outsider who has a point of view that might be useful to the insiders. And that is why I accepted the invitation to speak. I believe I know something about what technologies do to culture, and I know even more about what technologies undo in a culture. In fact, I might say, at the start, that what a technology undoes is a subject that computer experts apparently know very little about. I have heard many experts in computer technology speak about the advantages that computers will bring. With one exception – namely, Joseph Weizenbaum – I have never heard anyone speak seriously and comprehensively about the disadvantages of computer technology, which strikes me as odd, and makes me wonder if the profession is hiding something important. That is to say, what seems to be lacking among computer experts is a sense of technological modesty.

Michael P. Clough, Joanne K. Olson and Dale S. Niederhauser (Eds.), The Nature of Technology: Implications for Learning and Teaching, 7–14.

After all, anyone who has studied the history of technology knows that technological change is always a Faustian bargain: Technology giveth and technology taketh away, and not always in equal measure. A new technology sometimes creates more than it destroys. Sometimes, it destroys more than it creates. But it is never one-sided.

The invention of the printing press is an excellent example. Printing fostered the modern idea of individuality but it destroyed the medieval sense of community and social integration. Printing created prose but made poetry into an exotic and elitist form of expression. Printing made modern science possible but transformed religious sensibility into an exercise in superstition. Printing assisted in the growth of the nation-state but, in so doing, made patriotism into a sordid if not a murderous emotion.

Another way of saying this is that a new technology tends to favor some groups of people and harms other groups. School teachers, for example, will, in the long run, probably be made obsolete by television, as blacksmiths were made obsolete by the automobile, as balladeers were made obsolete by the printing press. Technological change, in other words, always results in winners and losers.

In the case of computer technology, there can be no disputing that the computer has increased the power of large-scale organizations like military establishments or airline companies or banks or tax collecting agencies. And it is equally clear that the computer is now indispensable to high-level researchers in physics and other natural sciences. But to what extent has computer technology been an advantage to the masses of people? To steel workers, vegetable store owners, teachers, automobile mechanics, musicians, bakers, brick layers, dentists and most of the rest into whose lives the computer now intrudes? These people have had their private matters made more accessible to powerful institutions. They are more easily tracked and controlled; they are subjected to more examinations, and are increasingly mystified by the decisions made about them. They are more often reduced to mere numerical objects. They are being buried by junk mail. They are easy targets for advertising agencies and political organizations. The schools teach their children to operate computerized systems instead of teaching things that are more valuable to children. In a word, almost nothing happens to the losers that they need, which is why they are losers.

It is to be expected that the winners – for example, most of the speakers at this conference – will encourage the losers to be enthusiastic about computer technology. That is the way of winners, and so they sometimes tell the losers that with personal computers the average person can balance a checkbook more neatly, keep better track of recipes, and make more logical shopping lists. They also tell them that they can vote at home, shop at home, get all the information they wish at home, and thus make community life unnecessary. They tell them that their lives will be conducted more efficiently, discreetly neglecting to say from whose point of view or what might be the costs of such efficiency.

Should the losers grow skeptical, the winners dazzle them with the wondrous feats of computers, many of which have only marginal relevance to the quality of the losers' lives but which are nonetheless impressive. Eventually, the losers succumb, in part because they believe that the specialized knowledge of the masters of a computer technology is a form of wisdom. The masters, of course, come to believe this as well. The result is that certain questions do not arise, such as, to whom will the computer give greater power and freedom, and whose power and freedom will be reduced?

Now, I have perhaps made all of this sound like a well-planned conspiracy, as if the winners know all too well what is being won and what lost. But this is not quite how it happens, for the winners do not always know what they are doing, and where it will all lead. The Benedictine monks who invented the mechanical clock in the 12th and 13th centuries believed that such a clock would provide a precise regularity to the seven periods of devotion they were required to observe during the course of the day. As a matter of fact, it did. But what the monks did not realize is that the clock is not merely a means of keeping track of the hours but also of synchronizing and controlling the actions of men. And so, by the middle of the 14th century, the clock had moved outside the walls of the monastery, and brought a new and precise regularity to the life of the workman and the merchant. The mechanical clock made possible the idea of regular production, regular working hours, and a standardized product. Without the clock, capitalism would have been quite impossible. And so, here is a great paradox: the clock was invented by men who wanted to devote themselves more rigorously to God; and it ended as the technology of greatest use to men who wished to devote themselves to the accumulation of money. Technology always has unforeseen consequences, and it is not always clear, at the beginning, who or what will win, and who or what will lose.

I might add, by way of another historical example, that Johann Gutenberg was by all accounts a devoted Christian who would have been horrified to hear Martin Luther, the accursed heretic, declare that printing is "God's highest act of grace, whereby the business of the Gospel is driven forward." Gutenberg thought his invention would advance the cause of the Holy Roman See, whereas in fact, it turned out to bring a revolution which destroyed the monopoly of the Church.

We may well ask ourselves, then, is there something that the masters of computer technology think they are doing for us which they and we may have reason to regret? I believe there is, and it is suggested by the title of my talk, "Informing Ourselves to Death." In the time remaining, I will try to explain what is dangerous about the computer, and why. And I trust you will be open enough to consider what I have to say. Now, I think I can begin to get at this by telling you of a small experiment I have been conducting, on and off, for the past several years. There are some people who describe the experiment as an exercise in deceit and exploitation but I will rely on your sense of humor to pull me through.

Here's how it works: It is best done in the morning when I see a colleague who appears not to be in possession of a copy of *The New York Times*. "Did you read The Times this morning?," I ask. If the colleague says yes, there is no experiment that day. But if the answer is no, the experiment can proceed. "You ought to look at Page 23," I say. "There's a fascinating article about a study done at Harvard University." "Really? What's it about?" is the usual reply. My choices at this point are limited only by my imagination. But I might say something like this: "Well, they did this study to find out what foods are best to eat for losing weight, and it turns out that a normal diet supplemented by chocolate eclairs, eaten six times a day, is the best approach. It seems that there's some special nutrient in the eclairs – encomial dioxin – that actually uses up calories at an incredible rate."

Another possibility, which I like to use with colleagues who are known to be health conscious is this one: "I think you'll want to know about this," I say. "The neuro-physiologists at the University of Stuttgart have uncovered a connection between jogging and reduced intelligence. They tested more than 1200 people over a period of five years, and found that as the number of hours people jogged increased, there was a corresponding decrease in their intelligence. They don't know exactly why but there it is."

I'm sure, by now, you understand what my role is in the experiment: to report something that is quite ridiculous – one might say, beyond belief. Let me tell you, then, some of my results: Unless this is the second or third time I've tried this on the same person, most people will believe or at least not disbelieve what I have told them. Sometimes they say: "Really? Is that possible?" Sometimes they do a double-take, and reply, "Where'd you say that study was done?" And sometimes they say, "You know, I've heard something like that."

Now, there are several conclusions that might be drawn from these results, one of which was expressed by H. L. Mencken fifty years ago when he said, there is no idea so stupid that you can't find a professor who will believe it. This is more of an accusation than an explanation but in any case I have tried this experiment on non-professors and get roughly the same results. Another possible conclusion is one expressed by George Orwell – also about 50 years ago – when he remarked that the average person today is about as naive as was the average person in the Middle Ages. In the Middle Ages people believed in the authority of their religion, no matter what. Today, we believe in the authority of our science, no matter what.

But I think there is still another and more important conclusion to be drawn, related to Orwell's point but rather off at a right angle to it. I am referring to the fact that the world in which we live is very nearly incomprehensible to most of us. There is almost no fact – whether actual or imagined – that will surprise us for very long, since we have no comprehensive and consistent picture of the world which would make the fact appear as an unacceptable contradiction. We believe because there is no reason not to believe. No social, political, historical, metaphysical, logical or spiritual reason. We live in a world that, for the most part, makes no sense to us. Not even technical sense. I don't mean to try my experiment on this audience, especially after having told you about it, but if I informed you that the seats you are presently occupying were actually made by a special process which uses the skin of a Bismark herring, on what grounds would you dispute me? For all you know – indeed, for all I know – the skin of a Bismark herring *could* have made the seats on which you sit. And if I could get an industrial chemist to confirm this fact by describing some incomprehensible process by which it was done, you would probably tell someone tomorrow that you spent the evening sitting on a Bismark herring.

Perhaps I can get a bit closer to the point I wish to make with an analogy: If you opened a brand-new deck of cards, and started turning the cards over, one by one, you would have a pretty good idea of what their order is. After you had gone from the ace of spades through the nine of spades, you would expect a ten of spades to come up next. And if a three of diamonds showed up instead, you would be surprised and wonder what kind of deck of cards this is. But if I gave you a deck that had been shuffled twenty times, and then asked you to turn the cards over, you would not expect any card in particular – a three of diamonds would be just as likely as a ten of spades. Having no basis for assuming a given order, you would have no reason to react with disbelief or even surprise to whatever card turns up.

The point is that, in a world without spiritual or intellectual order, nothing is unbelievable; nothing is predictable, and therefore, nothing comes as a particular surprise.

In fact, George Orwell was more than a little unfair to the average person in the Middle Ages. The belief system of the Middle Ages was rather like my brand-new deck of cards. There existed an ordered, comprehensible world-view, beginning with the idea that all knowledge and goodness come from God. What the priests had to say about the world was derived from the logic of their theology. There was nothing arbitrary about the things people were asked to believe, including the fact that the world itself was created at 9 AM on October 23 in the year 4004 B.C. That could be explained, and was, quite lucidly, to the satisfaction of anyone. So could the fact that 10,000 angels could dance on the head of a pin. It made quite good sense, if you believed that the Bible is the revealed word of God and that the universe is populated with angels. The medieval world was, to be sure, mysterious and filled with wonder, but it was not without a sense of order. Ordinary men and women might not clearly grasp how the harsh realities of their lives fit into the grand and benevolent design, but they had no doubt that there was such a design, and their priests were well able, by deduction from a handful of principles, to make it, if not rational, at least coherent.

The situation we are presently in is much different. And I should say, sadder and more confusing and certainly more mysterious. It is rather like the shuffled deck of cards I referred to. There is no consistent, integrated conception of the world which serves as the foundation on which our edifice of belief rests. And therefore, in a sense, we are more naive than those of the Middle Ages, and more frightened, for we can be made to believe almost anything. The skin of a Bismark herring makes about as much sense as a vinyl alloy or encomial dioxin.

Now, in a way, none of this is our fault. If I may turn the wisdom of Cassius on its head: the fault is not in ourselves but almost literally in the stars. When Galileo turned his telescope toward the heavens, and allowed Kepler to look as well, they found no enchantment or authorization in the stars, only geometric patterns and equations. God, it seemed, was less of a moral philosopher than a master mathematician. This discovery helped to give impetus to the development of physics but did nothing but harm to theology. Before Galileo and Kepler, it was possible to believe that the Earth was the stable center of the universe, and that God took a special interest in our affairs. Afterward, the Earth became a lonely wanderer in an obscure galaxy in a hidden corner of the universe, and we were left to wonder if God had any interest in us at all. The ordered, comprehensible world of the Middle Ages began to unravel because people no longer saw in the stars the face of a friend.

And something else, which once was our friend, turned against us, as well. I refer to information. There was a time when information was a resource that helped human beings to solve specific and urgent problems of their environment. It is true enough that in the Middle Ages, there was a scarcity of information but its very scarcity made it both important and usable. This began to change, as everyone knows, in the late 15th century when a goldsmith named Gutenberg, from Mainz, converted an old wine press into a printing machine, and in so doing, created what we now call an information explosion. Forty years after the invention of the press, there were printing machines in 110 cities in six different countries; 50 years after, more than eight million books had been printed, almost all of them filled with information that had previously not been available to the average person. Nothing could be more misleading than the idea that computer technology introduced the age of information. The printing press began that age, and we have not been free of it since.

But what started out as a liberating stream has turned into a deluge of chaos. If I may take my own country as an example, here is what we are faced with: In America, there are 260,000 billboards; 11,520 newspapers; 11,556 periodicals; 27,000 video outlets for renting tapes; 362 million TV sets; and over 400 million radios. There are 40,000 new book titles published every year (300,000 world-wide) and every day in America 41 million photographs are taken, and just for the record, over 60 billion pieces of advertising junk mail come into our mail boxes every year. Everything from telegraphy and photography in the 19th century to the silicon chip in the twentieth has amplified the din of information, until matters have reached such proportions today that for the average person, information no longer has any relation to the solution of problems.

The tie between information and action has been severed. Information is now a commodity that can be bought and sold, or used as a form of entertainment, or worn like a garment to enhance one's status. It comes indiscriminately, directed at no one in particular, disconnected from usefulness; we are glutted with information, drowning in information, have no control over it, don't know what to do with it.

And there are two reasons we do not know what to do with it. First, as I have said, we no longer have a coherent conception of ourselves, and our universe, and our relation to one another and our world. We no longer know, as the Middle Ages did, where we come from, and where we are going, or why. That is, we don't know what information is relevant, and what information is irrelevant to our lives. Second, we have directed all of our energies and intelligence to inventing machinery that does nothing but increase the supply of information. As a consequence, our defenses against information glut have broken down; our information immune system is inoperable. We don't know how to filter it out; we don't know how to reduce it; we don't know how to use it. We suffer from a kind of cultural AIDS.

Now, into this situation comes the computer. The computer, as we know, has a quality of universality, not only because its uses are almost infinitely various but also because computers are commonly integrated into the structure of other machines. Therefore it would be fatuous of me to warn against every conceivable use of a computer. But there is no denying that the most prominent uses of computers have to do with information. When people talk about "information sciences," they are talking about computers – how to store information, how to retrieve information, how to organize information. The computer is an answer to the questions, how can I get more information, faster, and in a more usable form? These would appear to be reasonable questions. But now I should like to put some other questions to you that seem to me more reasonable. Did Iraq invade Kuwait because of a lack of information? If a hideous war should ensue between Iraq and the U.S., will it happen because of a lack of information? If children die of starvation in Ethiopia, does it occur because of a lack of information? Does racism in South Africa exist because of a lack of information? If criminals roam the streets of New York City, do they do so because of a lack of information?

Or, let us come down to a more personal level: If you and your spouse are unhappy together, and end your marriage in divorce, will it happen because of a lack of information? If your children misbehave and bring shame to your family, does it happen because of a lack of information? If someone in your family has a mental breakdown, will it happen because of a lack of information?

I believe you will have to concede that what ails us, what causes us the most misery and pain – at both cultural and personal levels – has nothing to do with the sort of information made accessible by computers. The computer and its information cannot answer any of the fundamental questions we need to address to make our lives more meaningful and humane. The computer cannot provide an organizing moral framework. It cannot tell us what questions are worth asking. It cannot provide a means of understanding why we are here or why we fight each other or why decency eludes us so often, especially when we need it the most. The computer is, in a sense, a magnificent toy that distracts us from facing what we most needed to confront – spiritual emptiness, knowledge of ourselves, usable conceptions of the past and future. Does one blame the computer for this? Of course not. It is, after all, only a machine. But it is presented to us, with trumpets blaring, as at this conference, as a technological messiah.

Through the computer, the heralds say, we will make education better, religion better, politics better, our minds better – best of all, ourselves better. This is, of course, nonsense, and only the young or the ignorant or the foolish could believe it. I said a moment ago that computers are not to blame for this. And that is true, at least in the sense that we do not blame an elephant for its huge appetite or a stone for being hard or a cloud for hiding the sun. That is their nature, and we expect nothing different from them. But the computer has a nature, as well. True, it is only a machine but a machine designed to manipulate and generate information. That is what computers do, and therefore they have an agenda and an unmistakable message.

The message is that through more and more information, more conveniently packaged, more swiftly delivered, we will find solutions to our problems. And so all the brilliant young men and women, believing this, create ingenious things for the computer to do, hoping that in this way, we will become wiser and more decent and more noble. And who can blame them? By becoming masters of this wondrous technology, they will acquire prestige and power and some will even become famous. In a world populated by people who believe that through more and more information, paradise is attainable, the computer scientist is king. But I maintain that all of this is a monumental and dangerous waste of human talent and energy. Imagine what might be accomplished if this talent and energy were turned to philosophy, to theology, to the arts, to imaginative literature or to education? Who knows what we could learn from such people – perhaps why there are wars, and hunger, and homelessness and mental illness and anger.

As things stand now, the geniuses of computer technology will give us Star Wars, and tell us that is the answer to nuclear war. They will give us artificial intelligence, and tell us that this is the way to self-knowledge. They will give us instantaneous global communication, and tell us this is the way to mutual understanding. They will give us Virtual Reality and tell us this is the answer to spiritual poverty. But that is only the way of the technician, the fact-mongerer, the information junkie, and the technological idiot.

Here is what Henry David Thoreau told us: “All our inventions are but improved means to an unimproved end.” Here is what Goethe told us: “One should, each day, try to hear a little song, read a good poem, see a fine picture, and, if it is possible, speak a few reasonable words.” And here is what Socrates told us: “The unexamined life is not worth living.” And here is what the prophet Micah told us: “What does the Lord require of thee but to do justly, and to love mercy and to walk humbly with thy God?” And I can tell you – if I had the time (although you all know it well enough) – what Confucius, Isaiah, Jesus, Mohammed, the Buddha, Spinoza and Shakespeare told us. It is all the same: There is no escaping from ourselves. The human dilemma is as it has always been, and we solve nothing fundamental by cloaking ourselves in technological glory.

Even the humblest cartoon character knows this, and I shall close by quoting the wise old possum named Pogo, created by the cartoonist, Walt Kelley. I commend his words to all the technological utopians and messiahs present. “We have met the enemy,” Pogo said, “and he is us.”

NOTE

[1] A speech given at a meeting of the German Informatics Society (Gesellschaft fuer Informatik) on October 11, 1990 in Stuttgart, sponsored by IBM-Germany.
http://w2.eff.org/Net_culture/Criticisms/informing_ourselves_to_death.paper

CHAPTER 2

GORDON HULL

KNOW THY CYBORG-SELF

Thoughts on Socrates and Technological Literacy

"We are our own best artifacts, and always have been."

– Andy Clark (2003, p. 192)

There are no doubt many good reasons to encourage a general technological literacy. Here is one: U.S. society purports to aspire to democracy. Insofar as we live in a technologically-mediated society, if we want that democracy to amount to more than the rule of an ignorant mob, the "people" need to have some understanding of the technologies surrounding them. At one level, this is clearly a political question, and questions about the limits of popular knowledge, how much technical skill is necessary to qualify as technologically literate, what sorts of reference frames can and should be brought to one's understanding of technology, and so forth, immediately present themselves. But, as the term "democracy" suggests, it is also a question of values. Is it possible to give the question more teeth, and propose that the values in question are not just political, but moral? At the very least, is it possible to blur the boundary between political and moral questions in this case? Here I will argue that it is, particularly in the case of information technologies.

The essay proceeds as follows. In the first part, I discuss competing views of the relation between technology and human nature, with particular attention to recent theories to the effect that technology is fundamentally dehumanizing. In the next part, I suggest that both the utopic and dystopic variants of this story share the assumption that human nature is somehow detachable from its technological environment. I then provide evidence from recent discussions of human cognition that this assumption is unwarranted. In the final sections, I discuss three examples of how knowledge that we are tied to our technological environment should motivate concern about that environment: the practice of anonymous reading, the relative importance of amateur and professional culture, and the Platonic critique of books, which I interpret as an example of the problem of information glut. I will both start and end in the supposed birthplace of Western philosophy, ancient Athens.

Michael P. Clough, Joanne K. Olson and Dale S. Niederhauser (Eds.), The Nature of Technology: Implications for Learning and Teaching, 15–34.

1. WELCOME TO THE MACHINE

On trial for a variety of charges centered on the claim that he was corrupting the morals of the youth of Athens, Socrates offers three images of the philosopher: someone who does not fear death (29b),[1] someone who cares for his soul (29e), and a gadfly, *i.e.*, someone without whom the Athenians might very well "go on sleeping till the end of your days" (31a). The three images coalesce around the same thought: the philosopher is someone who, unlike his fellow citizens, will give "attention or thought to truth and understanding and the perfection of [one's] soul" (29e). Socrates and his fellow citizens agreed on only one point: he was not like them. After he failed to convince them that the presence of such a person as he was desirable, he famously remarked in sentencing that "examining both myself and others is really the very best thing that a man can do, and that life without this sort of examination is not worth living" (38a). Having further failed to convince the Athenians of either this point or that he should be rewarded by "free maintenance at the state's expense" for his efforts "to persuade each one of you not to think more of practical advantages than of his mental and moral well-being" (36d), Socrates was condemned to death.

I recount this well-known episode from the history of philosophy, and the context of the Socratic "know thyself" in order to provide a context with which to frame a contemporary question: what can philosophy tell us about the human aspects of technology? Admittedly, ancient Athens does not look like a promising place to start, as the Athenian distaste for the manual arts has been well-established. Aristotle even said that citizens should be kept away from the mechanical arts, lest their ability to govern themselves be corrupted (see Winner, 1995 for a quick summary). Aristotle also drew a fairly bright line between natural and artificial objects, proposing that art imitated nature. By the late medieval period, this line had hardened considerably, with a mainstream view insisting on the metaphysical priority of nature over art. Nonetheless, in a world that many think is fundamentally characterized by its abundance of technologies, the question is very much the Socratic one of an examined life and of the sort of political environment we want to live in.

How one interprets the injunction to know oneself depends on what one thinks it means to be human. Is there some sort of human nature that separates humans from other things in the world? Does this human nature exist independently of the environment it is in, such that some content can be given to "human nature," enough to ground ethical and other normative projects? The Greek concern about the corrupting influence of technology suggests precisely such a view of human nature. This sort of view would not deny that people in different contexts are in many ways different, but it would insist that there was some sort of core humanity present in every case, marking the human as authentically such. Of course, many people will not fulfill their nature, and their lives could then be criticized on ethical grounds, in the same way that Socrates criticized his fellow Athenians for failing to attend to what was most essential about themselves, instead frittering away their lives on idle amusements. Such a view would thus provide a clear way to evaluate the ethical implications of technology: does the presence of a certain technology fundamentally enhance or detract from our ability to fulfill our natures?

During the early part of the scientific revolution, which roughly coincided with the development of "modern" philosophy, the emergent view was that technology could remove many of the limitations placed on us by nature. Grounded in correct science, the capacity for such practical philosophy to enhance human life was essentially unlimited. Descartes can be taken as exemplary of this view. New principles in physics, he suggests "opened my eyes to the possibility of gaining knowledge which would be very useful in life, and of discovering a practical philosophy which might replace the speculative philosophy taught in the schools." By means of this philosophy, we might "make ourselves, as it were, the lords and masters of nature." He had particular hope for medicine: "we might free ourselves from innumerable diseases, both of the body and of the mind, and perhaps even from the infirmity of old age, if we had sufficient knowledge of their causes and of all the remedies that nature has provided" (1637/1985, pp. 142–3).[2]

I highlight early-modern optimism primarily to contrast it with more recent pessimism. Particularly in post-war Europe, a lot of recent work in the philosophy of technology views technology as fundamentally dehumanizing; the general claim is thus that a technologically saturated environment is, all things considered, a hindrance to being fully human. Adopting the worldview that living in such an environment encourages would thus be an ethical failing of the first order. The most philosophically significant exponent of this line of thought is probably Martin Heidegger. Heidegger, whose early work centered around Plato and Aristotle, and who explicitly pointed to the Aristotelian art/nature distinction, thought that there was something deeply and profoundly alienating about the technological ability to disrupt natural processes. Rather than let nature happen, technology treats nature as a "standing reserve," a set of resources reordered to provide energy on demand for the needs of an integrated technological system. As he puts it, technology "puts to nature the unreasonable demand that it supply energy which can be extracted and stored as such" (1977, p. 322). The integration of natural objects into the technological system changes what they are; thus, a dam on the Rhine makes the river "what [it] is now, namely, a water-power supplier, derives from the essence of the power station" (1977, p. 321). Of course, the river is still a river, but Heidegger's point is that we no longer think of it in any other way. Industrial agriculture provides another of his examples:

> The earth now reveals itself as a coal-mining district, the soil as a mineral deposit. The field that the peasant formerly cultivated and set in order appears differently than it did when to set in order still meant to take care of and maintain. The work of the peasant does not challenge the soil of the field. In sowing grain it places seed in the keeping of the forces of growth and watches over its increase. But meanwhile even the cultivation of the field has come under the grip of another kind of setting-in-order, which *sets upon* nature. It

> sets upon it in the sense of challenging it. Agriculture is now the mechanized food industry. Air is now set upon to yield nitrogen, the earth to yield ore, ore to yield uranium, for example; uranium is set upon to yield atomic energy, which can be unleashed either for destructive or for peaceful purposes (1977, p. 320).

In other words, the peasant initiates or occasions a process that is fundamentally a natural one. The process of industrial agriculture, on the other hand, is all about controlling nature.

At one level, all of this sounds either banal and obvious, or like the worst kind of luddism. Of *course* we extract energy from nature – we like to eat! One of Heidegger's main points – and this emerges more clearly in his other writings – is that we should find it odd that we do not find our relation to technology surprising. He claims that his essay is to prepare us for a "free relation to technology" (1977, p. 311), and a substantial part of that preparation is in making the current state of affairs seem abnormal. Most importantly, what this line suggests is that Heidegger's primary interest is ethical, and not in technology *per se*. Indeed, over the course of the essay, it becomes apparent that a central problem is that people treat themselves as technological resources; "the current talk about human resources, about the supply of patients for a clinic, gives evidence of this" (1977, p. 323). Maybe this is all for the better; Heidegger's point is that it ought to at least give us pause. As he puts it, the question is "whether we actually experience ourselves *as* the ones whose activities everywhere, public and private, are challenged forth by enframing" (1977, p. 329).

What emerges from a Heideggerian analysis of technology, is a general suspicion that immersion in technology-rich environments takes us away from something essentially human about ourselves. Even worse, we all go along for the ride, and so do not even recognize the problem as a problem.[3] The call to ethics, in this sense, is a call for people to return to themselves. It is at that level inherently Socratic. It also comes with a profound distrust of technology, or at least of the tendency of technologists and popular culture to make utopian claims about a world with more and more technology. As we will see, in this, too, it is profoundly Socratic.

2. THIS IS YOUR BRAIN ON TECHNOLOGY

So: people use tools. At some point, there are enough tools, and we are serious enough about them, that we cross a tipping point into modern technology. Embracing modern technology either puts us on the golden road to unlimited material progress, or takes us away from our essential humanity, whatever that is. Can a philosophical understanding of technological literacy do better than to vacillate between these alternatives? The problem, I would suggest, is an assumption common to both views: the assumption that "human nature" can be separated from technology. One effect of making this assumption is that the political and ethical questions raised by technology can be too quickly reduced to matters of personal ethics. On the utopian version, the only relevant questions are about efficiency: since technological progress is a good thing, we ought to adapt ourselves to whatever sets of policies will most efficiently bring about the greatest amount of that progress (these days, it is usually assumed to be free market competition). These decisions can best be left to experts in economics and the relevant technologies. On the dystopic version, we ought to turn away, as individuals, from using too much technology. Again, there is no need to inquire into the specifics of the technologies in question; since too much technology is a bad thing, the question is how best to minimize its diffusion. Langdon Winner takes the utopian version to dominate most American discourse about technology, and underscores that "it is important to note that as our society adopts one sociotechnical system after another it answers some of the most important questions that political philosophers have ever asked about the proper order of human affairs" (1986, p. 40). For example, should society favor large, bureaucratic organizations over smaller communities? Because of their economies of scale, most twentieth-century technologies apparently favored the former; many enthusiasts of the Internet think it favors the latter. If Winner's analysis is correct, and if either of these technologies is adapted for reasons of economic efficiency, then the calculus of "efficiency" effectively hides the fact that we've made some important ethical decisions about what-sized society we prefer without even knowing it.[4] The point I want to notice here is that the assumption that human nature is somehow independent of our technical systems generates an ethics, the most important questions for which have to do with how adopting a particular technology will affect our abilities to live according to our nature. If we reject this assumption as unsupported, then the ethical questions will be correspondingly transformed. As I will suggest, they will become both more fundamental and more political.

Let us return to Heidegger for some clues as to how this might be done. On the one hand, there does seem to be something profoundly correct in the way Heidegger puts matters: we do somehow "change" in the face of technology. Ordinary experience – of, for example, writing directly into a word processor versus writing by hand – and ordinary language – the way we say that "he is changed when he gets behind the wheel of a car" – both lend their support to the intuition. The problem is that Heidegger apparently couples this insight with an assumed "authentically human" baseline to generate what looks like an anti-technological position.

In response to an interview question, Jacques Derrida offers the following, which I think nicely captures an appropriate ambivalence about Heidegger:

> Heidegger's reaction was at once intelligible, traditional, and normative. The tradition of these norms is often respectable, and its reserve considerable when it remains vigilant in the face of technological mutations. But it also gives rise, sometimes in its least naïve form, to a confident dogmatism, an assurance that we have to interrogate. For instance, Heidegger deplores the fact that even personal letters are now typewritten and that the singular trace of the signatory

> is no longer recognizable through the shapes of the letters and the movements of the hand. But when we write 'by hand' we are not in the time before technology; there is already instrumentality, regular reproduction, mechanical iterability. So it is not legitimate to contrast writing by hand and 'mechanical' writing, like a pretechnological craft as opposed to technology. And then on the other side what we call 'typed' writing is also 'manual' (2005, pp. 20–1).

Derrida's point could be extended to Heidegger's other examples. For example, it is not like the peasant farmer works in a time before technology: agriculture is a profoundly technological activity, and always has been. Conversely, even modern agribusiness involves natural processes.

The constant here seems to be that humans have *always* supplemented their "natural" abilities with technology, or at least, they have done so throughout recorded history and for a long period before that. If this is the case, would we not be better off adding to our view of human nature the criterion that we are naturally tool-users? Such a strategy, arguably already in Heidegger, poses difficulties for the sort of philosophy undertaken by Descartes in particular. Having defined the mental as the authentically human, he deployed a variety of arguments to show that humans did have minds, and followed with arguments to show that other entities did not. Hence, he claims that animals are like machines in that they move only according to the arrangement of their organs; evidence for this is that they lack language (even if they can make language-like sounds, "they cannot show that they are thinking what they are saying" (1637/1985, p. 140)) and that they are incapable of adapting to new situations:

> Whereas reason is a universal instrument which can be used in all kinds of situations, these organs need some particular disposition for each particular action; hence it is for all practical purposes impossible for a machine to have enough different organs to make it act in all the contingencies of life in the way in which our reason makes us act (1637/1985, p. 140).[5]

That Descartes already uses the language of machines to specify the uniqueness of the human mind underscores the fragility of this strategy. Humans, says Descartes, do and should use tools. The human mind can also be usefully explained on the model of a universal tool. What prevents the mind/tool boundary from collapsing – why can we not say that the mind is somehow changed by its use of tools? Or that the use of tools is a constitutive part of our rationality? Descartes is opposed to those who would derive mind from such material things as brains, and draws a line in the sand: "the rational soul … cannot be derived in any way from the potentiality of matter, but must be specially created" by God (1637/1985, p. 141).

The Cartesian line in the sand sounds rather more like an empirical question for cognitive psychology, and cognitive psychology is squarely against Descartes' claim that there is anything metaphysically special about the human mind, at least insofar as we are speaking about cognition. If that is the case, then a further question arises: why stop at the brain? If all of the tool-language is apt, then should we say that somehow the *environment* is part of our cognitive processes? An affirmative answer to these questions forms the basis of Andy Clark's so-called "extended mind" hypothesis. According to this hypothesis, our cognition is in and through the environment. Clark's guiding intuition seems to be something like the following: we know from work in neuroscience that cognitive processes do not depend on specific neurons or arrangements of them to happen. Patients with substantial traumatic brain injuries, for example, are often able to re-learn to communicate using different parts of their brain. In this sense, one's exact neurological arrangement is a contingent feature of one's cognition. If that is the case, there is no particular reason to insist that all cognition be in the brain. For example, recent research into the way that people gesture strongly suggests that the act of gesturing itself does some cognitive work (Clark, 2007). Cognition, then, is an active, performative process (a point to which I will return in the context of fan fiction). But if all that is the case, then why should we stop at the limits of our body? Why can we not construe the environment as doing some of the cognitive work? Clark puts the intuition in a recent paper as follows:

> The cognitive scientist or philosopher of mind who chooses to treat the brain and central nervous system alone as the mechanistic supervenience base for mental states is rather like a neuroscientist who insists that neuroscience proper should not be concerned with the hippocampus or the cerebellum, because (they think) *all the real cognizing goes on in the cortex* (2008, p. 49).

Thus, just as "we need not care (within sensible limits) exactly where *within* the brain a given operation is performed, so too (it might be urged) we should not care whether … a certain operation occurs inside or outside some particular membrane or metabolic boundary" (2008, p. 50). In other words, given what we know about cognition within the body, the burden of proof ought to be on those who want to draw a cognitive boundary at the body's exterior.

Clark's argument is that, in fact, a great deal of our cognition can be said to happen in and through and with our environment. To think that cognition is somehow limited to our "skin bags" or our skulls is to miss the incredible richness of our interactions with the environment. He cites as an easy example: when someone asks you if you know what the time is, you first answer that you do, and then check your watch. Do you know the time? Yes, you do; "it is just that the 'you' that knows the time is no longer the bare biological organism but the hybrid biotechnological system that now includes the wristwatch as a proper part" (2003, p. 42). In an earlier paper, he and David Chalmers (1998, p. 12) propose the example of Otto, a patient with a mild case of Alzheimer's, who learns of an art show at the MOMA. Otto then looks into his notebook to learn where the MOMA is. How is this process, Clark asks, all that different from what happens when we pause for a moment to remember where the MOMA is? Or when we use mnemonic aids to recall things? Given the current advances in information technologies, this sort of cognitive enhancement is an increasingly obvious part of our lives. We do not normally notice much of this environmental cognitive work, because when we are sufficiently acclimated to the tool, "the tool itself fades into the background, becoming transparent in skilled use" (2003, p. 45). Deploying Heidegger for the distinction, Clark urges that such transparent technologies be considered part of our cognitive apparatus, as distinct from opaque tools, which remain "highly visible in use" and for which we "distinguish sharply and continuously between the user and the tool" (2003, p. 37)[6]

Clark's thesis is controversial from the point of view of philosophy of mind, and we need not pass judgment here on whether he is right in the strong sense (see Clark, 2007, for a defense of the stronger thesis as preferable to the weaker). In other words, we do not need to settle the question of the ultimate status of "mind" to push ethical questions to the fore. This is because a weaker version of the thesis will suffice: when any of us act as agents in the world, we carry our technological cognitive enhancements with us. From the point of view of my activities in the world, it does not matter if we conclude that my mind is contained in my skull but uses objects outside of it; or if my mind encompasses both the things in my skull and those outside. The ethically relevant point is that, in a very fundamental way, I require both sets of objects to be who I am. So too, when I encounter others in the world, they are carrying their cognitive enhancements along with them. Hence, neither for my sense of myself as an agent nor for my encounter with others as agents does the exact boundary of my mind matter all that much. Efforts to drive a wedge between the agent and her technological support system will tend to run afoul of the sort of reduction arguments intimated above. Suppose I were to take Otto's notebook from him. If Clark is right, then the injury is somehow morally on the same page as hitting him on the head and inducing amnesia, if that could be done without imposing collateral damage (in other words, any real-world bash on the head will have other physical and possibly psychological effects that make the act worse than simply stealing the notebook). Damage to either the extended system or its brain component can have the same effect, and either can be done with the intention to harm Otto's person.

The preceding moral hunch – and I intend it as a deliberately counterintuitive hunch – might turn out to be wrong, but that it is possible at all implies is that we should be deeply concerned with developing a sense of technological literacy and an understanding of how our cognitive environments are shaped and developed, if we want to have any purchase whatsoever on what it means to be an agent in the world. The "question concerning technology," then, would be a question concerning ourselves: it is a matter of making decisions about the sorts of people and agents we want to be, and (by consequence) about the sort of society we want to have; rather than talking about abstractions like "technology and society," we are compelled to talk about "technological society" (as Latour, 1992, proposes). In other words, if Socrates is right that "know thyself" is of paramount importance, then technological literacy is a necessary component. Clark concludes his *Natural Born Cyborgs* by noting that "if I am right ... the question is not whether" we modify ourselves with our technologies, but "in what ways we actively sculpt and shape it. By seeing ourselves as we truly are, we increase the chances that our future biotechnological unions will be good ones" (2003, p. 198). In other words, "technological education will be crucial if human-machine cooperation is to enrich and humanize rather than restrict and alienate. Once again, the lesson seems clear: *Know Thyself: Know Thy Technologies*" (2003, p. 183).

Clark is generally optimistic, and his work centers on new bodily enhancing technologies like telepresencing. I want to suggest here that the problem is more general. Before introducing examples, I want to illustrate how I think this generalization matters. In an appreciative but critical commentary on *Natural Born Cyborgs*, Adrian Mackenzie notes of the opaque/transparent distinction that "for any particular technology, we need to ask: opaque or transparent for whom?" This is because "transparency and opaqueness are not intrinsic to the technology. They relate to social, cultural, political and economic projects in which zones of opacity or transparency serve different functions. A relative opacity can be extremely useful in regulating who gets to play around with and alter the technology" (2004, p. 155). Langdon Winner makes the same general point in a slightly different context: ethicists like to talk about how "we" must make important ethical decisions, but the "important first task for the contemporary ethics of technology" is to assess "what is the identity of the moral communities that will make the crucial, world-altering judgments and take appropriate action as a result" (1995, p. 67).

A prosaic, and deliberately non-neuroscience example, drawn from Bruno Latour (1992), will serve to make the point. Consider the values that get embedded in the design of a door. In particular, the genius of hinge-pins is that they allow the door to effortlessly alternate between being open and closed: open, when we want to let things (like people) in or out, and closed when we want to keep the cold air out and the warm air in. Of course, people being people, they forget to close the door, and so the design temptation arises to outsource the opening and closing of the door to an automatic device. But here we must be clever: if the device closes too slowly, it is useless. If it closes too quickly, it discriminates against those who either move through the door slowly (small children, the elderly, the disabled) or who require it to remain open for other reasons (delivery people). Those who are able to do so, and who regularly walk through the door, will presumably develop habits such that they neither bonk their nose on the door waiting for it to open, nor have it pinch their heels before they finish getting through. The door becomes progressively more transparent, in Clark's terms. For those encountering the door the first time, it is relatively more opaque. For those who cannot alter their gait to fit the door, the device remains permanently opaque. This discrimination is of course not a conscious design decision, but it underscores that an automatic door that functions transparently for one set of individuals is opaque for another. Of course, a door that does *not* have an automatic opening and closing mechanism is orders of magnitude more opaque for the disabled than one that does, and so these sorts of questions need to be taken up in specific social contexts.

3. KNOW THY CYBORG-SELF

If the above analysis is correct, then it is easy to see why technological literacy is important: understanding our technological environment is a necessary condition for understanding ourselves; and taking responsibility for that environment is taking responsibility for ourselves. Advances in biotechnology clearly pose ethical questions in this sense, at least for the small percentage of humanity for whom those advances will be available. For example, consider the question of human cognitive enhancement: it is entirely possible that in the near(ish) future, it will be possible, through genetic means, to make more intelligent humans. Should we do this? Such questions represent, as it were, the tip of the biotechnological iceberg. Ethical and policy decisions made today about such technologies will also have clear downstream implications both for those for whom the technologies are available, and those for whom it is not. If there is a genetically enhanced "overclass," for example, that has clear implications for those who do not belong to it. Less obviously, and more immediately, resources spent developing medicines that will be purchased in relatively wealthier Western markets trade off with resources that could be spent developing treatments for diseases that predominantly strike in developing countries.[7] Here, I want to pursue the less spectacular topic of information technology. Ethical decisions about information technologies may seem mundane by comparison, but they more directly affect more people (although, again, issues of uneven distribution are both important and insufficiently discussed), and they more directly implicate the sorts of concerns that emerge from Clark's cyborg thesis. Here are two examples, specific to the U.S. context. The first concerns how technological literacy is important in relating to our social environment. The second is about how technological literacy requires at least a minimal competence with the legal systems regulating technologies.

The Personal: Reading Anonymously

Everyone knows that one's environment makes a difference in how one behaves, morally and otherwise. This is why no one wants his children to hang out with the Wrong Crowd, and why it is a bad idea for alcoholics to live upstairs from a bar. Technologies are analogously a part of that environment.[8] Consider the following. In the past, it was generally possible to read without anyone knowing what you read. A trip to the library, or a cash purchase at a bookstore, would secure a copy of a book that one could then take home and read privately, less than once (perhaps the book makes a nice doorstop!), once, or even several times, highlighting pen in hand. It is true that these initial transactions left a data trail: a library record, a cashier's memory of one's face, or the like. But there were several limitations to the ability of anyone to do anything with this data trail: librarians have a strong ethos of protecting their records; the memory of the cashier at the bookstore is far from perfect, and both privacy norms and physical barriers like curtains kept one's domestic reading behavior – even the fact that one was reading – thoroughly under the radar. Information technology looks to be changing that. Julie Cohen puts the fundamental problem as follows:

> The new information age is turning out to be as much an age of information about readers as an age of information for readers. The same technologies that have made vast amounts of information accessible in digital form are enabling information providers to amass an unprecedented wealth of data about who their customers are and what they like to read. In the new age of digitally transmitted information, the simple, formerly anonymous acts of reading, listening, and viewing – scanning an advertisement or a short news item, browsing through an online novel or a collection of video clips – can be made to speak volumes, including, quite possibly, information that the reader would prefer not to share (1996, p. 981).

We are all familiar with the harmless version of this: Amazon.com's uncanny ability to recommend titles based on our previous orders; most of us probably don't even mind it. We are also familiar with a perhaps less innocent version: the ability of the FBI under the Patriot Act to demand a patron's library records – and the legal requirement that patrons not be notified that their records have been obtained. Both uses of our reading data trail are enabled by technological advances in information storage, retrieval, and processing. They bring to salience a theoretical point about speech that perhaps had not been obvious before, the degree to which speaking presupposes reading. If that is true, then a robust right to free speech needs to include protection for reading and otherwise accessing information. The theoretical point is not new, and was put eloquently by the literary theorist Mikhail Bakhtin some time ago: "any speaker is himself a respondent to a greater or lesser degree. He is not, after all, the first speaker, the one who disturbs the eternal silence of the universe" (1986, p. 69). The legal point that follows about free speech is that, as Cohen argues "all speech responds to prior speech of some sort," and so the protection of people's speech should protect "the entire series of intellectual transactions through which they formed the opinions they ultimately chose to express. Any less protection would chill inquiry, and as a result, public discourse, concerning politically and socially controversial issues – precisely those areas where vigorous public debate is most needed, and most sacrosanct" (1996, p. 1006). Various information technologies, from Digital Rights Management to website tracking can be used to essentially eliminate the anonymity of reading (Cohen, 2003). The chilling effect should be obvious: I am less likely to read controversial materials, or materials that could be taken as evidence of my disloyalty or moral corruption, if I know that someone is likely tracking my reading habits.

The problem runs deeper than this. If there is any truth to the claim that "I" am my thoughts, then a change in the information environment that restricts or channels the information available to me effects a change in who I am. Developments in information technology again signal that such subtle changes are possible, and not just in the world of children. For example, the Supreme Court recently upheld Congressional legislation (essentially) mandating that public libraries install filtering programs on their Internet terminals (for a discussion and critique, see Hull, 2009). The idea was to protect children against pornography. But the filtering programs also tend to "overblock" a lot of non-pornographic material: medical research sites, breast cancer research sites, gay and lesbian social networking sites, amnesty international sites, and so forth. And they block it for all patrons, not just children. Patrons who want to view blocked information have to identify themselves to a librarian. Presumably, the librarian could be called upon to submit the record of who had been visiting what website at which computer. Should we have access to this information? Is the ability to anonymously digest information important to our sense of our political subjectivity? Yochai Benkler puts the underlying question in provocative terms: "the structure of our information environment is constitutive of our autonomy, not only functionally significant to it" (2006, p. 146). The point here is not to either attack or defend any of the previous developments; it is to make the case, on Socratic grounds, that we really ought to possess the technological literacy meaningfully to care. Furthermore, as the above should underscore, "technological literacy" is not about having the skillset to use a particular software package, or about knowing how to keep one's computer free from viruses. Technological literacy refers to the much more important need to reflect on the ways that the technologies we use, to a significant extent, make us who we are.

The Political: Whose Culture?

Consider the tradeoffs between "professional" and "amateur" culture. For most of the twentieth-century, as is well-known, most people got a lot of their culture from the mass media. This one-to-many model of communication had its virtues, among them that large media corporations could claim to operate as a "fourth estate" in keeping a watchful eye on government, and that the products of mass media were professionally produced. The development of more interactive technologies like the Internet seems poised to diminish the relative importance of mass media. Here, however, our knowledge of the technological capabilities of the Internet needs to be supplemented with an awareness of the legal and political environment within which it develops. As Benkler notes, different regulatory regimes "make institutional conditions more conducive for some approaches to information production than to other approaches." Strong intellectual property rights favor commercial speech, since commercial producers own large inventories of copyrighted material that they can exploit both for profit and as a resource base for developing new material. Weaker rights favor non-profit speech, as does, in theory, the decentralized architecture of the Internet

(2003, p. 181). As Benkler then shows, these regulatory regimes and technologies express interpretations of more fundamental moral values like democracy and autonomy. Thus, it is not that democracy is *per se* enhanced by one sociotechnical system or another – it is that the strong-rights version favors a consumer-choice model of democracy, and the weaker-rights system is more participatory (*op. cit.*).

We can notice the same complex relationship between values and technology and the regulatory environment at the level of human subjectivity. For example, what do we make of the fact that many people spend hours and hours writing fan fiction, and posting it on the Internet? Some of that fiction pays straightforward homage to the story on which it is based; other, "slash" fiction, reworks the function of gender within those narratives, usually by developing homoerotic relationships between masculine heroes. Slash is written largely by heterosexual women, for their own consumption, and the avowed aim is to imagine a world in which men are able to be more than military heroes. After gathering evidence in support of this reading of slash, Sonia Katyal concludes that "by taking a given (presumably heterosexual) text and reinscribing it with a largely homoerotic theme, slash acts to challenge the productive power of the author and offer a host of radically new political possibilities for a given narrative" (2006a, pp. 494–5). In this sense, slash presents a profound technological literacy, in that it enables those who write and read it to imagine a world different from the one they live in. However, fan fiction (and slash in particular), occupies very precarious legal territory. Since it uses existing characters from existing media productions, it is likely to be judged a "derivative work," and so would violative of the rights of the copyright owner of the original story.[9] Now, some copyright owners will care more about this state of affairs than others, but at least hypothetically, all could. A disapproving copyright owner could thus make a concerted effort to shut down fan sites; to the degree she succeeded, the fans would be deprived of that method of self-expression. With regard to fan sites, copyright owners have "chosen to undertake an approach that at once demonstrates lukewarm tolerance coupled with random, selected instances of control" (Katyal, 2006a, p. 479). It is often precisely the slash sites that are targeted for shut-down, because content owners find the material offensive.

Fan fiction may or may not matter in the larger scheme of things, but the example suggests a relationship between culturally available narratives and one's own self expression that does matter. Narratives (fictional or not) are an important part of how we understand ourselves: consider the way that people understand themselves through religious narratives ("WWJD?"). More broadly, as Cohen points out, fictional works are an important part of our cultural landscape. This process of self-constitution through cultural narrative begins in childhood, "when children imagine themselves into favorite fictional worlds or when they conclude, because they do not see characters resembling themselves, that those worlds have no place for them" (2007, p. 1202). Thus, "writing fan narratives carries forward these personal dialogues, and sharing them enables broader collective dialogues to take shape," and forms an important component of our cultural self-constitution (*ibid.*).

Other examples of this practice of amateur "glomming on" go beyond personal identification with narratives to include websites and blogs that quote mass media news and then comment on it (Balkin, 2004).

A legal regime could be designed to favor or disfavor such expressions of popular culture; the current regime generally disfavors them. Katyal summarizes one aspect of this disfavor: "copyright law's requirements of originality, tangibility, and fixation tend to minimize the contributions of non-market, amateur participants and often penalize them in the process" (2006a, p. 499). Another aspect has to do with the enforcement of copyright law, which the current regulatory regime essentially outsources to non-state actors. This outsourcing happens because the regime exempts internet service providers from liability for hosting infringing works if and only if they see to the removal of allegedly infringing material on receipt of a notice that posting the material violates copyright. Such a procedure only looks fair until one points out that the notices are sent by copyright owners, and that there is generally no remedy or appeal on the part of the individual posting the work. ISP's have no choice but to comply, and serve as a copyright enforcement police, since the alternative is to face substantial penalty if the work is in fact hosted in violation of the copyright statute.[10] Of course, there is no guarantee of an actual copyright violation, and no judicial process to answer that question. As a result, a great deal of legally protected, or protectable uses of works, are peremptorily taken from the public sphere. One recent study was even able to generate numerous spurious takedown letters for such non-infringing devices as network printers (Piatek, Kohno, and Krishnamurthy, 2008).

RIAA "take-down" letters allow us to see quite clearly the way that technological developments are driving the new salience of this problem. There has always been amateur culture, and it has always involved unauthorized use of copyrighted material by others. The "mix tape" is a very clear example; the relevant point here is that one did not mass-produce mix tapes. Fan fiction appeared in limited circulation, corner-stapled Zines. The legal system in turn tended to protect such unauthorized uses as "fair uses," if they were ever tested in court. However, with the exception of a few, very high profile examples, individual users did not need to know much about copyright, because their unauthorized uses fell well under the legal radar. Technological developments have changed this picture along at least two major axes. First, amateur culture is both easier to produce (in the sense that things like video cameras and the like are readily available), and much easier to distribute publicly. In economic terms, its costs have fallen dramatically.[11] In other words, high costs made the distribution of amateur culture very difficult outside of the local contexts in which it was produced. These difficulties in production and distribution in turn both limited the damage that amateur productions could do to the market for professional productions, and kept most people dependent on professional culture for their entertainment. Second, that reproduction is now increasingly digital means that almost every use involves making a copy (when I download a song to my computer, there's now a copy on both the host computer and mine). This means that a much larger percentage of our total cultural consumption is governed by copyright (Lessig, 2006).

The upshot is that copyright owners have both more to care about, and more reason to care. Now that technological developments have pushed down the costs of caring, we confront what legal theorist Lawrence Lessig has calls "latent ambiguities" in copyright law: is this amateur culture that has passed largely under the legal radar, engaging in activities that are legally impermissible but practically tolerated, something with independent normative value, or is it an unfortunate artifact of a time when we were technologically unable to enforce the legal regime? As Lessig puts it:

> Because of the changes in digital technology, it is now possible for the law to regulate every single use of creative work in a digital environment. As life increasingly moves into a digital environment, this means that the law will regulate more and more of the use of culture (2006, p. 196).

The question we confront, then, is the extent to which we think information ought to be commodified. This is a normative question for ethical and political reflection. Lessig thinks we need to be proactive in using the legal system to encourage a robust "commons" of freely available cultural material; not everyone agrees.

One important point to notice is that a normative decision to encourage such a commons might be expressed in a variety of regulatory structures.[12] One might try to enlarge the public domain directly, as for example by reducing the length of copyright terms, by re-instating a registration requirement to receive copyright protection, or by instituting some sort of "use it or lose it" requirement – to retain copyright protection, a publisher must keep works in print. One might also attempt to use contract licensing mechanisms to enable those who wish their works to be made generally available to do so without fear that future access to them might be limited (Lessig, 2006; for a sympathetic critique, see Elkin-Koren, 2005). Or, one might create more fair use exemptions, declaring, for example, that (non-commercial) fan fiction was definitionally fair use, and so is protected under existing copyright law (for this, see Cohen, 2007, p. 1202). One might even argue that strong intellectual property laws themselves do not decrease, but rather increase, the size of the cultural commons by generating spin-offs and other works unprotected by copyright (Wagner, 2003). The point is that, first, these are important questions made newly important by developments in information technologies; second, that they are not just policy questions, but are values-oriented; and, finally, they are the sorts of questions that everyone ought to care about, insofar as we all live in culture and become who we are in, and through, the cultural artifacts around us. Culture is us; if we are to know ourselves, we need to know the technological processes through which culture is created, and the ways the regulation of those processes modulates them.

4. BACK TO THE CAVE

As already noted, ancient societies did not concern themselves too much with technological questions – the general mood was that dealing with crafts was beneath the life of citizens. Still, Socrates was concerned with at least one kind of information technology: writing. In Plato's *Phaedrus*, Socrates explains that written discourse is inferior to oral. The most writing can do, he argues, is "remind one who knows that which the writing is concerned with" (275d). Written words are repetitive and therefore unable to instruct ("they seem to talk to you as though they were intelligent, but if you ask them anything about what they say, from a desire to be instructed, they just go on telling you the same thing forever" (*ibid.*)); they get into the hands of the wrong people ("not only of those who understand it, but equally of those who have no business with it" (275e)); and when misinterpreted, require help in the form of an authoritative interpreter. This multitude of sins is contrasted with the virtues of the "dialectical" (what we would call "Socratic") treatment of moral topics:

> The dialectician selects a soul of the right type, and in it he plants and sows his words founded on knowledge, words which can defend both themselves and him who planted them, words which instead of remaining barren contain a seed whence new words grow up in new characters, whereby the seed is vouchsafed immortality, and its possessor the fullest measure of blessedness that man can attain unto (276e-277a).

Books are bad; oral discourse is good. Oral discourse is interactive and participatory, so it educates its listener properly; books, on account of their inability to explain themselves, either do not educate at all, or miseducate.[13] But, as we already know from the *Apology*, not all oral discourse is good, either: only oral discourse aimed at finding truth, rather than advantage in argumentation, is good.

Such an argument will sound outlandish to current ears, but perhaps that reaction could be tempered by recalling the high value placed on the "Socratic method" in pedagogy, and by noticing frequent denunciations of new media – precisely for their tendency to get into the wrong hands and to damage those who handle them. Since we live in an "information age," it perhaps bears emphasis that writing, in a sense, was the first form of information. Socrates complains that writing detaches the communication from the agent who is communicating it; this is precisely the point. As Michael Hobart and Zachary Schiffman put it, "information is thus wedded to writing insofar as writing gives stability to the mental objects abstracted from the flow of experience, such that one can access them readily and repeatedly." They underscore the essential point succinctly: "the origin of writing therefore constitutes, at one and the same time, the first information technology and the birth of information itself" (1998, p. 34). The Socratic worry in the *Phaedrus*, then, is about the same kind of thing as denunciations of the supposedly corrupting influence of computer media on today's youth. Thus Plato inveighs against allowing youth to hear the wrong sort of poetry: "the young are not able to distinguish what is and is not allegory, but whatever opinions are taken into the mind at that age are wont to prove indelible and unalterable" (*Republic* 378d).[14]

Socrates, in short, was concerned with what we might today call "information glut," in particular the capacity of too much of the wrong sort of information to cause civic damage. Contemporary examples of this spring readily to hand: Lessig, for example, points to the Internet conspiracy theories surrounding TWA Flight 800, where the same erroneous information, circulated over and over, generated its own impression of truth. As Lessig puts it, "in a world where everyone can publish, it is very hard to know what to believe" (2006, p. 241). Too much information, in the wrong hands, was politically damaging. Plato's solution, as he makes clear in the *Republic* when he banishes most poets, is to regulate the content of information. "We must begin, then, it seems, by a censorship over our storymakers, and what they do well we must pass and what not, reject" (*Republic* 377c). If Plato were alive today, he would no doubt attempt to shut down any number of websites.

The point to emphasize in all of this is that a moral concern with our information environment follows naturally from the view that our information environment is essential to who we are and will become. More generally, a moral concern with our technological environment follows similarly from the view that our technological environment is essential to who we are and will become. In this sense, the Socratic call to "know thyself" includes a call to "know thy technological environment."

We thus return to where we began, the trial of Socrates. His worry that his fellow citizens were sleepwalking their way into fundamental moral corruption and civic ruin was perhaps overstated, but his concern with the information environment and its technologies turns out to be well-placed. When he said that books were morally corrupting, he was making a complaint that should be familiar to any of us that face a glut of information, much of it unwelcome. Still, it might have been possible on Socratic grounds to argue that one did not really need to know much about books: they were everywhere, but best avoided. I hope that the preceding has shown that the situation is more complicated than this. We do not need to fear books, but that is not because all technologies are good. Rather, it is because books are a constitutive part of who and what we are. We owe it to ourselves to understand this point, and to understand something of the technologies of which we are made. The core of the Socratic "know thyself" is as applicable today as it was in Socrates' time, and it includes at least two fundamental thoughts. First, we ought to understand something of the way that technology creates the environment from which we draw the information that makes us who we are, and the way that technological changes enable changes in the information environment. Second, we ought to notice that these technologies do not occur in a vacuum: understanding something of the political and legal environment in which they unfold are vitally important to understanding their effects on us.

NOTES

1 For all Plato References, I follow the standard, Stephanus page numbers, as these are consistent across editions and translations. I follow the text in Plato, 1961.

2 For a critique of this narrative, see Pippin, 1995. Pippin suggests that the Heideggerian critique (see below) tends to miss the early modern emphasis on mastery of nature.

3 For other examples of this approach, see the complaint in Dreyfus (1999), based on the existentialist Kierkegaard, that distance learning is inauthentic; or Jacques Ellul's general case against "technicist" society (1989).

4 Winner is perhaps too much of a determinist about the relation between certain technologies and corresponding political orders. In other words, he seems not to take enough account of the quite unexpected social effects of many, if not most, technologies. Certainly the Internet's trajectory from cold war survivability to ebay is the clearest, recent example. Late in the "*Techné*" essay, Winner poses as a question: "what forms of technology are compatible with the kind of society we want to build" (1986, p. 52)? Similarly, in his most famous paper (1980), he suggests that overpasses leading to the beaches on Long Island were deliberately low to keep buses (and the poor who ride them) off the beaches owned by the rich (That essay is criticized along the lines suggested here in Joerges, 1999). Four points to emphasize here: (1) what I want to retain is Winner's insistence that there are important political and moral questions to be asked; as will become apparent, I do think that certain socio-technical systems (interpreting the "system" to include regulatory and other environmental factors) favor some values over others, without necessarily binding us to them. (2) Andrew Feenberg (1995) usefully emphasizes the interdependence of social structures and technologies generally. (3) I am also sympathetic to Bruno Latour's (1992) argument that we would be better off dropping the "technology/society" classification in favor of the more neutral "technological society." A full assessment of Latour (and the literature discussing his work) is well beyond the scope of this essay, in particular whether adopting Latour's strategy enables or inhibits realization of democratic values. See Winner's (1991) critique of Latour and Latour's (2004) recent book length foray into political theory. (4) I would at all counts resist the cyberlibertarian assumption that the Internet is inherently liberating or inherently favors small-scale, deregulated communities; for a sustained critique of this assumption, see Lessig, 2006.

5 Those familiar with debates around so-called strong-AI will recognize both Descartes' anxiety about machines and his strategy for disqualifying them from human cognition.

6 Clark's engagement with, and debt to, "continental philosophy" seems to be underappreciated by most of his readers. The acknowledgements page of *Cyborgs*, however, is quite explicit.

7 For example, Margaret Chon notes that "99% of the global disease burden is concentrated in low and middle income countries. However, in 1992, less than 5% of the total global R&D was spent on their health problems. In 1996, only 0.5% of pharmaceutical patents related to tropical diseases such as malaria;" strengthening current IP law will only make matters worse: "economists agree that the global re-distributional effect of strengthening intellectual property laws will benefit the U.S. predominantly and only a handful of other developed countries in the short run, especially in the pharmaceuticals sector"(2006, p. 2884).

8 This is of course not an original point. As Benkler and Nissenbaum summarize, citing Winner, Lessig, Latour, and others, "the common idea [is] that technical systems and devices, in virtue of their properties, architecture, or functionality, have the capacity both to limit and to facilitate what individuals and collectivities are able to do" (2006, p. 416). Thus, "technical systems and devices are as much a part of political and moral life as practices, laws, regulations, institutions and norms that are more commonly seen as vehicles for moral and political values" (*op. cit.*, p. 417).

9 For the reasons why slash might or might not successfully mount a "fair use" defense, see Katyal, 2006a, pp. 497–517. Katyal generally thinks it is legally vulnerable, even though normatively it ought to be considered fair use. For the general theory of which slash is a paradigmatic case, see Katyal, 2006b. For the confusion surrounding predictions of "fair use," see Nimmer, 2003. For concerns about the social values embedded in fair use jurisprudence, see Tushnet, 2007.

10 For a general discussion how the collective actions of rational, risk-averse actors tend to inadvertently strengthen intellectual property regimes, see Gibson, 2007. It is worth pointing out that the copyright statute is very complicated and abstruse, and a rational actor will want to avoid litigation when it has little to gain from it. For the complexity, see Litman, 2001.

11 The general point is common enough. I develop it in the context of digital (vs. analog) reproduction and attempt to theorize the difficulties this poses for current copyright law in Hull, 2003.

12 The variety is meant to emphasize that a given normative commitment does not then bind one to a particular regulatory structure. For an elegant discussion of normative theories of intellectual property and the argument that one's theory of IP underdetermines policy even as having a theory facilitates policy conversation, see Fisher, 2001.

[13] Plato's stated preference for oral over written discourse is very difficult to sustain, as he depends on metaphors and vocabulary from writing to explain the priority of speech. This argument is famously made in Derrida, 1981.

[14] I will here suppress a question about the extent to which the speech/writing dichotomy may be too simplistic insofar as it does not attempt to theorize "code" as a third kind of information. That thought is pursued in Poster, 1990; and especially in Hayles, 2005, pp. 39–61.

REFERENCES

Bakhtin, M. M. (1986). The Problem of Speech Genres. (V. W. McGee, Trans.) In C. Emerson & M. Holquist (Eds.) *Speech Genres and other Late Essays* (pp. 60–102). Austin: U. Texas Press.

Balkin, J. (2004). Digital Speech and Democratic Culture: A Theory of Freedom of Expression for the Information Society. *New York University Law Review, 79*, 1–58.

Benkler, Y. (2003). Through the Looking Glass: Alice and the Constitutional Foundations of the Public Domain. *Law and Contemporary Problems, 66*, 173–224.

Benkler, Y. (2006). *The Wealth of Networks: How Social Production Transforms Markets and Freedom.* New Haven: Yale UP.

Benkler, Y. & Nissenbaum, H. (2006). Commons-based Peer Production and Virtue. *Journal of Political Philosophy, 14*, 394–419.

Clark, A. (2003). *Natural-Born Cyborgs: Minds, Technologies, and the Future of Human Intelligence.* Oxford: OUP.

Clark, A. (2007). Curing Cognitive Hiccups: A Defense of the Extended Mind. *Journal of Philosophy 104*, 163–92.

Clark, A. (2008). Pressing the Flesh: A Tension in the Study of the Embodied, Embedded Mind? *Philosophy and Phenomenological Research, 76*, 37–59.

Clark, A. & Chalmers, D. (1998). The Extended Mind. *Analysis, 58*, 10–23.

Cohen, J. E. (1996). A Right to Read Anonymously: A Closer Look at 'Copyright Management' in Cyberspace. *Connecticut Law Review, 28*, 981–1039.

Cohen, J. E. (2003). DRM and Privacy. *Berkeley Technology Law Journal, 18*, 575–617.

Cohen, J. E. (2007). Creativity and Culture in Copyright Theory. *University of California, Davis Law Review, 40*, 1151–1205.

Chon, M. (2006). Intellectual Property and the Development Divide. *Cardozo Law Review, 27*, 2814–2904.

Descartes, R. (1985). *Discourse on the Method* (J. Cottingham, R. Stoothoff & D. Murdoch, Trans.) In *The Philosophical Writings of Descartes* Vol. I (pp. 101–51). Cambridge: CUP. (Originally published 1637)

Derrida, J. (2005). *Paper Machine* (Rachel Bowlby, Trans.). Stanford: Stanford UP.

Derrida, J. (1981). Plato's Pharmacy. In *Dissemination* (Barbara Johnson, Trans.). Chicago: U. Chicago Press.

Elkin-Koren, N. (2005). What Contracts Can't Do: The Limits of Private Ordering in Facilitating an Information Commons. *Fordham Law Review, 74.*

Feenberg, A. (1995). Subversive Rationalization: Technology, Power, and Democracy. In A. Feenberg and A. Hannay (Eds.) *Technology and the Politics of Knowledge* (pp. 3–22). Indianapolis: Indiana UP.

Fisher, W. (2001). Theories of Intellectual Property. In S. R. Munzer (Ed.) *New Essays in the Legal and Political Theory of Property* (pp. 168–99). Cambridge: CUP, 2001.

Gibson, J. (2007). Risk Aversion and Rights Accretion in Intellectual Property Law. *Yale Law Journal, 116*, 882–951.

Hayles, N. K. (2005). *My Mother Was a Computer: Digital Subjects and Literary Texts.* Chicago: U. Chicago Press.

Heidegger, M. (1977). The Question Concerning Technology. In D. F. Krell (Ed.) *Basic Writings* (pp. 311–41). San Francisco: Harper Collins.

Hobart, M. E. & Schiffman, Z. S. (1998). *Information Ages: Literacy, Numeracy, and the Computer Revolution*. Baltimore: Johns Hopkins UP.
Hull, G. (2003). Digital Copyright and the Possibility of Pure Law. *qui parle, 14*, 21–47.
Hull, G. (2009). Overblocking Autonomy: The Case of Mandatory Library Filtering Software. *Continental Philosophy Review*. *42*, 81–100.
Joerges, B. (1999). Do Politics Have Artifacts? *Social Studies in Science, 29*, 411–31.
Katyal, S. K. (2006a). Performance, Property, and the Slashing of Gender in Fan Fiction. *Journal of Gender, Social Policy and the Law, 14*, 461–518.
Katyal, S. K. (2006b). Semiotic Disobedience. *Washington University Law Review, 84*, 489–571.
Latour, B. (1992). Where are the Missing Masses? The Sociology of a few Mundane Objects. In ed. W. E. Bijker & J. Law (Eds.) *Shaping Technology/Building Society* (pp. 225–58). Cambridge, Mass: MIT Press.
Latour, B. (2004). *Politics of Nature: How to Bring the Sciences into Democracy* (Catherine Porter, Trans.). Cambridge, MA: Harvard UP.
Lessig, L. (2006). *Code: Version 2.0*. New York: Basic Books.
Litman, J. (2001). *Digital Copyright*. New York: Prometheus Books.
Mackenzie, A. (2004). Contribution to "We Have always Been ... Cyborgs," Review Symposium for *Natural Born Cyborgs*. *Metascience, 13*, 153–63.
Nimmer, D. (2003). 'Fairest of them All' and other Fairy Tales of Fair Use. *Law and Contemporary Problems, 66*, 264–287.
Piatek, M., Kohno, T., & Krishnamurthy, A. (2008). Challenges and Directions for Monitoring P2P File Sharing Networks, or Why My Printer Received a DMCA Takedown Notice. *University of Washington Technical Report*, UW-CSE-08-06-01, at: http://dmca.cs.washington.edu/ (accessed 6/08).
Pippin, R. B. (1995). On the Notion of Technology as Ideology. In A. Feenberg and A. Hannay (Eds.) *Technology and the Politics of Knowledge* (pp. 43–61). Indianapolis: Indiana UP.
Plato (1961). *The Collected Dialgues* (E. Hamilton& H. Cairns, Eds.). Princeton: Princeton UP.
Poster, M. (1990). *The Mode of Information: Poststructuralism and Social Context*. Chicago: U. Chicago Press.
Tushnet, R. (2007). My Fair Ladies: Sex, Gender, and Fair Use in Copyright. *American University Journal of Gender and Social Policy, 15*, 273–303.
Wagner, R. P. (2003). Information Wants to Be Free: Intellectual Property and the Mythologies of Control. *Columbia Law Review, 103*, 995–1034.
Winner, L. (1980). Do Artifacts Have Politics? *Daedalus, 109*, 121–136.
Winner, L. (1986). *Techné* and *Politeia*. In *The Whale and the Reactor: A Search for Limits in Age of High Technology* (pp. 40–60). Chicago: U. Chicago Press.
Winner, L. (1991). Upon Opening the Black Box and Finding it Empty: Social Constructivism and the Philosophy of Technology. In J. C. Pitt and E. Lugo (Eds.) *The Technology of Discovery and the Discovery of Technology* (pp. 503–19). Blacksburg, VA: Society for Philosophy and Technology.
Winner, L. (1995). Citizen Virtues in a Technological Order. In A. Feenberg & A. Hannay (Eds.) *Technology and the Politics of Knowledge* (pp. 65–84). Indianapolis: Indiana UP.

Gordon Hull
Department of Philosophy,
University of North Carolina at Charlotte

CHAPTER 3

MICHAEL BUGEJA

THE INTERPERSONAL DIVIDE

Since the 1990s I have been investigating how higher education has invested in technology, often without any assessment. I have documented the cost of technology, not only in equipment, tuition and fees, but also in curricula, which has expanded dramatically to incorporate, maintain and support the software, hardware and wireless systems that enrich corporate entities whose restrictive service terms have altered our culture of transparency and due process. Also, I have studied how technology is perceived by digital natives, our students, who typically only know how to use the latest gadgets and applications as consumers rather than as critical thinkers. Worse, these changes occurred during an era of greed and convergence in news media that largely overlooked the current cultural devastation. I contend that all of these factors, and more to be explored in this chapter, have eroded interpersonal intelligence—knowing when, where and for what purpose technology use is appropriate or inappropriate—creating a divide that threatens to disenfranchise future generations tasked with resolving global issues that require self-knowledge and mutual understanding.

Like Neil Postman, the great social commentator and author, my primary expertise is cultural rather than technological. My research, again like his, focuses not on what technology can do for education, but also what it can undo. Moreover, I am a working journalist as well as a journalism director. I have great respect for my colleagues in the news media, especially now, with many reporters risking their lives in Iraq, Afghanistan and even in the United States. Many of us still realize that our job is to defend the Constitution by informing the public so that it can make intelligent decisions in the voting booth. However, because of the cost of technology, which also automates people—from telephone operators (voice mail) and postal workers (e-mail) to bank tellers (ATMs) to travel/airline personnel (online reservations), wreaking havoc in the service industry—there are fewer and fewer journalists in downsized newsrooms. Media owners focus on the bottom line, so much so, that we get a steady diet of celebrity and entertainment news augmented by "cost-effective" news, like health updates on diets or advice on personal finances. After more than a decade, we're no closer to monetizing the Internet for news because most traditional journalists believe that information still has value; Internet, as we will learn, believes

Michael P. Clough, Joanne K. Olson and Dale S. Niederhauser (Eds.), The Nature of Technology: Implications for Learning and Teaching, 35–50.

that content has little value and so is given away for free, enticing viewers to visit a news site so hidden computer technology can data-mine users and sell those data to e-marketers.

The goal of this chapter is to put all of these issues in perspective and then relate them to the interpersonal divide and what we can do about it. My research on the interpersonal divide—or the void that develops when we spend too much time using technology—is evident everywhere in society but more so on college campuses. Students are targeted for a lot of reasons, primarily because corporate venders can sell them real or virtual products at unprecedented rates. In the recent past, we never knew where students really lived. True, they would have campus or off-campus addresses, but were seldom there and rarely responded to direct mail advertisements. But even if you found them, you couldn't sell them much. They didn't have access to money or credit. They had weird consumer habits, too, like eating peanut butter for a month to purchase a stereo system. So businesses would throw as much paper at them as possible—usually in the form of controlled circulation magazines—distributed free of charge where marketers thought students might hang out on occasion, such as a residence hall or a campus building. It was important to sell things to students, even ones without much money, because they would be making brand decisions between the ages of 18 and 24 that they would keep into adulthood.

Because of portable handhelds, such as cell phones, all that has changed. We know now where to locate students any hour of the day because their digital gadgets have G.P.S. technology. Many students are posting their whereabouts on applications like Four Square, networks like Facebook and microblogs like Twitter. Moreover, the government has outsourced student loans to the banking industry, which allows students to borrow generous sums—so much, in fact, that students can live consumer lifestyles. Coincidentally, many of these banks also provide credit cards without which digital devices would be less entertaining and addictive.

Think about the implications of this at the typical college campus. We study and work in buildings whose architecture is meant to make us feel a part of something larger than ourselves. Typically students ignore those buildings and manicured grounds because they are texting, a carryover habit from high school, in which half of all teens send 50 or more text messages a day, or 1,500 texts a month, with one in three sending more than 100 texts a day, or more than 3,000 texts a month.[1] Academic institutions profess to celebrate diversity but students do not engage others who differ from themselves because they are too busy interacting digitally with like-minded individuals in their "affinity" group, a friendlier term for psychographics, or marketing niche. We hold hands with partners or spouses on the digital street while chatting with others on cell phones, using our free hands to multitask romance and socializing. We commit ourselves to environmental causes and preservation of species, ignoring both as we navigate the outdoors in a digital bubble of oblivion, listening to iTunes. *This was not supposed to happen.* We were supposed to inherit a global village that would enfranchise everyone with access. In the 1990s, Bill Clinton promised that advances in computer technology and the Internet would change the way America "works, learns, and communicates."[2] Rather, it has changed the way we waste time and escape from learning, communicating egotistically about who we are, where we're going, how we're going to get there, and what we want to do when we arrive. All the while we are networking, tweeting or texting, the content and pixels of our lives are being harvested according to terms of service of applications, sold to third-parties that spam us to death. When Bill Gates, founder of Microsoft, predicted that the Internet would be the town square of tomorrow, even he didn't realize that we would abandon or ignore the real town square in the digital street. Instead of a global village, we have inherited a global mall that promises to entertain and sell to us rather than enlighten and inform.

WE ARE AMUSING OURSELVES TO DEATH.

That also happens to be the title of the masterwork of Neil Postman. I often refer to Postman in my speeches at university campuses. Several people in the audience will have read his book, *Amusing Ourselves to Death,* published on the occasion of 1984. That year, of course, also is the title of the famous dystopian novel by George Orwell who predicted a Big Brother dictator using technology to surveil our every move in a police-state society. Postman, however, believed we would inherit a different type of dystopia—the one described in Aldous Huxley's *Brave New World*—as this passage from the introduction of his book illustrates:

> Orwell warns that we will be overcome by an externally imposed oppression. But in Huxley's vision, no Big Brother is required to deprive people of their autonomy, maturity and history. As he saw it, people will come to love their oppression, to adore the technologies that undo their capacities to think.
>
> What Orwell feared were those who would ban books. What Huxley feared was that there would be no reason to ban a book, for there would be no one who wanted to read one. Orwell feared those who would deprive us of information. Huxley feared those who would give us so much that we would be reduced to passivity and egoism. Orwell feared we would become a captive culture. Huxley feared we would become a trivial culture. ... As Huxley remarked in *Brave New World*, the civil libertarians and rationalists who are ever on the alert to oppose tyranny failed to take into account man's almost infinite appetite for distractions.[3]

Digital distractions threaten the quality of education as well as the fabric of society. Again, this assertion continues the work of Neil Postman who posited that there are winners and losers when it comes to technology. In a speech to technologists in 1990, he prophesied that students would be "easy targets for advertising agencies," especially when

> ... schools teach their children to operate computerized systems instead of teaching things that are more valuable to children. In a word, almost nothing

> happens to the losers that they need, which is why they are losers. ... Should the losers grow skeptical, the winners dazzle them with the wondrous feats of computers, many of which have only marginal relevance to the quality of the losers' lives but which are nonetheless impressive. Eventually, the losers succumb, in part because they believe that the specialized knowledge of the masters of a computer technology is a form of wisdom. The masters, of course, come to believe this as well. The result is that certain questions do not arise, such as, to whom will the computer give greater power and freedom, and whose power and freedom will be reduced?[4]

As you might imagine, Neil Postman, who taught communication at New York University, has influenced my research on the interpersonal divide. I often remark that professors still assigning his books may not even realize that Postman died in 2003. Perhaps some of you reading this are learning about his death for the first time. This is Postman's legacy. He had the misfortune of dying within weeks of the late entertainer Johnny Cash[5] and two days before Californians would elect former body builder and actor Arnold Schwarzenegger[6] governor in a year that featured "American Idol" as the top searched TV show online[7] and icons Paris Hilton and Britney Spears as the most sought after celebrities.[8]

Technology gives you the feeling that so much is changing so quickly that we can barely keep up with our lives. The fact is, not much has changed. In 2003, when Neil Postman died, the United States invaded Iraq. The Space Shuttle Columbia exploded. According to Lycos, the second top sought-after story online was the sex scandal of Kobe Bryant, the number four such top story was Michael Jackson's arrest, number nine was the Super Bowl, and number 10 was the Laci Peterson murder.[9] *Time magazine's* person of the year was "The American Soldier."

In 2006 we were still dealing with the Iraq war, American Idol, Britney Spears and Paris Hilton. NASA was in the news with astronauts gone wild concerning an illicit affair between two astronauts. *Time's* "Person of the Year" was not the physicist probing the edges of the universe, nor the chemist unraveling how DNA is copied to RNA, nor the economist creating wealth for the social underclass. No. *Time's* so-called person of the year was "You" in a tube, inspired by the sale of "YouTube" for $1.65 billion to the search engine Google, which grew an astounding 88 percent in 2006 with sales over $7.1 billion while employing a mere 6,790 people.[10]

At the start of 2007, I predicted top stories. Here is a passage from one of my speeches: "Top stories will be ... hmmm ... the Iraq War, a Presidential election whose outcome will be determined not by caucus but by YouTube and blog ... replete with nauseating follow-ups on American Idol, Paris Hilton and Britney Spears. This will be punctuated by scandals stereotyping black athletes and celebrities and beautiful women murdered or gone missing in exotic places."

Each time one of my predictions came true—and all of them had after OJ Simpson was arrested for robbery in September 2007—I would send an email to journalists and managing editors across the country, with the heading: "How am I doing so far?"

Here are some of their replies:

- "Spot on!" wrote Eli Flournoy, CNN senior international editor.
- "Sad but true" wrote Michael Redding, a reporter at WCNC in Charlotte, North Carolina.
- "The sad fact," wrote Mel Mencher, author of *News Reporting and Writing*, "is that NBC knows what people are interested in" offering Paris Hilton $1 million for her life behind bars *because of bars*. "The networks used to be proud of their commitment to important and meaningful news. Now, it's all ratings."
- "Yes! I thought of you when I heard about the NBC deal" to Paris Hilton, wrote Cheryl Mullenbach of Iowa Public Television. "How sad that our priorities are so distorted!"
- Janet Kornblum, who reports for *USA Today*, wrote, "Too bad you didn't take bets."

In an early draft of this essay, I wrote the following:[11]

> As I write, 2008 is a mere two weeks away. Top news stories will be, well, the Iraq War, the Presidential election, American Idol, Paris Hilton, Britney Spears, a sports event, a black athlete scandal, a black celebrity scandal, a beautiful woman gone missing who turns up dead with her husband as prime suspect (one version) or turns up in another city with a hitherto unknown lover (alternate version). Time magazine's "Person of the Year" will be the Democrat who wins the White House. Should a Republican win, the Person of the Year may be the adversary and/or gaffe that undid the Democratic contender.

As it turned out, revising the essay in 2010, the Iraq War was still top news in 2008 on most news sites and "American Idol" remained the most watched television show.[12] The presidential election was the second top news story after the failed economy, according to Time magazine; Britney Spears' "comeback" was the top celebrity story, according to UK's Telegraph; ESPN was so sure of Tiger Woods' being the top story, it compiled a list of his top 10 golf shots of 2008; Tiger Woods would become the top black celebrity and sports scandal of 2009, after confessing to myriad love affairs; Time's top crime story was former athlete O.J. Simpson's return to prison; so many beautiful white women went missing in 2008 that the term "Missing White Woman Syndrome" became popular.[13] Of course, Time Magazine's 2008 "Person of the Year" was Barack Obama.

Predicting these stories was no great feat. Nothing really changes year to year but the wars, the celebrities and the sport teams. But when we hear the news in a podcast or watch it on an iPhone, it all seems new and revolutionary as the gizmo in our pocket.

How did we get to this state of affairs? How did we escape the tyrannical Orwellian world only to inherit, as Neil Postman prophesized, the Huxleyan world with all its distractions?

Some of you reading this will remember how dangerously close we came to the oppressive, censored world of Orwell's Big Brother in the Joseph McCarthy era of the early 1950s. McCarthy, at the time, chaired the permanent Subcommittee on Investigations. Recently transcripts were released of secret interrogations of 395 Americans accused of subversive Communist activities. Here is how McCarthy addressed one person before his subcommittee:

> During the course of these hearings, I think up to this time we have some—this is just a rough guess—twenty cases we submitted to the grand jury, either for perjury or for contempt before this committee. Do not just assume that your name was pulled out of a hat. Before you were brought here, we make a fairly thorough and complete investigation. So I would like to strongly advise you to either tell the truth or, if you think the truth will incriminate you, then you are entitled to refuse to answer. I cannot urge that upon you too strongly. I have given that advice to other people here before the committee. They thought they were smarter than our investigators. They will end up in jail. This is not a threat; this is just friendly advice. ...[14]

The news media at the time understood what was going on but few had the courage to take on Sen. McCarthy. The person perhaps singularly responsible for rising to the challenge was Edward R. Murrow, the famous war correspondent and host of the CBS TV documentary series, *See it Now*, from which this excerpt is taken; again, I quote:

> [Sen. McCarthy's] primary achievement has been in confusing the public mind, as between the internal and the external threats of Communism. We must not confuse dissent with disloyalty. We must remember always that accusation is not proof and that conviction depends upon evidence and due process of law. We will not walk in fear, one of another. We will not be driven by fear into an age of unreason. ...[15]

I know what some of you are thinking. If we replaced the word "Communism" with the word "terrorism," a case could be made against certain rulers in the United States and elsewhere for realizing the Orwellian/McCarthy vision with unauthorized wiretaps and reports of clandestine interrogations and secret prisons. The historical situations differ, and so I cannot make that argument; but that is not the point. Consider again the ominous words of Neil Postman—that an amused society on information overload, much of it trivial, would hear about abuses of power and respond with passivity and disinterest.

Once again it was Edward R. Murrow who foretold our Huxleyan future in 1958, when U.S. society was sliding into the trivial with TV game shows rather than video games. Here is what Murrow said then about the state of the news media:

> We have currently a built-in allergy to unpleasant or disturbing information. Our mass media reflect this. But unless we get up off our fat surpluses and recognize that television in the main is being used to distract, delude, amuse and insulate us, then television and those who finance it, those who look at it and those who work at it, may see a totally different picture too late.[16]

More than a half century has passed since his speech, and we see ourselves in the carnival mirror of Murrow. In the darkest days of the McCarthy era only a few journalists like Murrow roused an aware but frightened public. In the digital era an equally small handful of journalists and scholars are trying to rouse an amused, distracted audience. Murrow tried to fight that phenomenon, but came to the conclusion that he could not defeat McCarthy and corporate journalism at the same time. The New McCarthyism is the media itself.

There are also fewer Murrows to take on deregulated media behemoths.[17] In 1983, 50 companies controlled most of the world's media. By 2003, there were six.[18] Annual profits often exceed 30 percent. Disney, one of the six behemoths, reported a 33 percent rise in profits in 2006 due to growth in media and entertainment.[19] In the past six years daily newspapers have lost more than 4,000 reporters and editors alone.[20] In 2007, newsroom cuts continue unabated at *The Atlanta Journal-Constitution* and *The Philadelphia Inquirer*, to name a few.[21] *The New York Times* has been buying out veteran journalists, as has the *Boston Globe,* including two Pulitzer Prize winners.[22] Since year 2000, network news has lost 10 percent of its staffs. Magazines have suffered, too. Time Warner, the largest, most powerful media company, has cut hundreds of workers from its magazines and closed bureaus of its flagship, *Time magazine.*[23]

We used to operate in the public interest. Many outlets, especially newspapers, operate on the concept of branding. Here is how one consultant discusses the public interest:

> Stewardship of the public trust remains central to the identity of most newsrooms. But perhaps newspapers are now structured on promises of value, such as independence and objectivity, which fail to substantially exist in the minds of consumers. Even if it were a perception that could be re-kindled, how much additional purchase intent would it generate?[24]

Purchase intent was not the constitutional intent of Thomas Jefferson and others who preferred newspapers over government *as long as the people could read.*[25] That caveat, reading, helped create public education under the theory that free speech *without* education was as meaningless as education without free speech. Let's ponder that with a bit more focus. Our nation was founded on the notion that an educated, informed public could discern truth from falsehood and so resist the manipulation of despots. Orwell's despot relied on education without free speech; Huxley's, on free speech without education and *that*, increasingly, may be our legacy as we accelerate socially from the death of print to the death of *reading.*

A key factor in that transition is the "allusion of enlightenment," which technology conjures along with revenue. We are moving from an era of social responsibility to fiduciary responsibility—so much so, that we have a new maxim in the media—"the public interest is what interests the public."[26] And what interests us more than ever are gizmos through which we amuse ourselves for much of the digital day. Perhaps the more important issue concerns how these distractions keep us from addressing the real issues of the day when during that day, on average, each of us consumes nine hours of media and technology with one third of us consuming two or more media at the same time, usually television and Internet.[27]

It seems the more technology we provide, the more incoming students say that they are bored in class. In 1985, according to national data, 26 percent of students were bored regularly. That figure now is 41 percent.[28] To test these data locally, though unscientifically, I conducted an unscientific interactive poll with 77 percent of those who responded claiming to be so bored in class that they just had to text a message or visit a social network or read email or watch movies or play a computer game or make an online purchase. More than a quarter of the 218 students who responded were taking my online survey *while they were in another class*.

A *New York Times* writer interviewed me on the subject of boredom in the classroom.

"I'm so tired of that excuse," I responded. "The idea that subject matter is boring is truly relative. Boring as opposed to what? Buying shoes on eBay? The fact is, we're not here to entertain. We're here to stimulate the life of the mind."[29] You cannot do that in a distracting environment. Distraction undermines education, namely, critical thinking.

At issue is who will be the bearer of truth in the digital age—the teacher and professor or the social network and the processor? This is what Postman calls the "Faustian bargain" that usually accompanies technological diffusion. According to Postman, "Technology giveth and technology taketh away, and not always in equal measure. A new technology sometimes creates more than it destroys. Sometimes, it destroys more than it creates. But it is never one-sided." [30] In the 30 plus years that I have worked in higher education I have witnessed technology being used to distribute content from a residential campus, then becoming content on residential campuses, and finally becoming the residential campus itself. Indeed, technology taketh away, so much so, we not only are willing to replace the professor with a processor but the residential campus with a virtual one.

At stake in these trade-offs with technology is our academic culture. Technology is its own autonomous system, independent of what it touches, radically altering whatever it touches with little change to itself.[31] Apply technology to the economy, and the economy henceforth is about technology. Apply it to journalism, and journalism is about technology. Apply it to education, and education is about technology. All must adapt, or else the application, network, computer program or server ceases to work or apply, pursuant to terms of service. Thus, we lose centuries of erudition because the interfaces of technology alter or undermine basic principles. Worse, because autonomous technology is independent of everything, *it cannot be blamed for anything.*

Even if you disagree with this and believe that technology is just a tool, then fathom the rhetoric of your conviction. The word "tool" is generic. A hammer and a hacksaw are tools. You don't pound nails with a hacksaw and you don't saw wood with a hammer. So you have to ask, a tool for what purpose? That is the central message of my book, *Interpersonal Divide*. In it, I challenge readers to list all of the gadgets in their homes and to recall why they bought the device and then assess how they are using it. If you don't ask and answer those questions, marketing will. Metaphorically, you could be hammering nails with hacksaws and sawing wood with hammers. You can't build a global village using tools like that.

Consumer technology promised us a global village. We found that village, all right, and it is peopled with as many jesters, peddlers and pickpockets as with wizards. For more than a decade we have been touting technology at great expense—not only to our pocketbooks—but also to our psyches. We were supposed to be citizens of a brave new media world that promised to enfranchise and enlighten everyone with universal access. Somehow we got hooked on vaudevillian devices that tell us we have mail, we have text, we have cell, we have friends, we have worth, we are hip, we are mobile. *We arrived.*

Where, exactly? According to the National Assessment of Educational Progress, the nation's report card, reading scores in 2005 were significantly worse than in 1992. And in math, only 23 percent of all 12th graders were proficient. Worse, these sinking scores occurred even though high school students averaged 360 more classroom hours in 2005 than in 1990.[32] The question is not whether one believes this is an accurate barometer of learning but whether anyone, by any measure, has weighed results against expenditures. For instance, have we compared scores of school districts lacking much technology against those that have invested heavily in it, adjusting for such factors as household income to see if the investment makes a difference? The Iowa Department of Education knew of no such assessment.

Then there is the issue of addiction. The effects of Internet addiction, which afflicts one in eight Americans, are being felt worldwide.[33] Administrators of India's top Institute of Engineering and Technology shut down Internet access from 11:30 at night to 12:30 in the afternoon in residence halls to prevent online gaming, social networking and downloads.[34] This, professors said, helps sleep-deprived students focus more on their studies. China's Internet addiction rates are staggering. An estimated 13 percent of that country's 18 million users *under* 18 are addicts.[35] Ironically, the distractions of Internet are afflicting education in countries hitherto identified as disenfranchised in past arguments about the digital divide.

In the mid 1990s we in education were warned that we would disenfranchise our own students unless we invested in technology. We responded by charging tech fees and inflating curricula and with it, tuition. Nonetheless, we continue to invest in consumer technology because of digital divide arguments that have yet to deliver. Renowned scholars like Nicholas Negroponte, co-founder of MIT's Media Laboratory, still operate on that assumption. In one of his columns for the online magazine *Wired*, Negroponte envisioned a global environment devoid of nationalism, a digital utopia so leisurely we would live rurally in our pajamas (no need for cities or office buildings). Better still, we would be free of the caste system of industrial society. In the new world order, he wrote, "landlords will be far less important than webmasters."[36] Landlords now collect rents online and provide or deny highspeed access as part of the housing agreement.

As much as anyone, I yearned for Negroponte's vision. He is a humane, ideal-driven innovator who believed technology would unite us globally. "As humans," he wrote in 1998, "we tend to be suspicious of those who do not look like us, dress like us, or act like us, because our immediate field of vision includes people more or less like us. In the future, communities formed by ideas will be as strong as those formed by the forces of physical proximity."[37]

The irony of that argument is the premise about the homogeneity of place where people look, dress and act similarly. If anything, Internet stereotypes according to consumer profiling while the physical world has become more diverse, even in rural American states historically homogeneous such as Iowa. According to the non-profit New York-based Population Council, America is becoming diverse so rapidly that by the year 2043, nearly one in four people will be Latino, with multiracial Americans common almost everywhere in every state.[38] The global village is outside our real front doors, not on the front doors of Internet sites where we brand ourselves by the companies—not the *human company*—that we keep.

Christine Rosen, a fellow at the Ethics and Public Policy Center, believes that loyalty to corporate brands is related to a concept she calls "egocasting." In a 2005 interview, she maintained that egocasting fosters "an on-demand attitude that infects consumers with fetish-like impulses."[39] According to Rosen, "It is ironic that the technologies we embrace and praise for the degree of control they give us individually also give marketers and advertisers the most direct window into our psyche and buying habits they've ever had." Rosen believes that universities have embraced technology to further education. "But," she says, "they have failed to realize that the younger generation views technology largely as a means of delivering entertainment—be it music, video games, Internet access or television—and secondarily as a means of communicating."

Technology also provides what Rosen calls "new and unusual ways to isolate oneself from opinions or ideas that make us uncomfortable, from people whom we would rather not have to know, from those often-awkward social interactions with strangers in public spaces. In the college context this is more worrisome," she says, "since liberal education is meant to expose students to ideas that challenge them to think in new ways about subjects or people that they had not encountered before." Such exposure, she says, fosters critical thinking.

Wireless technology fosters distraction. In 2004, only about a third of classrooms were wireless, according to the annual Campus Computing Survey. Wireless networks now cover more than half of all college classrooms.[40] Even technology experts such as Dennis Adams, chair of Information Sciences at the University of Houston, want to shut off the wireless in class. *You can't.* There are too many overlapping signals.

Adams worries about critical thinking in the wireless classroom, remembering "The Sesame Street Syndrome" by humanist Edna LeShan. Many current students were reared on that television show, brought to them by letters of the alphabet.

According to LeShan, those educational "commercials" taught students that there are right and wrong answers and that thinking and questions are immaterial … because adults do the asking and the answering.[41] Now we have moved from the "Sesame Street Syndrome" to the "Google Syndrome," which teaches Millennials that there are answers to questions but makes no claim as to the *accuracy* of those answers

Retired professor Theodore Roszak, author of *The Making of a Counter Culture*, advises that we should tell students "that thinking with your own naked wits is a pure animal joy that cannot be programmed, and that great culture begins with an imagination on fire. We should remind our [students] at every turn that more great literature and more great science were accomplished with the quill pen than by the fastest microchip that will ever be invented."[42] Roszak's greatest fear that is that technology "will reduce the mind to the level of the machine."

Many scientists share that concern. Physician David Ho is one of them. Dr. Ho, you may remember, is best known for his AIDS-related work using multiple drug therapies. What few people know is that Dr. Ho was trained in the physical sciences as well as biology. He had great respect for mathematics. As a boy, he even learned to "count cards" at blackjack.[43] Early in his research he would come upon treatments for AIDS that succeeded in the laboratory but failed in humans. But that's science, he said, because wrong answers help the discovery process. Soon he came to understand that the HIV virus mutates rapidly, resisting each single drug. That's when he and his team turned to mathematics, calculating probabilities of the virus mutating simultaneously around multiple therapies. Finally, the odds were in the patient's favor.

Computers can calculate those odds in a nanosecond, but they cannot formulate the question nor conceive the process by which to do so. Neither can Google. Earlier this year Dr. Ho expressed his concern to me about critical thinking in the wireless classroom. "We should be teaching our students to think creatively or to become innovators, not just test takers."[44]

David Skorton, another physician, formerly of the University of Iowa and now president of Cornell University, believes that students have been doodling since the days of chalk and slate. But he also wrote this to me in an e-mail: "The ability to check the weather or game scores or the headline news from their laptops during class puts an unprecedented barrier between the student and the instructor," impeding classroom discussion and the development of critical thinking.[45]

In sum, to restore critical thinking in the current generation, we must:

- *Inspire learners interpersonally as well as electronically in classrooms and advising sessions.* We must focus relentlessly on meaning and less on information. Again, it was Postman who noted that access to information is useless unless we know how to transform it into insight. The world remains incomprehensible now as it did in Postman's day. As he noted in 1990, "There is almost no fact—whether actual or imagined—that will surprise us for very long, since we have no comprehensive and consistent picture of the world which would make the fact appear as an unacceptable contradiction."[46] This may account for the pandemic of boredom reportedly suffered by Millennials in the traditional lecture. Inundated with information, they typically lack awe because, as Postman observed, "there is no reason not to believe" any ridiculous argument or hyperbolic thesis told to them in an environment of frivolous news. To counter that, we must emphasize the importance of discretion as an aspect of critical thought and teach students not only the alchemy of wisdom but also the ability to think independently, refine over-simplifications, exercise fair-mindedness, evaluate assumptions, listen and read analytically, and develop perseverance.[47] These are components of critical thinking that consumer technology typically undermines, as evidenced daily in the blogosphere.

- *Convey the importance of understanding technology philosophically and scientifically.* If we don't, students will only understand technology as consumers. Technology comes to us with Orwellian and Huxleyan motives mainly because it was developed by military or industry. The Orwellian is to *surveil*, the Huxleyan is to *sell*. Sometimes this happens simultaneously, which is why we surveil students on Facebook and why they order shoes in class on eBay. These motives are embedded in the interfaces and applications of technology, and we have an obligation to expose students to these data so that they can make independent decisions on everything from friendships to finances.
- *Explain the power of technology and the influence of marketing on everyday interactions.* We must remember why we purchased a product and then assess how we are using it, discerning whether it is advancing or hindering our priorities. Otherwise marketing will do that for us and give us new priorities associated with quarterly goals rather than with life goals. Many students are so debt-ridden that they have to take two jobs—one, to pay tuition, and another to pay the fees, upgrades and downloads associated with technology. They keep paying, too: to buy the device, to upgrade the device, to gain access for the device, to order products through the device, and to pay credit card interest because of the device. Multiple that by the number of devices that they own, usually a laptop, an iPod (or iPhone/iPad) and a cell phone. Multiply again by the number of media devices that they own, including interactive TV, DVD and gaming console, many of which are left on at home, school and work, padding the utility bill. As the national loan scandal illustrates, students are easy targets, addicted to the devices responsible in some part for the debt that will delete opportunities for real mobility—the financial freedom to establish how and where they will live.
- *Prepare the current generation to meet the challenges of a global environment.* To function effectively in that environment, they will have to interact face-to-face, especially when those faces are diverse culturally, ethnically, religiously, physically and attitudinally. They will have to be sensitive to other cultures and master discretion, diplomacy and self-awareness. Perhaps students can start by refusing to use consumer technology while strolling the grounds at their residential institutions or their own hometowns, paying attention to their surroundings and the

people in them rather than ignoring everything and everyone in their immediate presence, transforming them into an "it." This is the new objectification, and it is as demeaning as any sexist or racist one because ignoring others leads to ignorance. As such, we must model the behavior that we wish to see in our learners, opening our office doors and closing our laptops and cell phones, advising students in one-on-one sessions in our offices and conveying that they are worth our time and effort. Teachers also must be able to explicate the motive of interface or application and have the courage to know when technology detracts from content, devising other means to convey learning objectives to last a lifetime.

Otherwise the interpersonal divide will widen, delivering a dystopian world that will be mistaken as the path to self-knowledge and mutual understanding, as Postman prophesied, rather than as the path to ignorance and misunderstanding permeating and defining the current culture.

NOTE

1 Amanda Lenhart, "Teens, Cell Phones and Texting," Pew Internet & American Life Project, April 20, 2010; retrieved Nov. 11, 2010, from http://pewresearch.org/pubs/1572/teens-cell-phones-text-messages.

2 Bill Clinton, "President and First Lady," White House, retrieved Nov. 16, 2010 from http://clinton4.nara.gov/WH/EOP/html/principals.html.

3 Neil Postman, *Amusing Ourselves to Death*, New York: Penguin, 1985, vii-viii.

4 "Informing Ourselves To Death," by Neil Postman, speech, *the German Informatics Society, October 11, 1990, retrieved December 18, 2007 from* http://www.frostbytes.com/~jimf/informing.html

5 Johnny Cash died Sept. 12, 2003.

6 Arnold Schwarzenegger won the recall election on Oct. 7, 2003.

7 See "American Idol is 2003's #1 Television Show Online" at http://50.lycos.com/121103.asp.

8 See "Iraq War is 2003 #1 News Story Online" at http://50.lycos.com/121703.asp.

9 Ibid.

10 See "The InfoTech 100 Companies," Business Week, July 3, 2007, retrieved December 19, 2007 from http://www.businessweek.com/it100/2006/13.htm.

11 External reviewer of this chapter wrote in November 2010: "Again, do we want to date this chapter? ... Is there a way to make the point without being this specific with naming people or years?" Rather than delete dated material, the author chose to test his predictions in 2010.

12 "American Idol tops most-watched of 2008," by Tim Surette, TV.com, retrieved Nov. 16, 2010 from http://www.tv.com/american-idol-tops-most-watched-of-2008/story/12017.html.

13 Multiple sources: "The Top 10 of Everything in 2008," Time Magazine, http://www.time.com/time/specials/packages/completelist/0,29569,1855948,00.html; "Top 10 Showbiz and Celebrity Stories of 2008," The Telegraph, http://www.telegraph.co.uk/news/newstopics/celebritynews/3777498/Top-10-showbiz-and-celebrity-stories-of-2008.html; "For Tiger's best of 2008, it's not just about his triumph at Torrey," Bob Smiley, ESPN.com, http://sports.espn.go.com/golf/columns/story?id=3770145; "Missing White Woman Syndrome," Wikipedia, http://en.wikipedia.org/wiki/Missing_white_woman_syndrome.

14 Executive Sessions of the Senate's Permanent Subcommittee on Investigations of the Committee on Government Operations, 83rd Congress, First Session, Vol. 1, 1953, made public January 2003, available from http://a257.g.akamaitech.net/7/257/2422/06amay20030700/www.gpo.gov/congress/senate/mccarthy/83869.html Note from page 21 of the report: No witness was imprisoned for perjury, contempt, espionage, or subversion, although several were tried for contempt, with each case overturned on appeal.

15 See transcript, "A Report on Senator Joseph R. McCarthy," See it Now, March 9, 1954, retrieved December 19, 2007 from http://www.honors.umd.edu/HONR269J/archive/Murrow540309.html.

16 Edward R. Murrow, RTNDA speech, Chicago, October 15, 1958, retrieved December 19, 2007 from http://www.rtnda.org/resources/speeches/murrow.html.

17 Media chains grew more powerful during deregulation of the Reagan era and later via the Telecommunications Act of 1996, which eliminated many ownership rules.

18 Viacom, Time Warner, Disney, Bertelsmann, Murdoch's News Corporation, General Electric. See Ben Bagdikian's *The New Media Monopoly*. Excerpt available from http://www.benbagdikian.com/Docs/excerpt2.htm.

19 "Disney annual profits jump 33%," RTE Business, November 10, 2006, retrieved December 19, 2007 from http://www.rte.ie/business/2006/1110/disney.html?rss.

20 See "The State of the News Media 2007," retrieved December 19, 2007 from http://www.mediachannel.org/wordpress/2007/03/13/the-state-of-the-news-media-2007.

21 Eric Sass, "Updated: Southern Discomfort: Atlanta Journal-Constitution Cuts Newsroom," February 21, 2007, *Media Daily News*, retrieved December 19, 2007 from http://publications.mediapost.com/index.cfm?fuseaction=Articles.showArticleHomePage&art_aid=55841.

22 "Boston Globe cuts 24; Pulitzer Prize winners take buyouts," March 22, 2007, the Associated Press, retrieved December 19, 2007 from http://www.chron.com/disp/story.mpl/headline/biz/4653042.html.

23 See "The State of the News Media 2007."

24 David Miller, "Finding the Reason to Be: Creating a Durable Brand Identity," Ideas: The Magazine of Newspaper Marketing, p. 12.

25 "Excerpts from the Correspondence of Thomas Jefferson," Letter to Edward Carrington, Jan. 16, 1787; retrieved December 19, 2007 from http://www.cooperativeindividualism.org/jefferson_d_01.html#D10.

26 M.S. Fowler & D.L. Brenner, "A marketplace approach to broadcast regulation," *Texas Law Review*, 1982, Vol. 60, No. 2, p. 10.

27 Tobi Elkin, "Media Usage Tops Consumers' Daily Activities," Media Daily News, Sept. 27, 2005, retrieved December 19, 2007 from http://publications.mediapost.com/index.cfm?fuseaction=Articles.san&s=34518&Nid=15710&p=314136.

28 See "The American Freshman: National Norms for 1997" available from http://www.gseis.ucla.edu/heri/norms_pr_97.html and "The Chronicle Almanac" available from http://chronicle.com/weekly/almanac/2006/nation/0101901.htm#aster.

29 "New Class(room) War: Teacher vs. Technology," by Samuel G. Freedman, The New York Times, November 7, 2007, retrieved December 19, 2007 from http://www.nytimes.com/2007/11/07/education/07education.html?_r=1&oref=slogin.

30 "Informing Ourselves To Death."

31 Jacques Ellul, "The "Autonomy" of the Technological Phenomenon," in Scharff, Robert C.& Dusek, V. (Eds.), *Philosophy of Technology: The Technological Condition* (Malden, Mass: Blackwell), p. 386.

32 Diana Jean Schemo, "Grades Rise, But Reading Skills Do Not," The New York Times, Feb. 23, 2007, retrieved December 19, 2007 from http://www.nytimes.com/2007/02/23/education/23tests.html?_r=1&ref=us&pagewanted=print&oref=slogin.

33 "Scientists Study Internet Addiction," UPI report in *Science Daily*, Oct. 17, 2006.

34 Ramola Talwar Badam, "India tech institute curbs Internet use," Associated Press via Yahoo News, March 21, 2007, available from http://news.yahoo.com/s/ap/20070321/ap_on_hi_te/india_internet.

35 "Worried China bans new Internet cafes for a year," Reuters, March 7, 2007, available from http://www.reuters.com/articlePrint?articleId=USPEK24407720070307.

36 Nicholas Negroponte, "Beyond Digital," *Wired*, December 1998, available from http://www.wired.com/wired/archive/6.12/negroponte.html.

37 Ibid.

38 Erin Texiera, "U.S. Will Be More Diverse at 400 Million," *USA Today*, Oct. 21, 2006 available from http://www.washingtonpost.com/wp-dyn/content/article/2006/10/21/AR2006102100402.html.

39 Christine Rosen, email interview, personal correspondence, November 11, 2005. Other quotations are taken from the same interview.

40 "Wireless Networks Reach Half of College Classrooms," The 2006 National Survey of Information Technology in US Higher Education, October 2006, available from http://www.campuscomputing.net/.

41 See Edna LeShan's "The Sesame Street Syndrome," *The Humanist*, Nov.-Dec 1972, pp. 9–11.

42 Theodore Roszak, email interview, personal correspondence, October 11, 2006. Other quotations are taken from the same interview.

43 See "David Ho Interview," Academy of Achievement, May 23, 1998, retrieved December 19, 2007 from http://www.achievement.org/autodoc/page/hoa0int-2.

44 David Ho, email interview, personal correspondence, November 21, 2006.

45 David Skorton, email interview, personal correspondence, November 22, 2006.

46 "Informing Ourselves to Death."

47 See Jill Geisler, "Critical Thinking: What Do You Mean By That?" Poynteronline.com, available from http://www.poynter.org/column.asp?id=34&aid=81581

Michael Bugeja
Greenlee School of Journalism and Communication,
Iowa State University

CHAPTER 4

SUVI TALA

THE NATURE OF TECHNOSCIENCE (NOTS)

INTRODUCTION

In science education an important component of understanding exists that encompasses not only the content of science but also the broader issues related to the production and justification of scientific knowledge, as well as understanding the impact of science in society, and *vice versa*. These aspects are often referred to as the Nature of Science (NOS). Recently, in addition to NOS there has been both a parallel interest in the Nature of Technology (NOT)[1] as well as suggestions to deal with this in education (e.g., ITEA 2000/2002/ 2007; Project 2061; NRC 2012, see also Jones et al. 2011). Curiously, these issues, although closely related, have not yet been seen from a common viewpoint, nor as having an equal footing, although such a unifying viewpoint is quite obvious, considering that the interaction between NOS and NOT would support understanding both NOS and NOT content and thus help with problems met in implementing them in education. Irrespective of the century-long history of the place of NOS in many curricula, the NOS objective has not yet been reached by students of various ages, teachers, or the general public (see e.g., Driver et al. 1996; Khishfe & Abd-El-Khalick 2002; Lederman et al. 2002; Lederman 1992; McComas et al. 1998; Osborne et al. 2003 and their references) and similar difficulties have been encountered when implementing the newcomer, NOT (see Bame et al 1993; Jones 1997; Daugherty & Wicklein 1993, see also Compton & Compton 2011).

The difficulties in learning about NOS and NOT is here suggested to be accelerated by the fragmentation of education and a taken-for-granted abstract view of NOS and NOT. NOS and NOT are often taught in separated science and technology lessons – and rarely in integrative settings with history and philosophy teachers, for example. Indeed, NOS and NOT contents should be integrated within the scientific and technological content under study, because the abstract NOT and NOS can be understood and the relation between them seen, when those are contextualized. Such contextualized NOS and NOT understanding would also support understanding of the scientific and technological content under study and the application of the knowledge learned. Thus, what is needed is a concretization of the general, abstract NOS and

Michael P. Clough, Joanne K. Olson and Dale S. Niederhauser (Eds.), The Nature of Technology: Implications for Learning and Teaching, 51–84.

NOT ideas in the concrete practices of science and technology for education, and to study the relationship between these ideas in practice. Development in this direction is supported also by the recent curricula frameworks (NRC 2012).

The kind of new context-sensitive view suggested here can be reached by using science and technology studies to support the design of didactic solutions (Clough 2011; deVries 1996; Matthews 1994; Metz et al. 2007; Sandoval 2005). The present chapter scrutinizes the relationship between NOS and NOT from the viewpoint of recent philosophical and historical analyses of the actual activities and practices of science, scientists, and engineers (e.g., Chang 2004; Ihde 1979; Kroes 2003; Mitcham 1994; Rothbart 2007; Tala 2011; Vincenti 1994), where a close – even intertwined – relationship between technology and science is admitted and taken as a starting point. This kind of scrutiny reveals the connections between scientific and technological progress in the development of conceptual and theoretical structures as well as the material resources available for experimentation and instrumentation in science and engineering, thus bringing forward a new view of *technoscience*, the term suggested by Latour (1987). The previous view has widely taken into account the material, economic, social and psychological dimensions of the unification of science and technology. However, its epistemological impact needs to be studied further (Tala 2009) – and in this way its scope extended for science and technology education.

The intertwined nature of the development of science and technology is easiest to observe in the science of the 21st century. Nowadays a large field of research and development exists that cannot be considered pure science or pure technology, but rather an area in which the two are merged. It is apparent that "Big Science" – such as, for example, high energy physics – is so closely tied to big technological machines that one can meaningfully speak of a single, complex phenomenon which is at the same time science, scientific technology, and technological science: *technoscience*. "Big Science," of which large research groups involved with the scientific design and testing of experimental machines, accelerators, and detectors (see Baird 2004; Galison & Thomson 1999) serves as a good example, and is very technology driven; here pure science and technology are nearly inseparable. Consider, for instance, the recent research and design efforts in connection with the Large Hadron Collider (LHC). It is not easy, or even worth trying, to say which roles are played by technology and which by science when producing and filtering experimental data or when designing, constructing, and running the computer models of the expected mechanisms underlying the phenomena for comparison. While the LHC technology is large, the same intimate relationship is also known in the science where the subject of study is the very small; the research projects in nanoscale, in particular, are very often interdisciplinary, transdisciplinary or even super-interdisciplinary (see Schummer 2004). Indeed, new methods developed by "nanophysicists" and "engineers," for example, are used by chemists or biologists, or in integrative projects where one cannot say who advances biology, who chemistry, who physics, and who the computer sciences (Tala 2011). Further, consider such fields as genetic engineering, cybernetics, food and environmental sciences or cognitive science. Similar examples can be found also in diverse areas of "applied physics and chemistry," such as in geo- and biophysics and biochemistry, as well as atmospheric sciences, research on soft and hard condensed matter, low temperature research, brain research and nuclear engineering, to mention only a few examples, where (in traditional terms) natural science is motivated by and merged with technological research or – to put it the other way around – technological research is advanced and developed by natural science.

But it is not only at the level of 'big science' and 'big technology' and in the new fields of research and development, where the unification happens. The need for co-operation between natural scientists in different fields and engineers can be seen also in the areas of development that individuals deal with in everyday life, such as communication technologies or the web- and mobile-based technologies, automotive technology, and aeronautics; in its original usage the term *technoscience* frequently referred to the hybrid that resulted from the fusion of science and industry. The merger of science and technology can be seen to take place also in everyday research, where simpler instruments or experimental set-ups are used: In the history of modern laboratories, scientific knowledge was not simply "discovered" from nature as obvious facts, but has been painstakingly constructed through careful and well-planned experimentation and the accompanying interpretation of the events, which is nowadays often supported by computer modeling and simulations. In that process, the available technological knowledge and practice for the design of laboratory or virtual set-ups and their manipulation for control are crucial. This kind of experimentally "based" natural science is the focus of this chapter, used to analyze the connections of NOS and NOT. Of course, the natural sciences also aim to answer research questions where experimentation is limited – if even needed (e.g. taxonomy) – and where theoretical considerations play a fundamental role. Respectively there are also such tasks in the wide field of technology, which are not directly advised or advanced by science, and the nature of which is discussed in other articles.

At the concrete level it is unquestionable that parts of our physical and chemical world are instrumentally and technologically revealed in experimentation and modeling, and *vice versa*: even in daily news we can learn of examples of how technological development is frequently improved by scientific research. Nevertheless, in scientific knowledge construction through experimentation and modeling, technology and design do not only gain a methodological means of support, but they gain, rather, a highly cognitive role. In this way, technology not only promotes scientific progress, but also unknowingly affects and changes scientific reality. The role of technology at higher levels of all scientific theorizing is often not as apparent, and few philosophers have focused on this question. An early exception is Martin Heidegger (1927), for whom technology is humanity's way of being in the world; it is not a means to an end, but rather a mode of human existence. Thus, Heidegger extends the idea of technology-driven science probably most radically, by holding that what we take to be science, even at its most theoretical core, is an effect of a technological way of apprehending things, of "revealing a World" technologically. Finally, technology also figures out the ways of informing public and the information and views shared in such popularization about scientific and technological research (cf. Postman 1985) – and then the discussion about it, which increasingly takes place in the social media embodied in technoscientific artifacts.[2]

Nevertheless, science and technology education, both as distinct and integrated subjects (Layton 1993) still widely rely on and teach a traditional conception of science and technology as two fundamentally different and distinct enterprises[3]: as a consequence, NOS and NOT are explained and taught separately. It is time to consider the bi-directional, comprehensive *technoscientific* relation between science and technology, in technology and science education as well. A natural place to start this revision is in scrutinizing the recent NOS and NOT components of education, because the nature of the subject under study is a natural starting point of educational planning. This chapter thus considers the views from the worlds of current science education and present design education as two different perspectives on the same concerns, the unification of which would benefit both. This *technoscientific* view, where technology acquires a crucial role in scientific progress, and conceptual and theoretical structures an essential role in technological progress, guides embodying NOT and NOS themes in the practices of science and technology for education in a fruitfully unifying way. Such unification has the potential to simultaneously support learning about science and technology and about the nature of science and technology.

THE RELATIONSHIP BETWEEN SCIENCE AND TECHNOLOGY IN EDUCATION

There have thus been two parallel objectives in liberal education, both capturing the basis of scientific technological development, with one aiming to disseminate understanding about NOS and the other to disseminate understanding about NOT. In the past decades, numerous attempts have been made to define both NOS and NOT, for education. Typically what is meant by NOS are the features of scientific knowledge, values, and assumptions assumed to be inherent to science, to conceptual inventions in science, or in general to epistemology of science (see, e.g., AAAS 1993; Driver et al. 1996; Lederman 1992; Lederman et al. 2002; Matthews 1994; McComas & Olson 1998; NRC 1996). Respectively, in the field of education, the nature of (engineering) technology refers to features of technological action or design, values inherent to the design process, and to skills; it may also refer to practical knowledge employed in design processes (e.g., AAAS 1993; Arthur 2009; Cajas 2002; ITEA 2000; Jones 1997; NRC 2012; NAE/NRC 2002; NRC 1996). Indeed, NOT emphasizes the special relationship between technology and society and the diversity of technology's fields (see, e.g., AAAS 1993; ITEA 2000; Rasinen 2003, Shaw 2002).[4]

Understanding these issues has been considered an important part of scientific and technological literacy (STL, see e.g. NRC 1996; Hodson 2008 and the reviews by Cajas 2001; Jones 2011; Miller 1983, and Thomas & Durant 1987). The objective is also defended in terms of instrumental reason – understanding what one is learning or teaching about helps to learn and teach it better (see Abd-El-Khalick 2012; Jones 2011) – but even more in terms of democratic, moral, cultural, and economic reasons (e.g., AAAS 1993; Arthur 2009; Dewey 1916; Driver et al. 1996; Hansen 1997; NRC 1996; NAE/NRC 2002; McComas et al. 1998; Rudolph 2005). The main motivation of this chapter is that deep conceptualization and conceptual understanding – the traditionally valued goal of all science and a newcomer in technology education – includes not only remembering the scientific ideas (product of the process) but also understanding, which constitutes the area of application of this knowledge and, thus how we acquired that knowledge or how it can be justified. How the knowledge is produced often also defines the area of the application of that knowledge.

Since there have been a variety of views about why NOS and NOT should be included in education, there has also been a diversity of views concerning what should be included as NOS and NOT content and how this should be taught. In most countries, science education and technology education are organized separately, as independent subjects, but promising integrative projects also occur. Nevertheless, the views promoted as regards the relationship between science and technology (or NOS and NOT) in education have been (at least) one-sided.

NOS and NOT themes

For numerous interest groups, the discussion about what we should teach youth about science and technology has been lively. In order to dissociate NOS and NOT from metaphysical or political commitments, the focus has been on listing *typical* characters of science or technology as NOS and NOT *themes* appropriate for overall teaching,[5] which has been done in multidisciplinary co-operation. The dividing of NOS and NOT into themes helps to focus the discussion concerning different views on the nature of science and technology. At best, the teaching and discussion around NOS and NOT themes helps students to recognize the viewpoints they have adopted before and may also help them to revise their views in the light of new experience. Nowadays there is a kind of consensus among science education experts on the basic, simplified set of formal NOS themes – and some kind of consensus on NOT themes – that students should understand upon leaving school (Driver et al. 1996; Lederman et al. 2002; Matthews 1994; Osborne et al. 2003; Sandoval 2005, and cf. AAAS 1993; ITEA 2000, 2006; NAE/NRC 2002; NRC 1996; NRC 2012). These themes are included in many curricula and various science education reform documents worldwide (for exemplary comparisons, see Cajas 2002; McComas & Olson 1998; Osborne et al. 2003; Rasinen 2003).

Basically, most recent lists of NOS themes mention *the importance of the empirical basis of scientific knowledge* (Lederman 1992; Lederman et al. 2002; McComas

& Olson 1998; Osborne et al. 2003; Sandoval 2005), highlighting thus the role of empirical experience and experimentation in science. Natural sciences can even be seen as experimentally-laden (Chang 2004; Hacking 1983; Ihde 1979). On the contrary, the basis of technology does not lie in "nature" but in human creativity, and, indeed, the aim of the technological process is to change "nature": one NOT theme states that *technology extends our abilities to change the world* by manipulating and controlling nature *to meet human needs* (see, e.g., AAAS 1993; NRC 1996; ITEA 2000, 2006; NAE/NRC 2002). Thus, indeed, the essence of engineering is seen to be *the design of artifacts under constraint* (material, economic etc., NOT theme).

The importance of methodology is mentioned in both NOS and NOT lists, but neither science nor technology education practices reflect the picture portrayed by NOS and NOT lists: Contrary to previous NOS views and the practices of science education, which promote the myth of the scientific method (cf. Chinn & Malhotra 2002; Hodson 1996; Metz et al. 2007), in present NOS lists it is said that there is actually a variety of methods (cf. Feyerabend 1975; Nola 1999). Likewise in the methods of science, a variety of methods are used in technology: in NOT descriptions different design methodologies are highlighted[6] (AAS 1993; ITEA 2000, 2006; Mawson 2003; NAE/NRC 2002; Rasinen 2003). Anyway, in technology education, design has often been taught as algorithmic problem solving (for examples see e.g., ITEA 2000; Johnsey 1995; Zubrowski 2002); this step-by-step approach has been much criticized (see deVries 1997; Jones 1997; Johnsey 1995; Johnson 1997; Mawson 2003; McCormick 2004 and the references therein).

Technology is often seen as a third kind of culture, placed between the natural sciences and the practical arts. Then in NOT it is emphasized, that human *creativity is at the heart of technological design*. The same can be said about science: Against the romantic view of science describing human-independent Nature, the enterprise called science, as well, and the nature of science is inherently bound to us as human beings. Among NOS themes, *scientists' creativity* is often mentioned. Indeed, in NOT descriptions, the innovative nature of design is emphasized; inventions and *innovations* are being seen as *the results of special, goal-oriented research* (AAAS 1993; ITEA 2000, 2006; NAE/NRC 2002). Thus, the development of technological designs is not a history of trial and error but something frequently *advanced by technological and scientific knowledge* (AAAS 1993; ITEA 2000, 2006; NAE/NRC 2002).

Understanding the scientific knowledge has traditionally been at the focus of science education, while technology education has focused more on doing. Thus, among NOS themes many ideas linked to scientific knowledge have been suggested, such as *theories differ from laws*. The respective categories in technological knowledge – and their relations – seem not to be recognized or discussed among NOT themes. Nevertheless, recent educational literature notes the conceptual objectives of technological processes; namely understanding some combination of technical and technological knowledge has been highlighted (see Johnson 1997; Jones 1997; Jones et al. 2011; McCormick 2004; Mitcham 1994). However, it is not clear what technological knowledge should be taught in all-round education. No particular discipline underlies the content of technology education, but rather a multiplicity of disciplines, ranging from auto mechanics and automatics to the other fields of technological research and development and even to sociology and philosophy of science. Assistance in selecting the knowledge base of technology education has been sought, for example, from the rather young field of philosophy and history of technology (deVries & Tamir 1997; deVries 1997; Jones et al. 2011; Mitcham 1994). Later in this article, moreover, knowledge in engineering technology is examined.

In science, the relation between theory and practice is not as apparent for an outsider as it is in engineering. Thus the *theory-ladenness of experimenting* is listed as a NOS theme (Lederman 1992; Lederman et al. 2002; McComas & Olson 1998; Osborne et al. 2003): scientific action – from theorizing to modeling and experimentation – always takes place in a theoretical framework. The conceptual frameworks guiding the scientific processes are not ready, but under development. This is referred to in NOS themes by the *certainty and different forms of scientific knowledge* (Lederman 1992; Lederman et al. 2002; Osborne et al. 2003; Sandoval 2005). Among NOT themes, a need for a similar theme has not been taken up. Instead, this idea can be seen in the background of the theme *all technological systems can fail* (a NOT theme; e.g. AAAS 1993; ITEA 2000; NAE/NRC 2002): when a system fails, the underlying knowledge of its development has to be developed. The justification of technological knowledge lies in its practical functionality: the technological claim is 'true' if it guides one in successful action (cf. deVries 2011).

Finally, enterprises called science and technology do not take place in a vacuum, but are deeply rooted in societies. Both *technology and science are essentially global phenomena* having both global and local influences (e.g., NAE/NRC 2002). Social, psychological, economic, and cultural forces shape technological development, and *vice versa*: technology shapes the modes of human existence, including thought and action; these issues concerning *the special interaction between technology and society* have been a focus of NOT (AAAS 1993; ITEA 2000/2002/2007, 2006; NAE/NRC 2002; Rasinen 2003). In fact, technology not only shapes the processes of society, but also shapes the ways individuals see the world; thus it also influences the ways individuals think about and act in the world. Respectively, the practitioners of science are the products of the culture(s) in which the enterprise is embedded.[7] Hence, *the creative role of human beings, societies and cultures in* constructing *science* and, in turn, *the impact of science on cultures and societies* is frequently mentioned among NOS themes in research reports and in curricula (Lederman 1992; Lederman et al. 2002; McComas & Olson 1998; Osborne et al. 2003; Sandoval 2005).

The above introduced lists of NOS and NOT themes portray quite an abstract view of science and its relations to society, because it is the consensus view. Beyond these general characterizations, science educators and philosophers, historians, and sociologists of science, as well as scientists themselves, are quick to disagree on specific explanations of the NOS themes. What one understands a particular NOS or NOT theme to mean depends upon the epistemological views (s)he has adopted or the scientific context under consideration.[8] In any case, science and engineering as such do not need to commit to any particular epistemology; the answers to 'why' and 'what' questions of science and engineering are reached by answering an iterative series of 'how' questions under the guidance of flexible epistemological ideas (for example, see Tala 2011). Thus, any particular definition of a NOS or NOT theme is as tentative, if not more tentative, and context-dependent than scientific and technological knowledge themselves. Moreover, teachers' metaphysical commitments typically differ from students' views (see Chinn & Malhotra 2002; Lederman 1992). Thus, the lists of NOS and NOT themes can provide a fruitful basis for educational discussions, while those portray different aspects of scientific and technological process. But then we need practical frameworks guiding understanding of NOS and NOT themes, which are indeed sensitive to different, shared viewpoints and contexts of sciences and engineering. The concrete scientific and technological practices, for example, are rarely considered in NOS and NOT materials; even the practices figure out the nature of these enterprises. Moreover, the school-lab often does not reflect the NOS and NOT ideas that are being taught (e.g. Chinn & Malthora 2002; Clough 2006). Thus, what is the nature of the relations between NOS and NOT, as well as science- and technology-based elements in present education?

The Relation Between Abstract Themes and Educational Practices

It has been articulated that by recent ways of studying science and technology, students have not reached the aspired level of knowledge about the nature of science and technology. Thus, at the heart of many arguments, which defend the place of NOS and NOT in the curricula, is the idea that students need to be explicitly taught about these issues. A challenge of this kind of approach is the explanation of the NOS and NOT themes for understanding (e.g., Clough 2006). The NOS and NOT frameworks cannot be understood without contextualizing – and thus, historical and also some contemporary science stories have been developed in order to help teachers explain the abstract ideas (e.g., Clough 2011; Begoray & Stinner 2005). In explaining by stories, one has to be careful not to turn teaching into indoctrination (see Feyerabend 1975) or give a student the role of a spectator instead of somehow being immersed in the scenario.[9] This kind of explicit pedagogical approach to teaching NOS and NOT can, however, provide a fruitful context for reflective, inquiry-based learning (e.g., Dagher et al. 2004). When NOS and NOT issues are closely linked to the conceptual contexts under study and to the hands-on activities, studying NOS and NOT contents decrease the fragmentary character of science education. Respectively, different results are reached when students' NOS and NOT views are studied as contextualized: Instead of asking definitions of general concepts, by such questions as "what is science?", "what is technology?", "what is an experiment?" and "what is the relationship between science and technology?" (cf. Lederman 2007), one might ask the role of certain kind of knowledge in the task at hand, for example.[10]

In addition to the explicit approach to NOS and NOT, among science and technology educators an implicit view emerges, which proposes that the NOS and NOT views are best absorbed through the practices of inquiry and design, which indeed is the primary means to express and investigate them (for examples, see Khishfe & Abd-El-Khalick 2002 and also Jones 1997; Sandoval 2005). That view supports the development of experimental or design activities where students "imitate" scientists and engineers – consider, for example, the "discovery science" approach (Hodson 1996). Such an approach develops the analytical skills for scientific and technological literature: STL does not imply only knowing that 'science is theory-laden' and 'innovations are results of special, goal-oriented research,' but being able to assess particular claims encountered in everyday life and in public discourse. That being said, however, the epistemological authenticity of science and technology education is often poor: most scientific inquiry or design tasks given to students in schools reflect neither the core attributes of authentic scientific or technological reasoning (e.g., Abd-El-Khalick 2012; Chinn & Malhotra 2002; deVries 1997) nor the contextual NOS and NOT themes of education. For example, in the formal lists of NOS themes it is frequently mentioned that science is a creative and theory-laden enterprise constructed by collaborating human beings through employment of a variety of methods; nevertheless, the practices of science education typically support a view of science as an algorithmic activity employing a universal scientific method and neutral instruments, which transform the facts awaiting us in nature (Chinn & Malthora 2002; Hacking 1983; Hodson 1996). Respectively, when the step-by-step design or general problem solving schemes are used as the basis for hands-on tasks in technology education, the teaching does not reflect the authentic practices and formal NOT themes (e.g., deVries 1997, 2011; McCormick 2004).[11] By revising the practices in science and technology classes, students' views concerning NOS and NOT may be improved. Indeed, students cannot be little scientists and engineers creating new knowledge, but instead they are studying science and technology, which has been developed during the past centuries by numerous experts. Empirical research indicates that students' conceptions cannot be changed through practice only (e.g., Abd-El-Khalick & Ledermann 2000; Jones 1997; Ledermann 1992; Richard et al. 1998).

In sum, at best, the objectives of NOS and NOT education guide approaches and methods used in teaching and evaluating. By employing the explicit or implicit approach alone we cannot reach the typical level of NOT and NOS objectives, which imply that students not only remember abstract NOS and NOT ideas or are competent to act in particular school tasks, but require also understanding the ideas in a way which supports conceptual understanding and application of that knowledge (cf. Ledermann 2007). The relationship between practical and formal NOS views as well as teaching and learning NOS and NOT have been found to be very complex; the empirical evidence argues for the middle ground (see, e.g., Abd-El-Khalick 2012; Akerson et al. 2000; Clough 2006; Johson 1997; Lederman et al. 1992; Sandoval 2005; Schwartz 2004; Palmquist & Finley 1997). Moreover, several arguments emphasize an understanding of the social and political nature of scientific and technological progress (e.g., AAAS 1993; Bijker et al. 1989; Latour 1987; Hansen 1997; Rudolph 2005). In this article the presented context-sensitive framework, the nature of *technoscience*, is a view of practicing scientists and engineers, which is simultaneously a bridge to both formal and practical epistemologies.

It is through the practices of knowledge construction that NOS and NOT become inherently bound together in research fields and in education. Thus, the students' picture of the nature of science and of technology and the relationship between these disciplines is influenced by reflecting on both the educational content and study methods employed. Indeed, an inherent part of such a process is guiding students in recognizing their own views (e.g., Clough 2007) about NOS and NOT and the views highlighted by different kinds of educational material used. The view of recent education as regards the relationship between NOS and NOT is to be discussed next.

The Relationship Between NOT and NOS in Recent Education

Science and technology studies have basically discussed two parallel approaches of the science-technology relationship: science-driven technology, in which 'pure' science is seen to be 'applied' to practical solutions, and the converse, technology-driven science, where science is seen to advance through technological progress. In the latter view, technology is mostly considered as a concrete methodological means, as tools and apparatus, from telescopes, compound microscopes and electron microscopes to devices used for constructing the special laboratory environments for phenomena, such as Boyle's air pumps and their more advanced successors (nowadays, we have vacuum science), to quantitative measurement-enabling instruments such as clocks, thermometers, the galvanometer and electrometers, as well as to a very special class of technological measuring devices used in sensor technology, including advanced products like the detectors in the LHC at CERN. In addition, the NOT and NOS descriptions mention the role of technology in science by highlighting the role of these types of instruments (e.g., AAAS 1993); the role of instruments would be a natural place to consider the role of technology also in school laboratory practices.

In education, the role of science in technological development is seen mostly as a "cathedral of knowledge" (Layton 1993; Fensham & Gardner 1994) to be applied in practical solutions.[12] This view has been opposed the most vigorously by technology educators; however, as described below, it is the view of a typical integration project. The opposite view, "the materialist," where science is seen to be nothing more than a field of technology (e.g., Latour 1987; Janich 1978), is rarely presented in educational literature. Indeed, quite a few authors of science and technology studies have developed interpretations in which science is conceptualized as basically similar to technology. This view does not occur in education. Although these positions could be seen as complementary rather than exclusive, it is somewhat surprising to find that this is not the situation in the educational literature.

Alongside the hierarchical views, educators have been encouraged to omit "the interactionist" view, where science and technology are seen as working together like "company stores" (Layton 1993; Fensham & Gardner 1994). This view is easily supported by examples from the recent past, such as gene technology, bioengineering and the other examples mentioned in the introduction, but here again the concrete integrative educational tasks oversimplify the situation. In favor of the interactionist view are established projects in which students are asked to develop and improve items of everyday technology, such as miniature bridges or moving toys (for examples, see Roth 2001; Sadler et al. 2000; Zubrowski 2002). In this way they are supposed, by applying scientific knowledge, to enhance their understanding of that knowledge, or to learn more by testing and developing these material artifacts. These projects bring forward an important emphasis on the interaction between science and technology and are touted as motivating students (Layton 1993). Nevertheless, in reality, the path from scientific knowledge to practical devices and from the skilful action of practitioners to scientific knowledge is "non-linear" and in authentic situations it takes a long time. Thus, despite the promise of this kind of approach, the teachers attempting it have been faced with many challenges (e.g., Kolodner et al. 2003; Sidawi 2007).

To promote a more authentic view of the relationship between NOT and NOS, we need contextualized views and educational approaches reflecting these views. Looking at science and technology as practices, indeed as formalized and creative practice, links science to culture, because culture is a set of socially accepted practices and values: on one hand, it is part of the scientific culture and on the other, part of entire societies. Internationally, contextualized approaches and science-rich classroom-based studies showing how teachers and students represent the NOS propositions and support them to embody NOS at a particular level of education are only now emerging; this chapter continues in this direction.

A closer look at the following reveals an unexpectedly deep, bi-directional relationship between the development of science and of technology: The two interact and grow together, whereby science promotes technology and technology promotes science. This interaction is so active that, finally, in the practices of science and technology it is often difficult to separate part that is science from that which is technology. Since this has been widely studied in the sociological literature, the views therein are already considered in the field of education from sociological viewpoints, for example in such movements as STS (science-technology-society), STSE (science-technology-society-environment) and SSI (social scientific issues).[13] In education, it is time to consider especially epistemological and cognitive views of *technoscience*[14] as well.

PHYSICS AND ENGINEERING AS FIELDS OF TECHNOSCIENCE

Traditionally, it is thought that scientists discover plain facts, universal laws which are already waiting for them in nature and engineers create new technical artifacts.

However, in laboratories we do not find scientists "observing" pure nature, not more so now than in early experimental laboratories and institutions. Instead we see them actively and intentionally creating and designing experimental settings, instruments and machines, which produce or isolate interesting phenomena, which do not exist outside the instruments and machines as such. Thus, science has many times progressed through producing artifacts – both material and conceptual set-ups and instruments: experimental scientists create phenomena by using the scientific instruments of very special design towards that purpose (Hacking 1983) and in interaction with material artifacts they produce also conceptual artifacts. The reliance on the development of scientific technology is nowadays easy to see also in computer modeling, which is increasingly used in natural sciences as a research method (for examples, see Sundberg 2006; Tala 2011).

Also, as the employed technological knowledge becomes developed, the products of technological design are not only technical artifacts but also new methods and technological knowledge. Through experimental design, engineers develop our practical capabilities to construct, manipulate, and control the material systems and knowledge of these capabilities. The knowledge of the capabilities of a particular system, indeed, develops in interaction with more general scientific and technological knowledge. On a concrete level, on the one hand, the capacities of technological capability lead to the productive capacities of experimental design and these capacities become developed in these experimental processes of science and technology, and, on the other hand, the capacities of the experimental systems under consideration are revealed by technological instruments. Furthermore, such technological instruments are designed to measure scientific quantities and thus it can be said that the scientific understanding is established and developed in them as integrated with the technological ability and limitations. This is in contrast to the traditional view of education seeing technology as an application of ready scientific theories. It is also in contrast to an oversimplified empiricist view on the role of technology in science, where experimental technology is only seen as a means for collecting new data from Nature; that is to say, once the new data has been produced, technology has played its part and the real scientific work – theorizing – may begin. In regard to the knowledge and skills developed in the design processes of science or engineering, it is often useless to try to say which of those parts are technological and which parts are scientific.

These notions suggest looking for unifying views from technology and science, which take into account and help to better understand the inevitable intertwined nature of these fields. This can be provided by *technoscience*, which acknowledges the most experimental, natural, and engineering science as an amalgamation of science and technology. A product of *technoscience* is not only an understanding about the regularities of the physical and chemical phenomena of the natural and artificial environment, but that also necessitates and provides the capability to create phenomena and design ways to control and manipulate them. Through experimental technology, a "technological window," scientists gain access to the parts of nature that would otherwise remain hidden due to either the human being's shortcomings or to contingent boundary conditions prevailing in our universe. Through experimental design and research, a "scientific window," engineers advance the development of material and methodological artifacts better than what otherwise would have been possible. Thus, by studying scientific and technological reality and simultaneously producing it, the *technoscientific* process embodies the dialectic between the two trajectories of the science-technology relationship: science-driven technology and technology-driven science.

In such a view, technological devices and the phenomena they produce are also part of scientific research and of scientific interest; moreover scientific design also interests engineers. The relationship between science and technology is more closely studied in the following in the limited context of the practices of natural, experimental sciences and engineering science, and by concentrating on the epistemological issues. These epistemological viewpoints, namely how the new knowledge under study is reasoned, cannot be overlooked in education and it is a central question among researchers (e.g., Tala 2011). In a *technoscientific* view, the dialogical tension between the two mentioned trajectories – that of science-driven technology and technology-driven science – is considered to be a primary motor of scientific and technological progress for most parts of experimental natural sciences, as well as a part of engineering research, but in some areas it is more visible than in others.

Technology not only encompasses a methodological role in experimental natural sciences, but in addition encompasses an epistemological and cognitive role, even to the extent that it affects our ontological positions. It is easiest to see how technology affects our ontological positions by considering scientific phenomena or artifacts, which do not exist outside of laboratories as such, for example Hall's effect, pure chemicals, or many nanostructures. Indeed, this new perspective shifts the object and processes of scientific research nearer to the ones of technological research; the challenge of scientific research is not only in highly theoretical considerations but even more often in the ability to construct and control the technologically staged laboratory-phenomena. This also supports understanding about the cognitive aspects of design and the variety of knowledge employed and developed in mutual interaction with material innovations. In such a manner, the *technoscientific* view improves both technology and science education (studied more closely in the last subsection).

DESIGNING TECHNOSCIENTIFIC REALITY

Science and technology studies have quite recently paid increasing attention to experimentation and design and, in consequence, they have highlighted the dialectic between the history and philosophy of science and the history and philosophy of technology.[15] Experimentation is used by natural scientists as a means to actively interfere with the material world, in order to acquire knowledge, which is then used to support the more hypothetical generalizations. If these generalizations can be used as a basis for successful predictions, through consequential justification, the circle is then closed and new scientific knowledge is acquired. Design, the central process of engineering, includes management of constraints, the product of which is not only the material ability but also knowledge concerning these processes of construction, control, and manipulation. A design process starts from the analysis of objectives and constraints and aims to satisfy the non-negotiable constraints and optimizing those that are negotiable.[16] During the design process of engineering, both the knowledge about the natural order underlining the functioning of the device and the knowledge about constraints are sharpened. Also the very success of the experimentation lies in the control and manipulation of material laboratory phenomena. Then, experimental design is used by engineers and scientists as a means to actively interfere with the material world, in order to reach new solutions in the form of methods and knowledge. It is these types of conceptions of experiments and design, which provides the *technoscientific* view.

In actual laboratories, we find researchers or engineers not following a stepwise-sequenced problem-solving or universal scientific method, as proposed in the traditional views, and as often displayed in science and technology education. There is no universal method to follow. Instead, the different methods of experimentation and design are developed and become frequently improved in the experimental design, when scientists and engineers tailor laboratories to the needs of their research projects. In the same process they develop the scientific and technological understanding. Thus, the experimental design work aims to develop methods as well as individual scientific or technological claims in a cyclic and iterative *technoscientific* design, which provides creative and critical planning, construction, and development of material settings in interaction with the developing understanding of it (figure 1). In the *technoscientific* design, knowledge become practically and socially justified: the justification of the designed experimental systems and instruments lies in the pragmatic, scientific determination that they work. Finally, the *technoscientific* design process intertwines the development of science and technology – considered both as theories and as action.

Constructing Scientific and Technological Knowledge

Scientific knowledge is conceptual and often hierarchical in nature: Concepts are defined in relation to other concepts in the same conceptual system – and the scientific principles are expressed by using these concepts – or by the measurable counterparts, quantities. Indeed, the knowledge can be organized hierarchically from specifics to generalities. Textbooks concentrate typically on the established scientific knowledge structures, which can be applied to numerous contexts, such as scientific principles and laws of different kinds. Abstract technological knowledge, however, especially abstract descriptive technological knowledge, and its hierarchical nature, are less known. The following stages of engineering knowledge have been found through case studies (e.g., Vincenti 1990) and philosophical analysis: 1) technical

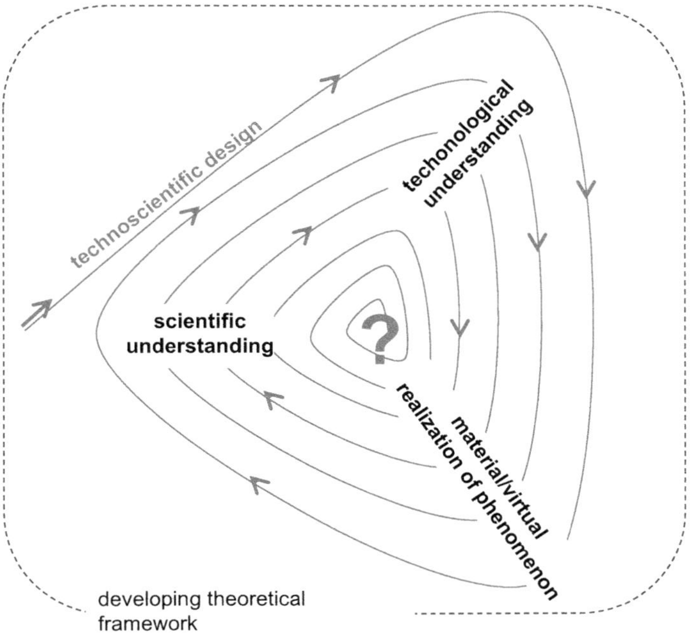

Figure 1. An iterative design process aims to develop methods, technoscientific artifacts and knowledge about it.

know-how including sensorimotor skills or technemes, 2) functional rules or "rules of thumb" and structural rules, (Mitcham 1994; Vincenti 1990), 3) technological laws, and 4) technological theories.[17] This is a quite imprecise categorization, and sublevels naturally exist; for example, some technological laws are similar to scientific empirical laws and the ones derived from theory.

In the case of technological knowledge and thinking it may be more apparent than with scientific knowledge that this knowledge is not only conceptual, but can also be symbolic and embodied in material artifacts and techniques (cf. Baird 2004; Rothbart 2007); we have become familiar with the vocabulary of points, lines and shapes used in diagrammatic reasoning through the different sources including engineering sketches and final blueprints. Additionally, just like in science, implicit, "tacit knowledge" (Polanyi 1966, 1958) also exists in engineering, of which the case of the TEA laser is a good *technoscientific* example. Researchers could not transfer from Canada to Britain the expertise needed to build a TEA laser through plain text; instead, they needed the concrete experience gained through visiting in the group to be able to build it. Successful experimentation and design requires special knowledge and training in special experimental techniques. In this particular case, the successful transfer of "tacit knowledge" necessitated personal contact with an accomplished practitioner, and as the knowledge was "invisible;" the scientists and engineers did not know whether they had the appropriate ability to build a laser until they tried (e.g., Collins 1985). In addition, both scientific and technological processes require a kind of sociotechnological understanding.

Engineering knowledge, the kind of technological knowledge concentrated on here, is by and large constructed and developed in the same and corresponding experimental and design schemes as scientific knowledge. Engineering knowledge can thereby be seen to be constructed in a corresponding hierarchical structure. The development of both scientific and technological knowledge can be considered from the perspective of constructive, model-based views. In model-based views on science, such as Ronald Giere's (1988, 1999) and Bas van Fraassen's (1980) views, theoretical knowledge is seen to be constituted of the hierarchy of conceptual models – and this view is here applied also to understanding technological knowledge.[18] Giere (1988) describes scientific knowledge as a hierarchy of models, at the top of which are theoretical models guiding the construction of the models at the next lower level (level of general laws) and developing in this manner. These guide construction of the models at the next lower level etc. In these interactive processes knowledge of the different levels are iteratively brought in line with our cognitive structures and shared scientific and technological understanding of the success of current scientific or technological theories in explaining the materialization of experimentation and design.

The lowest level of the system of models suggested by Giere (1988) constitutes visual models. From the viewpoint of models, the epistemological significance of technology in science – and the nature of technological knowledge – becomes more apparent, if we were to go further. Namely, experimental set-ups and devices can be seen as the concrete material models, which carry the "thing knowledge" (Baird 2004). Apparent examples of such material models are the cloud-chamber in its original usage in meteorology (Galison & Assmus 1989), Francis Crick and James Watson's metallic model of the structure of a DNA double helix, didactical scale-models of the solar system, etc. But every experimental system, instrument, or specimen embody the experimenters' epistemic and methodological ideas, of which material models they are. When the cloud-chamber is used to study particles, there is still thing knowledge embodied in the experimental success, but the epistemic part of it does not concern cloud formation or optical phenomenon. Indeed, the material models carry knowledge about designs, materials and practical conditions needed to construct them or complete successful experiments, namely the technical know-how and functional rules needed for experimental success, for example (cf. Mitcham 1994; Norman 1998; Vincenti 1990). These scientific and technological ideas are primarily present in experimenters' skillful practices, and developed in design plans and experimental inquiry, which prepare scientists for action. The experimenters' and designers' epistemic ideas about idealized relationships between experimenters, instruments, and laboratory phenomena, and indeed their functioning in various conditions, can be read in design plans (see Rothbart 2007) and in laboratory notebooks. Thus, these documents establish the bridge from abstract knowledge structures to the concrete physical world. It is through the design and development of material models that concepts and the quantities of theories and laws get empirical meanings and are frequently defined and developed. That is done, to cite a central example, by using and designing instruments, which have the purpose of making the concepts mutually measurable and materially controllable (cf. Chang 2004). Nevertheless, these features of modeling embodied in the material models at the lower levels are totally missing in the descriptions of knowledge within the educational literature. It could be highlighted in the school laboratory, for example, by deeply considering the reasons for the purification of phenomena for the experiments performed or experimental failure met.

In consequence, also the structural understanding of relations, connections, and interactions of the physical and chemical world, both the natural and artificial parts of it, is linked to technology, to our capacity to control processes in designed experiments through intelligence. This *technoscientific* base of scientific and technological understanding and progress is worth introducing in education.

What is Technoscience About?

Technological knowledge concerns the world as it is constructed creatively by human beings. As seen above, also the phenomena studied in experimental natural sciences are "created." And they are created in at least two ways, materially and cognitively: On the one hand, the materialization of the experiment is the creation and controlling of an experimental setting that produce or isolate the phenomena under study and the instruments that reduce and modify the events into experimental measurements. On the other hand, the simultaneous conceptualization of nature takes place in the minds of the intentional researchers; it is neither produced by the objects under manipulation, nor does it arise directly from the observation of "Nature." Should we then give up the traditional conception of natural sciences as describing human-independent, constant features and causal regularities of "Nature"? Does it make sense to talk about scientific knowledge as different from technological knowledge?

Social Factors Shape Technoscience

In full-fledged philosophical constructivism, the previous notion has led to such views, according to which the experimental objects and processes are nothing but artifacts and science as a whole is a branch of technology rather than an equal field of culture producing universal knowledge (see, e.g., Latour 1987; Janich 1978). For example, for Latour (1987), the experimental success has nothing to do with accessing the world's real structures or capacities. Practitioners of *technoscience* can consider Latour's (1987) view that Nature does not exist in the laboratory to be right in many senses: the sociology of scientific knowledge reveals that the hardness of scientific fact is a human creation, not Nature's direct effect (MacKenzie 1989). Moreover, sociology reminds that these creation processes swallow up a lot of money, for example, and thus these enterprises become figured out by sponsors. Nevertheless, in the face of the strong influence of scientific and technological progress on our everyday lives the attempt to reduce science and technology to sociology and power games should be refuted in education.

To accept the above-presented ideas of the creative knowledge construction that phenomena studied in science and engineering are created, does not imply however some sort of subjectivism or relativism in the sense that everything is possible to reason as a true claim in science (cf. Hacking 1983). An experimenter (or designer) can neither create anything he or she can imagine nor can he or she make the entities and species of the world that are captured in a laboratory behave according to his or her own will (cf. Hacking 1983), but humans experience all kinds of constraints when intervening in the world. Technological, economic, political, ethical, social among other constraints of experimentation, which are linked to us as human beings, can change over time and place: for the limited financial resources, for example, experimenters can try out only part of the interesting ideas and for the limited technological ability, modelers' simulations have to be reduced in order to make it possible to run those on limited computational power (see Tala 2011). The natural constraints differ from technological (and other) essential constraints in the sense that they cannot be overcome – and that is how scientists reveal them. In consequence, scientific designs become shaped by innumerable external (social, economic, legal, political, cultural, material, and perhaps also esthetic) factors; the products of design, whatever they are, must be optimized not only in the material context but also in the ecological and psycho-social dimensions. These external social, political, and economic forces strongly influence the choosing of what technologies or scientific research will be undertaken, given attention in the media, and thus invested in and probably also used (see, e.g., Bijker et al. 1989).

Moreover, the products of *technoscience* have greatly influenced the course of history and the nature of society, and continue to do so. Mainly, it is these factors extrinsic to the experimentation and design processes that determine the available options between different research topics and methods that occurs before settling on specific techoscientific research projects. Financial policy also influences the speed of development of mature fields of research. Naturally, the objectives of research and development, having a more direct link to or even taking place in industrial laboratories, is more directly ascertained by extrinsic needs. Thus the kind of result reached through *technoscientific* research depends also on many external factors, including political institutions, culture, trends, the economy and the context and objectives of the research projects. The role of these extrinsic social forces differs substantially from the role of the intrinsic social and psychological aspects of the *technoscientific* construction and justification process, which make up the jury deciding whether success is achieved in the experimental design process. Thus when the social aspects of science and engineering are discussed in education, the role of the communities of practitioners of *technoscientific* processes should be discussed as separate from the interaction between *technoscientific* research and society.

In the process of accepting new knowledge, the Kuhnian idea seems to hold; there is no standard higher than the assent of the relevant community[19](Kuhn 1962). The decisions of relevant communities are not necessarily correct; nevertheless, that is the court which judges whether the research questions and methodology are relevant or not. Without such a community structure, the justification process would result in endless regression without coming to any conclusive views: the material setting itself cannot tell whether our conceptual models correspond to the experimental process (further explained in the following subsections, see also Collins 1985; Nickles 1989; cf. MacKenzie 1989). Thus, the social dimension of epistemology can thus be seen to be positive, as supporting the processes of science (see e.g., Collins 1985; Hacking 1983; Nickles 1989 and Rothbart 2007). The proponents of this view acknowledge the rather indisputable fact that scientific or technological inquiry is a social process and reasoned judgment is itself socially defined. Therefore it is natural and necessary that the logic of science has a certain social basis, but no primary role needs to be attached to sociology. This kind of moderate sociological view serves as a valuable guide to understanding the social background of scientific and technological research, which remains recognizable to their practitioners. It is this moderate view which is proposed here to be adopted as the basis of the *technoscientific* justification processes. The social nature is considered only to the extent in which it bears on epistemology, as supporting the progress of *technoscience.*

The Functional Nature of Engineering and Scientific Laws

Through providing both scaffoldings and limits of scientific reality accessible to us, technology necessarily also affects our conception of scientific reality. What kind of aspects of "natural" phenomena are we studying when we study, for example, the interaction between short-lived entities produced by particle accelerators, the spectra of pure chemicals, or the objects of genetic engineering? What do we "see" through an electron or scanning tunneling microscope or sonic probing? Or what do we reach by employing human-made measurement instruments, such as thermometers and voltmeters, producing quantities, which are not properties of nature as such, but rather what have been called "phenomenological profiles" (Ihde 1979) of instruments? These questions are not problems in the field of technology, which by definition concern the human-made world.

Since ancient times, nature itself has been compared metaphorically to a machine. In modern science, the natural order is also assumed to underline an experimental specimen or system in a laboratory; thus, those function as one of the world's machines with capacities to generate a "natural" change when sufficiently agitated (Cartwright 1999). Indeed, as discussed above the material realization of instruments, an experimental apparatus or system, is a kind of material model of the ideas the experimenter has of the "functioning" of the "natural" phenomena under study. To this material model is linked a large amount of technological knowledge, which is iteratively developed in the design process. Hence, finally, both the scientific and technological laws obtain upon the capacities of the experimental system: for example, in classical mechanics, repulsion, attraction, resistance, pressure and stress capacities are "observed" when the experiment is running properly (cf. Cartwright 1999). This idea is most apparent in the development of laboratory phenomena in the tradition of mimetic experimentation in the 18th and 19th centuries, which aimed to replicate a visible piece of nature in laboratory settings [e.g., Aitkens's miniature cloud-building in a cloud chamber (Galison & Assmus 1989), Van Marum's experiments with artificial clouds, Cavendish's model of electric fish (Hackmann 1989), Theodoric's study on rainbow geometry in water-filled flasks, etc.].

The situation is much more complicated on the level unattainable by our senses. In nanoscience and with nanolevel phenomena, there is no better way to secure that things are actually working as we imagine them to work than to wait and see whether the observable outcome fits our ideas of the supposed actions at the nanolevel. Indeed, when we cannot see, there remains a question about whether the entities expected in theoretical ideas really exist. In practice it is supposed that, if the nanostructures under changing macroscopically controlled conditions behave as expected, that is, if the nanoconveyors and minimachines manage to cause expected macroscopic effects, then the entities behind these causes must be real. Nowadays the data are increasingly produced through agitations, manipulations and inferences rather than through human "observations of nature." Consider, for example, the differing methods of spectrometry including a signal (e.g., electromagnetic radiation) generator, detector, signal processor, and readout device, or scanning electron microscopes, field ion microscopes, and scanning tunneling microscopes and so on (see Rothbart 2007). Often also computer simulations are developed and used in close relation to experimentation and to compensate it as well (Tala 2011).

Indeed, science cannot distinguish between the artificial and the natural (Kroes 2003), since all the objects involved are physical. Then, the focus sharpens the "back-inference" from laboratory phenomena to the world outside laboratories and the relation of measurements taken to the factual features of the world – and this relation is important for science. Even if the origin of the phenomena under study is more or less artificial, in practice it is assumed that "Nature" can have some decisive role in the outcome of the experiment and, thus, the underlining causal relationships are expected to be natural.

Nevertheless, in an experimental invention, certain states of affairs are intentionally brought about, which would not have arisen without the interference, and, we could also have chosen to realize another state (Bohr 1958; Janich 1978). For example, by running the original Bohrian apparatus of modern physics we can cause either particle phenomena or a wave phenomenon to occur (see Bohr 1958), depending on which hypotheses or model the apparatus is designed to support. A similar example is the cloud chamber, by the running of which we can either select to study meteorological phenomena or elementary particles. Thus, what experimental laws of science – like the ones of engineering – describe are the ways by which we may interact with our physical and chemical environment in idealized situations (Woodward 2003): we cannot prepare a system which offends these laws, such a system in which objects in a vacuum will not fall at the same speed (the laws of free fall), the resistance in a metallic conductor in an electric circuit will not be constant in constant temperature (Ohm's law), or two thermodynamic systems, which are separately in thermal equilibrium with a third, would not be in thermal equilibrium also with each other (the zeroth law of thermodynamics). In sum, our conceptions of human-made systems give rise not only to the engineering laws but also to scientific laws, which (are thought to) describe more general regularities of the world.[20]

The predictive power of engineering, physics, and chemistry lies in the laws, which define the relations between different measurable physical features, quantities. Working scientists and engineers aspire to accurate measurements, for example, but they do not reach "absolute truth values." Measurement devices are assumed to detect states of an experimental system by a regular causal relationship occurring between the states of the "world" and the state of the instrument. This ability is built into the instruments: the measurement devices are often developed in interaction with the theoretical understanding of what has been measured (theory-ladenness). For example, this is easy to see in how the understanding about thermal phenomena developed in close interaction with the development of thermometers (see Chang 2004; Middleton 1964). This lead to a situation, where by deciding the criteria for the accepted method to measure the quantity, the character of the theory underlining the measuring and the quantity is decided (Collin 1985; MacKenzie 1989). Thus, even the final scientific "truth" is defined by the methods by which it was reached.

The Iterative Hunt for Technoscientific Reality

The experimental design takes place within a theoretical framework and the objective is that the frameworks themselves become developed in the iterative design processes: the product of the original design of a measurement device, for example, is not only more exact measurement devices but also the sharpened meaning of the quantity measured. In the technoscientific design process the independent, yet interactive, stages of "know-how" and "know-that" knowledge develop iteratively. In that sense, engineers and scientists adopt at every moment the existing systems of "know-how" and "know-that" knowledge and existing material abilities to control experimental systems or the relation between the states of this system and the state of the instrument (without any firm assurance of the correctness and accurateness) and, moreover, they aim to sharpen and correct them both (cf. Chang 2004).

Thus, "know-how" and "know-that" develop in mutual interaction; the design process combines the stage of "knowing" and "doing" into a continuum between the abstract level of theory and the material level of action. This, on the other hand, creates the ability to construct, manipulate and control the dimension of the phenomenon that is being pursued. On the other hand, it creates knowledge of the functioning of the phenomena, in a special form of quantities and laws, a very special product of experimental process made possible by the instruments and machines developed for that purpose in the theoretical framework. The aim of the *technoscientific* design is to reach the greatest technically feasible accuracy, and then extend the local and technical limits for progress. Progress means the crossing of limits of particular material settings, which is reproducing the ideas in different material settings,[21] in close interaction with the developing theoretical framework.

Through such kinds of self-corrective processes of experimental design, conceptual and material models of science and technology are developed in mutual interaction, and closure is achieved when they fit together adequately. It is just this end result that van Fraassen refers to as "empirical adequacy" and Giere refers to as "similarity." However, there remains a question – when can we be satisfied that such closure is achieved? In many cases just this problem is at the core of scientific disputes and controversies. The problem seems to be that there is no objective, sociologically neutral or unambiguous method to settle this question (e.g., Nola 1999). Rather, as described above, any methodology that manages to demonstrate control over phenomena, the ability to intervene and manipulate or capabilities for creating phenomena by using technological devices, is accepted. In this process, social aspects cannot be simply bypassed – also epistemology becomes intertwined with sociology.[22]

SUMMING UP: THE TECHNOSCIENTIFIC VIEWS ON EDUCATION

In science education, experimental work has long been considered to be an integral part of learning science (Hodson 1996), whereby design education is widely omitted as a basis of technology education. In addition to acquiring certain skills, these activities are important because they support conceptualization, aid in learning about the scientific and technological process, and teach about the empirical foundation of science and challenges met in the practical foundation of technology (e.g., Jones 1997; Koponen & Mäntylä 2006; Matthews 1994; NAE/NRC 2002). For a couple of decades there has been an active discussion in technology education about the technological knowledge appropriate to instill in all-round education (AAAS 1993; ITEA 2000; Johnson 1997; Jones 1997; McCormick 2004 deVries & Tamir 1997; deVries 1997; Mitcham 1994; NAE/NRC 2002). Indeed, the relationship between science and technology is considered to some degree in many curricula and a diversity of practical projects have integrated science and technology education. Such conditions could bring forward the intimate connection between NOT and NOS, but as yet they have not. Many new ideas discussed above connecting NOS and NOT can be implemented in the already existing framework by re-organizing the science- and technology-based elements of education in a new and fruitfully unifying way.

Technoscience Improves Understanding of Both NOT and NOS

As an educational approach, *technoscience* embodies a moderate constructivist view of science and technology, but rejects the radical constructivist views as well as radical constructivist epistemologies.[23] In the *technoscientific* view, experimentation and design play a central or even crucial role, but give experiments a very different and more authentic role than they have in discovery learning, or in learning by inquiry. The view differs also from step-by-step design education and views that reduce the design process to power games. The advantage of *technoscience* is that it approaches education from the perspective of the working engineer and scientist, and emphasizes the constructive nature of knowledge building, avoiding fixing upon any specific philosophical orthodoxy or fashionable theme (be it empiricism, realism or social-constructivism). Thus it can enhance understanding of NOS and NOT regardless of the metaphysical orientation of students or teachers.

Basically, on the one hand, *technoscience* adds to our understanding of what "the empirical nature" of science – which is often a slogan in textbooks and in the core of NOS – means in the case of the experimental natural sciences; indeed, it sheds light on the fact that this NOS notion cannot be understood correctly without paying close attention to the role of technology. On the other hand, *technoscience* helps us to understand the epistemic and cognitive side of technology and how engineering combines these with practical and social values (a NOT theme). Technology, as a design science, is based on the knowledge constructed in science or in parallel research activities as scientific knowledge. By considering these two sides more closely, the conception of scientific knowledge also becomes extended; it is primarily through the lower levels, the material and visual models, where the epistemology and ontology of science and technology are intertwined. These models become iteratively developed in interaction with the models of higher levels operating in the *technoscientific* design process.

The products of *technoscientific* design are not only those having industrial value but also those that are scientific, such as having the capability to construct, manipulate and control the material environment in experimental systems as well as instruments and knowledge connected to these capabilities (cf. NOS theme *certainty and forms of scientific knowledge*). In this way, physics and chemistry, as experimentally "based" sciences, and to some degree biology, geology and other natural sciences, become inexorably linked to technological productivity and capability. A process in which science makes progress and where technological advantage is central employs *a variety of methodologies* (a NOS theme) and *technoscience* actually shows how these methodologies of science and technology became constructed and developed in the design process. A variety of methods and means are combined and developed creatively also in design processes. Since *technoscience* supports the conception that not only technological reality but also physical and chemical reality is revealed through, as well as modified by, technological action, it differs from the flawed views of traditional science education where an experimental method is necessary only for the verification of theories, and instead deepens understanding of the knowledge generative nature of experimentation in theory formation (cf. Koponen & Mäntylä 2006). Indeed, it reveals that the technological design process has also the objective to construct and justify knowledge. In short, contrary to traditional views, *technoscience* enforces the view that we design or construct our conceptions of what exists in the world and how it does so through active manipulation and intervention of designed experimental settings. Thus it happens that experimental design plays a central role in science and in (engineering) technology, and scientific design processes intertwines the two. Finally, both science and technology progress through constructing and developing artifacts – material as well as conceptual.

Technoscience also embodies, in the concrete practices of both science and technology, the other frequently mentioned general notions of NOS and NOT; the NOT statements can be used to understand the practical meaning of the NOS statements and *vice versa*. For example, the statements, *scientific knowledge is theory-laden and tentative*, *at the heart of engineering is design under constraint*, and *engineering is improved through scientific knowledge* can be compared:[24] Then, we extract the point that was examined above: that at the heart of *technoscience* lies creative design under constraint, which is theory-laden in the sense that (to be meaningful) it takes place in a theoretical framework and at same time it also happens that the theoretical framework is itself thereby developed. For a special example of theory-ladenness, which is quite a complex theme, considered above were the crucial design of experimental systems and instruments. In these processes the scientific and technological quantities and laws used to describe the world around us are designed also. This is an iterative, self-correcting development of independent but interactive stages of "know-how" and "know-that" knowledge: experts adopt at every moment the existing systems of know-how and know-that knowledge and existing material abilities to control experimental systems, and aim to sharpen and improve them both. This kind of process requires total immersion in the previous theories and scientific worlds and in the tacit and explicit knowledge and skills of the experimental and design work; this is the practical meaning of the *theory-ladenness of scientific* – and technological – *knowledge*. It thus reveals how also technological knowledge is theory-laden.

Scientific knowledge is likewise experimentally-laden and thus founded on functionality: this constitutes its empirical nature in *technoscientific* practice. To maintain the vitality of science, the controlling and manipulation of laboratory phenomena is the very basis of experimental success and functions independently of the theoretical interpretations (Hacking 1983). The capacities of human-made experimental systems give rise to the regular behavior of the "world" that we express in our scientific and technological laws. *Technoscience* links the abstract concepts to each other and to the world in the cognitive-material medium of experimental design, which, indeed, provides students with a concrete view of modeling as a means of producing scientific and engineering knowledge. In this manner, *technoscience* facilitates the understanding and application of ideal scientific laws. Moreover, *technoscience* guides understanding in the area of application of the scientific and technological knowledge, and shows how we have succeeded in "knowing."

By considering the technological side of scientific process, it is easy to realize that *scientists are creative* (a NOS theme). To design something is not to follow recipes, but rather to think and act creatively and critically (Johson 1995; Layton 1993; Mitcham 1994; Rothbart 2007; Vincenti 1990), in order to realize cognitive goals which can be achieved only through, and merged with, technological devices. Nevertheless, every *technological system can fail* (a NOT theme, see also Cajas 2001), whereby experimental systems and instruments can also fail. In the *technoscientific* process this is understood as a lack of control (NOT theme) and is to be rectified by developing both the material and conceptual control over the phenomenon under study. Knowledge about technological constraints for control is also often developed in this way. Indeed, this is also what the NOS themes *tentativeness, certainty and forms of scientific knowledge* mean in the practices of science.

In addition, the social system, external to the knowledge-constructing practices of science and engineering, place limitations on *technoscientific* processes in terms of economics and ethics. Typically these limitations force choices between the research questions and methods before the projects are launched, by approving and supporting some and refusing or not supporting others. As such, the abstract NOT and NOS themes referring to the *strong interaction between scientific or technological systems on the one hand, and social systems, on the other*, and to the certainty and forms of scientific knowledge, can indicate any commitment, from radical constructivism and radical relativism to more moderate and realistic views. *Technoscience* helps us to understand that the motivational view of scientists and engineers must be that the constraints underpinning the world order are ontologically independent of human power games, inquiry, perception and action. Nevertheless, it reveals that the justification of iterative design processes lies in the practical functionality of materialized ideas in the human-made systems; indeed, the practical success is itself socially defined. Ultimately, the *technoscientific* view leads us to see technology not only as a means to more reliable and accurate knowledge, but also as a part of scientific objects and value as an object of scientific research itself.

Technoscience in Educational Practice

In practice, highlighting the technological nature of scientific reality and the scientific nature of technological process can be supported by moderately implementing the ideas of knowledge construction through *technoscientific* design in education as methodological means and contents. Indeed, reflection is needed. In hands-on education this means, for example, the design and development of simple measuring instruments (instruments to measure length, time, tensile strength, stream velocity or temperature, for example), and simple experimental design in interaction with knowledge concerning these in contexts of science and technology, or *technoscience*. This can also mean becoming acquainted with particular sketches, laboratory notebooks, and historical accounts of this kind of design work and its role in the development of science and technology. In science education, it guides at least paying attention to the idealization and purification process performed and the technoscientific know-how needed when designing the material settings of simple school laboratory experiments. Indeed, paying students' attention to the laboratory devices is a short way to study the very nature of the phenomena studied: as discussed above, the data in experimentation tell typically much more about the measuring instruments and experimental settings used than about the phenomenon under study (cf. Postman 1985).

Research suggests that content courses, where it is possible to talk about particular theories or particular pieces of historical evidence, provide a fruitful venue for addressing and developing learners' views on the nature of the discipline under study (see, e.g., Dagher et al. 2004; Hanuscin et al. 2006; Lederman 1992; Matthews 1994; Schwartz et al. 2004). Success has been frequently achieved when the explicit, implicit and social approaches to NOS are combined, for example, by reflective exercises. Thus, the perspectives of NOS and NOT can be reflected on, for example, in particular hands-on-activities (Lederman 1992, 1998; Matthews 1994; Metz et al. 2007). When explicitly reflecting on the practices of a particular science class, it is easier to get students immersed in the discussion about the nature of the field(s) of knowledge that they are studying. For example, the above mentioned scientific design activities can follow historical storylines (Metz et al. 2007) reconstructed from the perspectives of NOTS objectives, which guide understanding of both scientific and technological contents and practices. It is worth noting that also an experimental failure in traditional school laboratory tasks is a place to learn about the variety of scientific and technological knowledge and skills needed in order to demonstrate the desired phenomena.

Reflective *technoscientific* design tasks, involve not only the usual motivational factors, but become guided by the human ambition to understand and control the environment – and understand the basis of this understanding. In these tasks students participate in collaborative projects in which they are able to construct and control design solutions that are both concrete and conceptual, which is interrupted by reflection. Hence design tasks merge the procedural, conceptual, and NOS and NOT objectives and, additionally, they increase the authenticity of the content. In such a manner, *technoscience* promotes the connection between "doing" and "knowing" and supports the students' own construction of knowledge.

In sum, a *technoscientific view* supports learning in experimental and design tasks. It improves the authenticity of and coherence between NOS- and NOT-based as well as conceptual and procedural elements of education. It does this by learning from the coherence between the respective elements – theoretical, methodological and epistemological – in the practices of science and engineering. This strong confluence of experimentation, and design in education is thus more than the sum of their respective parts, in the sense that their combination genuinely improves learning.

NOTES

1 The slogan 'nature of technology' is not as frequent as the slogan 'nature of science.' The views of nature of technology in educational literature are frequently referred to by such terms as 'image of technology,' 'attitudes toward technology', 'perception of technology' and 'concept of technology.'

2 The ideas of *technoscience* can be profitably compared to Neil Postman's (1985) idea of an epistemological shift toward public knowledge rooted in the different forms of media. In his book "Amusing ourselves to death," Postman describes how changes in media bring about a corollary change in the structure of people's mind and cognitive capacities. The example of television as a medium defines the limits and forms of communication, and finally also directs knowledge in the ways of knowing. Respectively, in a *technoscientific* process, by designing experimental systems and measuring instruments our knowledge structures of "the world" are also configured, a process that is further discussed in sections 3 and 4.

3 Naturally, focusing, in an introductory manner, on the differences helps in identifying science, technology and applied sciences, for example, but more could be achieved, if the fields could be considered also from a common viewpoint.

4 There has been an active discussion about how to limit the constructs. This is an especially difficult question in the case of NOT, because technology education is not defined by any particular field of expertise. Nevertheless, often it is identified with engineering. Among the general population there is a tendency to perceive technology as a narrow, restricted field confined to automobiles, televisions, and the web- and mobile-based technologies, which suggests that the present treatment of technology in the school curriculum may be too fragmented or too abstract.

5 There is on-going discussion about the construct which should be listed, whether it is NOS 'ideas,' 'themes,' 'tenets,' or ' statements,' for example. Furthermore, Irzik and Nola (2011) have suggested that we should discuss 'family resemblances,' instead, because there may be no set of purposeful NOS themes shared with every field of science. From the viewpoint of this article, it is natural to see the various lists of constructs as aiming toward the same objective, to list the typical characters of science to be taught to every citizen.

6 In engineering philosophy of technology (see Mitcham 1994), which supports the discussion here, the technological process is described by four "ends": design, construction, operation (Vincenti 1990) and production, of which design is considered as the central mission of engineering.

7 Some authors also list the subjective nature of the scientific process among NOS themes, because in addition to scientists' previous knowledge and theoretical commitments, also their beliefs, training, experience and expectation figure in the results.

8 The possibility of reaching an agreement about NOS or NOT can be easily denied. For example, a comparison between the works of Dewey, Popper, Kuhn, Lakatos, Feyerabend, Heidegger, Bijker and Giere, or between the metaphysical commitments of scientists or engineers, for example, such as H. V. Regnault and Kelvin, both of whom developed thermo physics (Chang 2004), makes it clear that there can be no meaningful agreement about the innermost character of the nature of science in relation to the world. However, such a consensus is not even needed when NOS and NOT is studied as linked to practices.

9 However, an accurate understanding of NOS does not necessarily lead to instruction reflecting that understanding (Lederman 1992) and participating in authentic instruction does not necessarily lead to accurate understanding of NOS and NOT.

[10] See the tests VNOS- A, VNOS- B and VNOS-form C, summarized in Lederman et al. (2002).

[11] Both scientific and technological processes are context-sensitive. This idea may be easier to understand in the context of technology: for example, when a designer has gained expertise in one area of the design field, say bridge design, (s)he naturally cannot necessarily operate in another design field, say refrigeration device design. So, in terms of hands-on activities in science and technology education, the experimental process and design process have been simplified.

[12] In education, also both hierarchical views have been reasoned in terms of reconstructed, simplified examples of the development of science and technology. The demarcationist view is supported by such examples as 'scientific optics was a basis for the development of eyeglasses,' 'the research of Hertz and other scientists was a basis for radio and TV' and by stating that many newcomers, such as atomic energy, DNA engineering, microchips, computer design, and many fundamental principles mirrored by them, have been considered to be applications of science. The opposition has reminded us that many technical inventions – such as the wheel, scissors, bridges, and sailboats – were developed earlier without recourse to institutionalized science. In addition, present-day engineers and scientists have their own libraries and training programs.

[13] Early examples of the projects promoting this view are SATIS, Chemcom, Science at work and Physics Plus, Biology Plus, Chemistry Plus, Salter's science and Salters' chemistry.

[14] Technoscience is introduced here from the viewpoint how it combines and improves understanding about NOT and NOS, and relation of NOT and NOS. A deeper discussion about epistemology of seeing physics as technoscience is in Tala 2009, which discusses many ideas mentioned here about technoscience.

[15] In fact, already Pierre Duhem (1914) discussed advanced views about the role of technology in science, which are in line with the views presented here, but not many contemporary scientists of his day were interested in speculations of that kind.

[16] Nevertheless, experimentation or design does not (necessarily) require hands-on working; in it the functioning of a device, e.g., experimental setting or instruments, can be tested and to some extent also judged in the symbolic world, through thought experiments or in computer modeling (see Rothbart 2007) without deploying metals, plastic wires, components, apparatuses and the likes in any particular laboratory.

[17] From this list, we see that technological knowledge has several dimensions: there is functional knowledge concerning, for example, how specimens and systems are made, maintained, and function. In addition, there is theoretical, descriptive knowledge about the principles underlying the technological capacities. But this is also the case with scientific knowledge; we have no other way to learn about many aspects of nature than to construct an experimental system and run it to see if how it functions (as discussed later on). Thus, in addition to theoretical knowledge, both the scientific and engineering processes require and develop a special kind of experimental knowledge which is functional or operational: knowledge about how to construct, develop, and control experimental systems, specimens, and instruments. (For a wider consideration, see Mitcham 1994; Vincenti 1990.)

[18] Naturally, a theory as well as the technological capability for experimentation and design can also develop independently (for examples see Hacking 1983), but in practice, a theory typically, sooner or later, meets with experimental testing.

[19] However, there are a variety of views on what kind of community it is, and what aims and views the individuals of the community need to share, in order to be counted as members of the community. Many Kuhnian and post-Kuhnian philosophers refer to 'the scientific community' as a quite homogenous group of equal experts who take part in defining the truth within that community (see Kuhn 1962; Latour 1987). Many others, who have studied the practices of science (e.g., Van Fraasen 1980; Nickles 1989; Hacking 1989; Harré 2003) perceive the (techno)scientific community as a heterogeneous group of practitioners, who may have different views and methods, but who all have a shared objective.

[20] If considering from a constructivist viewpoint, it is unnecessary to see the scientific laws as universal and true generalizations of nature (like logical empiricists, for example) (Giere 1988, 1999).

[21] The simultaneous development of technoscientific measurement devices and quantities can be considered as an iterative "hunt for the reality" (Chang 2004), where scientific "truth" is defined by

the methods by which it was reached (Chang 2004, see also Tala 2009): for example of the mutual development of conceptual understanding of thermodynamical phenomena and thermometrical means to measure, see Chang (2004) and Middleton (1964).

[22] This is how epistemology becomes intertwined with sociology in *technoscience*. Postman (1985) discussed the same phenomenon in a different context, in the context of the epistemology of public discussion.

[23] To read about the problems met while these extreme views have been applied in education see e.g., Matthews (1998), or Mayer (2004).

[24] Naturally, the nature of *technoscience* does not encompass all the [great] variety of meanings that the general notions of NOT and NOS may encompass.

REFERENCES

AAAS (American Association for the Advancement of science) (1993). *Benchmarks for science literacy*. Oxford: Oxford University Press.

Abd-El-Khalick, F. (2000). Improving science teachers' conceptions of nature of science: A critical review of the literature. *Internatinal Journal of Science Education, 22*(7), 665–701.

Abd-El-Khalick, F. (2012). Teaching With and About Nature of Science, and Science Teacher Knowledge Domains, *Science & Education*, online first August 2012.

Akerson, V.L., Abd-El-Khalick, F., & Lederman, N.G. (2000). Influence of a reflective explicit activity-based approach on elementary teachers' conceptions of nature of science. *Journal of Research in Science Teaching, 37*(4), 295–317.

Arthur, W. B. (2009). The Nature of technology: What it is and how it evolves? New York: Free Press.

Baird, D. (2004). Thing knowledge. Berkeley and Los Angeles: University of California Press.

Bame, E., Dugger, W., Jr., de Vries, M., & McBee, J. (1993). Pupils' attitudes toward technology -PATT-USA. *Journal of Technology Studies, 19*(1), 40–48.

Begoray, D. & Stinner, A. (2005). Representing science through historical drama: Lord Kelvin and the age of the earth debate. *Science & Education, 14*(3–5), 457–471.

Bijker, W.E., Hughes, T. P. & Pinch, T. (Eds.) (1989). *The Social construction of technological systems. New directions in the sociology and history of technology*. Cambridge MA: MIT Press.

Bohr, N. (1958). *Atomic physics and human knowledge*. New York: Wiley.

Cajas, F. (2001). The science/technology interaction: Implications for science literacy. *Journal of Research in Science Teaching, 38*(7), 715–729.

Cajas, F. (2002). The role of research in improving learning technological concepts and skills:
The context of technological literacy. *International Journal of Technology and Design Education, 12*(3), 175–188.

Cartwright, N. (1999). *The dappled world: A study of the boundaries of science*. Cambridge: University Press.

Chang, H. (2004). *Inventing temperature: Measurement and scientific progress*. Oxford Studies in Philosophy of Science, Oxford.

Chinn C. A & Malhotra B. A. (2002). Epistemologically authentic inquiry in Schools: A theoretical framework for evaluating inquiry tasks. *Science Education, 86*(2), 175–218.

Clough, M. P. (2006). Learners' responses to the demands of conceptual change: Considerations for effective nature of science instruction. *Science & Education, 15*(5), 463–494.

Clough, M. P. (2007). Teaching the nature of science to secondary and post-secondary students: Questions rather than tenets. *The pantaneto forum 25*, http://www.pantaneto.co.uk/issue25/front25.htm.

Clough, M. P. (2011). The Story behind the science: Bringing science and scientists to life in post-secondary science education. *Science & Education, 20*(7–8), 701–711.

Compton, V.J. & Compton, A. D. (2011). Teaching the nature of technology: determining and supporting student learning of the philosophy of technology. *International Journal of Science Education,* online first August 2011.

Collins, H. (1985). Changing order: Replication and induction in scientific practice. London: Sage.

Dagher Z., Brickhouse N., Shipman H., & Letts W. (2004). How some college students represent their understandings of the nature of scientific theories?. *International Journal of Science Education, 26*(6), 735–755.
Daugherty, M. K. & Wicklein, R. C. (1993). Mathematics, science, and technology teachers' perceptions of technology education. *Journal of Technology Education, 4*(2), 30–45.
de Vries, M. J. & Tamir, A. (1997). Shaping concepts of technology: What concepts and how to shape them. *International Journal of Technology and Design Education, 7*(1–2), 3–10.
de Vries, M. J. (1996). Technology education: Beyond the "technology is applied science" paradigm. *Journal of Technology Education, 8*(1), 7–15.
deVries, M. J. (1997). Science, technology and society: A methodological perspective. *International Journal of Technology and Design Education, 7*(1), 21–32.
deVries, M. J. (2011). The Nature of Technological Knowledge: Philosophical Reflections and Educational Consequences *15*(2), 149–154.
Dewey, J. (1916). Democracy and education. New York: Macmillan.
Driver R, Leach J., Millar R., & Scott P. (1996). *Young people images of science*. Buckingham: Open University Press.
Duhem, P. (1914/1954). *The aim and structure of physical theory* (translation of La Théorie Physique: Son Objet, Sa Structure. 2nd edn. 1914 Paris). Princeton: Princeton University Press.
Fensham, P., & Gardner, P. (1994). Technology education and science education: A new relationship?. In Layton, D. (Ed.) *Innovations in science and technology education*. Paris: UNESCO.
Galison, P. & Assmus, A. (1989). Artificial clouds, real particles. In Gooding, D., Pinch, T. & Schaffer, S. (Eds.). *The uses of experiment: Studies in the natural sciences*. Cambridge: Cambridge University Press.
Galison, P. & Thompson, E. (Eds.) (1999). *The Architecture of science*. Cambridge: The MIT Press.
Giere, R. (1988). *Explaining science. A cognitive approach*. Chicago: The University of Chicago Press.
Giere, R. (1999). *Science without laws*. Chicago: University of Chicago.
Hacking, I. (1983). *Representing and inventing: Introductory topics in the philosophy of natural science*. Cambridge: Cambridge University Press.
Hackmann, W. D. (1989). Scientific instruments. In Gooding, D., Pinch, T. & Schaffer, S. (Eds.). *The uses of experiment: Studies in the natural sciences*. Cambridge: Cambridge University Press.
Hansen, K.-H. (1997). Science and technology as social relations towards a philosophy of technology for liberal education. *International Journal of Technology and Design Education, 7*(1–2), 49–63.
Hanuscin, D. L., Akerson, V. L. & Phillipson-Mower, T. (2006). Integrating nature of science instruction into a physical science content course for preservice elementary teachers: NOS views of teaching assistants. *Science Education, 90*(5), 912–935.
Harré, R. (2003). The materiality of instruments in a metaphysics for experiments. In Raddér, H. (Ed.). *The philosophy of scientific experimentation*. Pittsburgh: University of Pittsburgh press.
Heidegger, M. (1927/1978). *Being andtime* (translation of Sein und Zeit. Robinson, E. & Macquarrie, J. (trans.)). Oxford: Blackwell.
Hodson, D. (1996). Laboratory work as scientific method. *Journal of Curriculum Studies, 28*(2), 115–135.
Hodson, D. (2008). Towards Scientific Literacy A Teachers' Guide to the History, Philosophy and Sociology of Science, Rotterdam/Tapei: Sense Publishers. Ihde, D. (1979). *Technics and praxis*, D. Boston: Reidel Publishing Company.
Irzik, G. & Nola, R. (2011). A Family resemblance approach to the nature of science for science education. *Science & Education, 20*(7–8), 591–607.
ITEA (The International Technology Education Association) (2000/2002/2007). *Standards for technological literacy – content for the study of technology*. Reston, VA: Author.
ITEA (2006). *Technological literacy for all: Rationale and structure for the study of technology*. Reston, VA: Author.
Janich, P. (1978). Physics – natural science or technology?. In Krohn, W., Layton, E.T., & Weingart, P. (Eds.). *The dynamics of science and technology*. The Netherlands: Dordrecht, Reider.
Johnsey, R. (1995). The design process — does it exist?. *International Journal of Technology and Design Education, 5*(3), 199–217.

Johson, S.D. (1997). Learning intellectual concepts and developing intellectual skills. *International Journal of Technology and Design Education, 7*(1–2), 161–180.
Jones, A. (1997). Recent research in learning technological concepts and processes. *International Journal of Technology and Design Education, 7*(1–2), 83–96.
Jones, A., Buntting, C.M. & de Vries, M. (2011). The developing field of technology education: a review to look forward. *International Journal of Technology and Design Education*, online-first June 2011.
Khishfe, R. & Abd-El-Khalick, F. (2002). Influence of explicit and reflective versus implicit inquiry-oriented instruction on sixth graders' views of nature of science. *Journal of Research in Science Teaching, 39*(7), 551–578.
Kolodner, J. L., Camp, P. J., Crismond, D., Fasse, B., Gray, J., Holbrook, J., Puntambekar, S., & Ryan, M. (2003). Problem-based learning meets case-based reasoning in the middle-school science classroom: Putting learning by design(tm) into practice. *Journal of the Learning Sciences, 12*(4), 495–547.
Koponen, I. T. & Mäntylä, T. (2006). Generative role of experiments in physics and in teaching physics: A suggestion for epistemological reconstruction. *Science & Education, 15*(1), 31–54.
Kroes, P. (2003). Physics, experiments, and the concept of nature. In Raddér, H. (Ed.). *The Philosophy of Scientific Experimentation.* Pittsburgh: University of Pittsburgh press.
Kuhn, T.S. (1962). *The structure of scientific revolutions.* Chicago: University of Chicago Press.
Latour, B. (1987). *Science in action: How to follow scientists and engineers through society.* Milton Keynes: Open University Press.
Layton, D. (1993). *Technology's challenge to science education: Cathedral, quarry or company store?.* Buckingham and Philadelphia: Open University Press.
Lederman, N. G. (1992). Students' and teachers' conceptions of the nature of science: A review of the research. *Journal of Research of Science Teaching, 29*(4), 331–359.
Lederman, G., Abd-El-Khalick, F., Bell, R. L. & Schwartz, R. S. (2002). Views of nature of science questionnaire: Toward valid and meaningful assessment of learners' conceptions of nature of science. *Journal of Research in Science Teaching, 39*(6), 497–521.
Lederman, G. (2007). Nature of Science: Past, Present, and Future. N. G. Lederman 2007 In Abell, S.K. & Lederman, N.G. (Eds.), Handbook of Research on Science Education, 831–880.
MacKenzie, D. (1989). From Kwajelein to Armageddon? Testing and the social construction of missile accuracy. In Gooding, D., Pinch, T., & Schaffer, S. (Eds.). *The uses of experiment: Studies in the natural sciences.* Cambridge: Cambridge University Press.
Matthews, M. (Ed.) (1998). *Constructivism in science education: A philosophical examination.* New York: Kluwer Academic Publishers.
Matthews, M. R. (1994). *Science teaching: The role of history and philosophy of science.* New York: Routledge.
Mawson, B. (2003). Beyond 'the design process': An alternative pedagogy for technology education. *International Journal of Technology and Design Education, 13*(2), 117–128.
Mayer, R. (2004). Should there be a three-strikes rule against pure discovery learning? The case for guided methods of instruction. *American Psychologist, 59*(1), 14–19.
McComas, W. F. & Olson, J. K. (1998). The nature of science in international science education standard documents. In W. F. McComas (Ed.) *The Nature of Science in Science Education: Rationales and Strategies.* Dordrecht: Kluwer.
McComas, W. F., Almazroa, H., & Clough, M. P. (1998). The nature of science in science education: An introduction. *Science & Education, 7*(6), 511–532.
Metz, D., Klassen, S., McMillan, B., Clough, M., & Olson, J. (2007). Building a foundation for the use of historical narratives. *Science & Education, 16*(3), 313–334.
Middleton, W. E. K. (1964). *The history of barometer.* Baltimore: The Johns Hopkins Press.
Miller, J. D. (1983). Scientific literacy: A conceptual and empirical review. *Daedalus*, 112(2), 29–48.
Mitcham, C. (1994). *Thinking through technology: The path between engineering and philosophy.* Chicago: The University of Chicago Press.
NAE/NRC (National Academy of Engineering/National Research Council) (2002). *Technically speaking: Why all Americans need to know more about technology.* Washington, D.C.: National Academies Press.

Nickles, T. (1989). Justification and experiment. In Gooding, D., Pinch, T. & Schaffer, S. (Eds.). *The uses of experiment: Studies in the natural sciences.* Cambridge University Press, Cambridge.

Norman, E. (1998). The Nature of Technology for Design. *International Journal of Technology and Design Education* 8(1), 67–87.NRC (National Research Council) (1996). *National science education standards.* Washington, DC: National Academy Press.

NRC/Committee on Conceptual Framework for the New K-12 Science Education Standards (2012). *A Framework for K-12 Science Education: Practices, Crosscutting Concepts, and Core Ideas.* The National Academies Press.Nola, R. (1999). On the possibility of scientific theory of scientific knowledge. *Science & Education, 8*(4), 427–439.

Osborne, J, Collins, S., Ratcliffe, M., Millar, R., & Duschl, R. (2003). What "ideas-about-science" should be taught in school science? A delphi study of the expert community. *Journal of Research in Science Teaching, 39*(5), 497–521.

Palmquist, B. C., & Finley, F. N. (1997). Preservice teachers' views of the nature of science during a postbaccalaureate science teaching program. *Journal of Research in Science Teaching, 34*(6), 595–615.

Polanyi, M. (1966). The Tacit Dimension. New York: Doubleday.

Polanyi, M. (1958). Personal Knowledge: Towards a Post-Critical Philosophy. London: Routledge.

Postman, N. (1985). *Amusing ourselves to death. Public discourse in the age of show business.* New York: Viking.

Rasinen, A. (2003). An analysis of the technology education curriculum of six countries. *Journal of Technology Education, 15*(1), 31–47.

Roth, W.-M. (2001). Learning science through technological design. *Journal of Research in Science Teaching, 38*(7), 768–790.

Rothbart, D. (2007). *Philosophical instruments. Minds and tools at work.* Urbana and Chicago: University of Illinois Press.

Rudolph, J. L. (2005). Inquiry, instrumentalism, and the public understanding of science. *Science Education, 89*(5), 803–821.

Sadler, P. M., Coyle, H.P., & Schwartz, M. (2000). Engineering competitions in the middle school classroom: Key elements in developing effective design challenges. *The Journal of the Learning Sciences, 9*(3), 299–327.

Sandoval, W. (2005). Understanding students' practical epistemologies and their influence on learning through inquiry. *Science Education, 89*(4), 634–656.

Schummer, J. (2004). Interdisciplinary issues in nanoscale research. In Baird, D. Nordmann, A., & Schummer, J. (Eds.). *Discovering the nanoscale.* Amsterdam: IOS Press.

Schwartz, R., Ledermann, N. G. & Craftfors, B.A. (2004). Developing views of nature of science in an authentic context: An explicit approach to bridging the gap between nature of science and scientific inquiry. *Science Education, 88*(4), 610–645.

Shaw K. E. (2002). Education and technological capability building in the Gulf. *International Journal of Technology and Design Education, 12* (1), 77–91.

Sidawi, M. M. (2007). Teaching science through designing technology, *International Journal of Technology and Design Education,* online first: 11 December 2007. DOI 10.1007/s10798-007-9045-1.

Sundberg, M. (2006). Credulous modellers and suspicious experimentalists? Comparison of model output and data in meteorological simulation modelling. *Science Studies, 19*(1), 52–68.

Tala, S. (2009). Unified view of science and technology for education: Technoscience and technoscience education. *Science & Education18* (3–4), 275–298.

Tala, S. (2011). Enculturation into technoscience: Analysis of the views of novices and experts on modelling and learning in nanophysics. *Science & Education 20*(7–8), 733–760.

Thomas, G., & Durant, J. (1987). Why should we promote the public understanding of science?. In Sortland, M. H. (Ed.). *Scientific literacy papers.* Oxford: Department of External Studies.

Van Fraassen, B. C. (1980). *Scientific image*, Oxford: Clarendon Press.

Vincenti, W. G. (1990). *What engineers know and how they know it: Analytical studies from aeronautical history.* Baltimore: The Johns Hopkins University Press.

Woodward, J. (2003). Experimentation, causal inference, and instrumental realism. In Raddér, H. (Ed.), *The philosophy of scientific experimentation.* Pittsburgh: University of Pittsburgh press.
Zubrowski, B. (2002). Integrating science into design technology projects: Using a standard model in the design process. *Journal of Technology Education, 13*(2), 48–67.

Suvi Tala
Department of Physics,
University of Helsinki

CHAPTER 5

TERESA J. SHUME

COMPUTER SAVVY BUT TECHNOLOGICALLY ILLITERATE

Rethinking Technology Literacy

INTRODUCTION

We humans are toolmakers. Since the time of our ancient ancestors, human ingenuity has created tools that alter the natural world in ways that deeply affect our lives. Imagining a sector of modern life that is not influenced by machines and devices of human design is difficult. Agriculture, medicine, transportation, communication, and entertainment continue to undergo profound changes in light of technological advancements. Each advancement spawns new solutions as well as new problems in a spiral of ever-increasing complexity. To make a meaningful difference in tomorrow's technological world, our future citizenry will need more than the skills of reading and writing; children will also need the ability to understand the nature of technology, and to apply this understanding to wisely use and manage technology – so that technology does not use us.

Like all sectors of modern life, education is undergoing profound changes because of technology. Technological developments in education have changed how students undertake many academic tasks, such as exchanging information and producing visual representations of data. Some technological changes encourage student to interact in ways that were not previously possible, such as real-time video links between distant classrooms and participation in virtual environments. Students are immersed in a world drenched in information, a state of affairs for which computer technology can be both blamed and celebrated. Gargantuan portions of educational resources are being funneled towards computer technology in schools and we are scarcely aware of what is being left behind (Oppenheimer, 2003; Cuban, 2001). For better and worse, computer technology is changing education in profound ways.

Alongside scientific literacy, ecological literacy, media literacy, visual literacy, religious literacy, and a whole host of other forms of literacy aimed at designating proficiency of knowledge and skills in particular fields, technology literacy has become a principal concern of educators across the world. This chapter will trace

Michael P. Clough, Joanne K. Olson and Dale S. Niederhauser (Eds.), The Nature of Technology: Implications for Learning and Teaching, 85–100.

the recent history of the dominant perspective on technology literacy and will argue that a technologically literate citizenry will not result. Further, this chapter will explore some key aspects of robust technology literacy including a foundation in a broader definition of technology, explicit rejection of technological determinism, technological instrumentalism, and technological fundamentalism, as well as congruence with democracy and ecological sustainability.

DOMINANT CONCEPTION OF TECHNOLOGY LITERACY

During the past decade, substantial energy has been channeled towards the creation of standards and technology plans at the national, state, and local levels. An array of national and international technology literacy documents have been created with the aim of shaping state and local technology literacy standards, guiding the integration of technology into school curriculum, as well as impacting teacher preparation. Some of these undertakings include *United Nations Educational, Scientific and Cultural Organization's Information and Communication Technology Competency Standards for Teachers* (UNESCO, 2008), *Standards for Technological Literacy* (International Technology Education Association, 2007), *Technology for All Americans: A Rationale and Structure for the Study of Technology* (International Technology Education Association, 1996), as well as standards for students, teachers, and administrators produced by the National Educational Technology Standards Project (International Society for Technology in Education, 2011). Every state in the U.S. has developed technology standards (Education World, 2008), and local school agencies seeking technology funds from federal programs such as the *E-Rate Program* (Universal Service Administrative Company, 2008) and the *Enhancing Education Through Technology (Ed-Tech) State Program* (U.S. Department of Education, 2007) are required to create and implement technology plans. The ultimate aim of these efforts is to help schools produce technologically literate citizens.

Perhaps the most influential and widely used set of documents impacting technology literacy in the United States today is the International Society for Technology Education (ISTE)'s *National Educational Technology Standards (NETS)* (Alliance for Childhood, 2004; ISTE, 2011). ISTE released its current standards for students, *NETS for Students: The Next Generation* in 2007, after producing the first set of national technology standards for students in 1998. Each version includes sets of performance indicators for preK-2, grades 3–5, grades 6–8, and grades 9–12. Additionally, ISTE released revised *NETS for Teachers* in 2008 after producing original standards for teachers in 2000, as well as revised *NETS for Administrators* in 2009 based on the original 2001 standards for administrators. By 2003, 48 of 50 U.S. states had adopted, adapted, aligned with or referenced the NETS standards in their own department of education documents pertaining to technology; these included technology plans, curriculum plans, assessment plans, certification, licensure or other such documents (ISTE, 2003). Today, NETS standards have not only been broadly adopted in the United States but are actively being adapted by schools in several other countries, including Norway, Costa Rica, Malaysia, Japan, Australia, Philippines, Micronesia, Korea, and Turkey. (ISTE, 2011). Established in 1979 and operating with a budget over $14 million U.S. dollars in 2010, ISTE's membership includes 20,000 individual members, 64 corporate members, 76 affiliate organizations, 86 individual member countries, and encompasses five affiliate regions including Australia, Canada, India, the United Kingdom and the United States (ISTE, 2011).

ISTE's first set of standards, *Technology Foundation Standards for Students* was focused heavily (but not exclusively) on computer skills and included these six standards: (a) Basic Operations and Concepts; (b) Social, Ethical, and Human Issues; (c) Technology Productivity Tools; (d) Technology Communication Tools; (e) Technology Problem-Solving and Decision-Making Tools; and (f) Technology Operations and Concepts. An examination of the standards and related performance indicators reveals that most of these standards focused on abilities to manipulate computer technologies to perform particular tasks. The standard on social, ethical, and human issues was the exception to this pattern, however it was widely ignored in K-12 classrooms (Neiderhauser, Lindstrom, & Strobel, 2007). The computer skills included in the standards did not simply lists types of software students should learn to control (e.g. students will create PowerPoint presentations, will author web pages, will master desktop publishing, and so on). Instead, skills and knowledge were grouped by purpose (e.g. technology for productivity, for communication, for research, and so on). Nonetheless, the essence of the skills and knowledge sought by these early national standards clearly centered on learning to use computers effectively for certain purposes, in short: computer skills.

National and international documents, including the ubiquitous NETS standards, are shifting beyond technical computer skills and towards integration of critical, cognitive and problem-solving skills with digital technology and communication tools, an approach seeking literacy in "Information and Communication Technologies" or ICT. *The United Nations World Youth Report 2005* (United Nations Department of Economic and Social Affairs, 2005, p. 77) defines ICT as "all technologies that enable the handling of information and facilitate different forms of communication." Another international document that demonstrates this shift away from pure technical skills and towards critical and cognitive skills is *Digital Transformation: A Framework for ICT Literacy* (International ICT Literacy Panel, 2007). Convened by Educational Testing Service (ETS), the panel, comprised of educators, technology experts, and scholars from five countries including the United States, defines information and communications technologies literacy this way, "ICT literacy is using digital technology, communications tools, and/or networks to access, manage, integrate, evaluate and create information in order to function in a knowledge society" (International ICT Literacy Panel, 2007, p. 1). The panel goes on to report that, "ICT literacy cannot be defined primarily as the mastery of technical skills. The panel concludes that the concept of ICT literacy should be broadened to include both critical and cognitive skills as well as the application of technical skills and knowledge" (International ICT Literacy Panel, 2007, p. 1). This statement infers that the previous focus was primarily on technical computer skills. It also demonstrates that ICT is grounded firmly in the realm of computer technology whose chief purposes include information management and communication.

The organization and content of ISTE's newest educational technology standards for students, *NETS for Students: The Next Generation* also embodies an ICT approach (ISTE, 2007). This document is now divided into six different standards: (a) Creativity and Innovation; (b) Communication and Collaboration; (c) Research and Information Fluency; (d) Critical Thinking, Problem-Solving and Decision-Making; (e) Digital Citizenship; and (f) Technology Operations and Concepts. The insertion of the term "ICT" into the title of the *NETS for Students* performance indicators that accompany the standards is the most obvious signal of ISTE's shift toward ICT. In 1998, the performance indicators were entitled *Profiles for Technology Literate Students: Performance Indicators*. In 2007, the new title for the performance indicators was *Profile for Technology (ICT) Literate Students.*

Worthy of note is that ISTE's first version of *NETS for Students* included the nature of technology as part of the standard on basic operations and concepts, while the new ISTE standards omit any reference to the nature of technology. In 1998, *NETS for Students* stated, "Students demonstrate a sound understanding of the nature and operations of technology systems" (ISTE, 1998, p.14), while the 2007 version indicates, "Students understand and use technology systems" (ISTE, 2007). In the 1998 version, this standard was widely interpreted by K-12 teachers to focus exclusively on computer skills and not the nature of technology (Neiderhauser, Lindstrom, & Strobel, 2007), so no significant shift has occurred in practical terms. Nonetheless, an interesting observation is that the 2007 ISTE standards are devoid of both explicit and implicit references to the nature of technology. Even the "Digital Citizenship" component eschews any hint of critically examining the nature of technology and its impact on society, an egregious and unfortunate missed opportunity. The dominant trend towards ICT in educational technology has failed to incorporate analysis of the nature of technology, and remains grounded in a conception of technology limited to the usage of computers and other digital technologies.

EDUCATIONAL TECHNOLOGY OR TECHNOLOGY EDUCATION?

Educational technology seeks to use "multimedia technologies or audiovisual aids as a tool to enhance the teaching and learning process" (ITEA, 2007, p. 238). It has also been defined by the Association for Educational Communications and Technology (in Januszewski & Molenda, 2008, p.1) as "the study and ethical practice of facilitating learning and improving performance by creating, using and managing appropriate technological processes and resources." In classrooms across the country, educational technology is synonymous with the use of computers, telecommunications and other digital technology as well as electronic networks to support learning. The restricted focus on computer technology drives more than the dominant national standards for producing technology literate students. Indeed, the *U.S. Department of Education's National Education Technology Plan* (2004) and the portion of the *No Child Left Behind Act* known as *Enhancing Education Through Technology Act of 2001* (U.S. Department of Education, 2001) also equate technology literacy with an information technology literacy approach, as does the United Nations Scientific and Cultural Organization (UNESCO, 2008).

The central argument of this chapter is that while ICT literacy is widely regarded as the best path to technology literacy, it is not authentic technology literacy. In spite of the vast resources we are spilling into ICT literacy across the United States (Oppenheimer, 2003; Cuban, 2001), this approach will not result in a technologically literate citizenry. Technology literacy is aptly described as "a study of technology, which provides an opportunity for students to learn about the processes and knowledge related to technology that are needed to solve problems and extend human capabilities" (ITEA, 2007, p. 242). It stems from a conception of technology that expands vastly beyond computers and related digital media, encompasses technology in all its forms, and delves into the realm of critical analysis of technology._

Widespread confusion exists between educational technology and technology education (ITEA, 2003, 2007). Even though every state has technology standards in place (Education World, 2008), very few students across the nation receive any exposure to the study of technology resonant with the definition of authentic technology education described above (ITEA, 2007). Disparate conceptions of technology are one of the key differences that distinguish authentic technology education from educational (ICT) technology.

CONCEPTIONS OF TECHNOLOGY

A Gallup poll conducted in the spring of 2001 revealed that the American public holds a very narrow view of technology, defining it as primarily computers and the Internet (ITEA, 2003). An ICT literacy approach operates on a broader conception of technology that includes not only computers and the Internet, but an array of other digital devices and electronic environments. A richer conception of technology recognizes the products of human invention, such as cell phones, vehicles, pesticides, antibiotics, magnetic resonance imaging, microwave ovens, fire retardant fabrics, and a whole host of other artifacts of technological know-how that inhabit our lives.

A still deeper understanding of technology recognizes that at its core, technology is about humans modifying the natural world to meet human needs and to extend human capacities. Technology includes not only artifacts of human invention, but also the knowledge, processes, and systems that produce those artifacts. Consider the following definition of technology from ITEA (2007, p.242) that, though very concise, offers a vastly expanded conception of technology as "the innovation, change, or modification of the natural environment to satisfy perceived human needs and wants." Another rich perspective on technology is offered by the American Association for the Advancement of Science (AAAS, 1990, p. 25):

> In the broadest sense, technology extends our abilities to change the world: to cut, shape, or put together materials, to move things from one place to another, to reach farther with our hands, voices, and senses. We use technology to try to change the world to suit us better. The changes may relate to survival needs such as food, shelter, or defense, or they may relate to human aspirations such as knowledge, art, or control.

The history of technology is as old as the history of humans. Indeed, discoveries of the earliest ancient tools are widely accepted by archaeologists as signs of the beginnings of human culture (Nye, 2006). From the times of chipping flint and harnessing fire to today, technology has been a potent force in the unfolding of civilization over the past several thousand years. It stands beside language, commerce, ritual, and the arts as core elements of cultural systems. Technology both influences and echoes a cultural system's values (AAAS, 1990).

These expanded definitions not only envelop the impact of computer technology, but also a much broader perspective on the role of technology in civilization. A cogent perspective on technology literacy must stem from a broad and encompassing conception of technology, one that recognizes the role of technology in civilization and not simply a narrow focus on computers and telecommunications technologies used for learning.

Further, insightful conceptions of technology encompass an awareness of the nature of technology and its impacts on our lives, our society, and our ecosphere. In *The End of Education* (1995), for example, Postman enumerates ten principles about the nature of technology that he posits all K-12 students should understand. For example, he points out that technological change is a Faustian bargain because new technologies bring not only advantages, but corresponding disadvantages. Postman also describes how all forms of technology carry intellectual, emotional, sensory, social, and content biases. A conception of technology that underpins authentic technological literacy must expand beyond computers and digital media, beyond the artifacts of technology that populate our lives, beyond an understanding of technology as a force that shapes the unfolding of civilization, and delve into critical analysis of the nature of technology itself.

TECHNOLOGICAL DETERMINISM AND INSTRUMENTALISM

In addition to stemming from an astute definition of technology, robust conceptions of technology literacy should explicitly equip students to think critically about technological determinism, a broad cultural assumption that "technologies forge ahead under their own steam, independent of human influence" (Alliance for Childhood, 2004, p. 12). Technological determinism holds that technological advances and their corresponding social changes are inevitable and cannot be stopped. This perspective can be contrasted with an instrumentalist view of technology that asserts technology is benign, neutral, and entirely subservient to human purposes (Carr, 2010).

From medicine to transportation to communication, overwhelming evidence exists that technologies can and do drive social change. In *The Disappearance of Childhood*, Postman (1982) constructs a cogent line of reasoning for the role of the printing press in the emergence of childhood and the role of television and other media in the erosion of childhood innocence, profound social changes that technologies have induced. More recently, Carr (2010) offers a compelling thesis for how the Internet is altering our ability to read and eroding our capacity for deep thought. Instrumentalist perspectives lack awareness of the ways that technologies can use us if we embrace technology without carefully examining larger social and cultural questions that critics such as Postman and Carr raise. Does this mean that determinism is the more compelling perspective? Clearly, technology can initiate social and cultural transformations that transcend time and geography. Acknowledging that potent social changes can stem from technological advances, however, is not necessarily congruent with ceding that such changes are ineluctable and beyond the control of humans.

Carr (2010, p.211) offers a particularly useful concept, technological momentum, that can bring clarity to this seemingly dichotomous debate.

> Technological momentum...acknowledges that once a system such a railroad or an electrical grid has been designed to certain specifications and put in place, it has a rigidity and direction that can seem deterministic to those who use them.

In other words, while technologies can and do drive profound social change, they do not *have* to. Because technology is a human enterprise, humans have the capacity to control the invention, dissemination, and use (or rejection) of technology. Indeed, the gun would seem an irresistible form of technology once introduced, but the Japanese samurai class rejected guns and Japan banned guns for a span of three centuries. Further, Mennonites and Amish in the United States have long been highly selective about adopting any new technologies, demonstrating that cultures and communities can make deliberate choices about the use of technology. Resisting technological determinism does not mean denying that technologies can shape society in tacit and powerful ways, but rather becoming aware of such social impacts and striving for deliberate and cognizant responses to them, which may include shunning certain technologies all together.

Insightful conceptions of technology literacy prepare students to tease apart concepts such as technological determinism and technological instrumentalism, in order to recognize the serious flaws in both of these perspectives. For example, ITEA (1996, p.26) includes linkages between technology, society, and the environment in its knowledge base for technologically literate citizens.

> Decisions concerning the development and the use of technology cannot be made by today's world without an understanding of how technology influences and is affected by society and the environment. Individuals, societies, and academic disciplines all affect technology and, in turn, are changed by new

> technological developments. These influences and impacts can be positive or negative, anticipated or unanticipated, depending on the situation.

A counter example (Alliance for Childhood, 2004) can be found in one of the grades 9–12 performance indicators of the original *NETS for Students* (ISTE, 1998, p. 24), "Make informed choices *among* technology systems, resources, and services" [emphasis added]. This performance indicator tacitly silences any voice wishing to discuss *if* a particular technology system, resource or service should be utilized or not. Students need to be encouraged and equipped to explore knotty questions about human agency and cultural assumptions in the development of technology in order to resist both technological determinism and instrumentalism.

TECHNOLOGICAL FUNDAMENTALISM

A separate but related concept is David Orr's idea of technological fundamentalism, tenacious belief in the assumed value of the explosive juggernaut of technology without questioning basic assumptions about how tools relate to larger purposes or prospects[1] (Orr, 2002). Any robust conception of technology literacy would explicitly eschew technological fundamentalism because unbridled enthusiasm for technology brings an unbalanced and myopic perspective to decision-making about technology. ITEA (2007, p. 9–10) offers this insightful remark about technology literacy, "A technologically literate person is comfortable with and objective about the use of technology, is neither scared of it nor infatuated with it."

The organization that produced the dominant educational technology standards, however, is steeped in technological fundamentalism as it pertains to educational technology. ISTE's NETS project promotes an unquestioned assumption that immersing children and teachers in educational technology is overwhelmingly positive and that society should enthusiastically welcome computer technology's ceaseless and ineluctable permeation into our lives. Consider, for example, this quotation entitled "Envisioning the Future of Education and Technology" from the "Educator Resources" section of ISTE's web site (ISTE, 2008).

> Click each blue link to transport yourself into the future of School 2.0. Every click of the mouse will lead you on a different journey. Travel full speed ahead with cutting edge tech tools. The future is about classroom redesign, innovation in student creativity, Second Life. Design your lesson plans to join in on the avatar-action yourself. Don't think that the students have all the fun. Give yourself a little MySpace and take a stroll through Facebook. Research in a new dimension and just Wikipedia it. Set your standards high with Classroom Blogging. Upgrade your school to version 2.0. Buckle up. Hold on. And begin your Learning Journey!

Though research is underway into whether or not computer technology results in sustained positive gains in student learning, this question has not been answered conclusively through independent research from unbiased funding sources (Alliance for Childhood, 2004). Nonetheless, ISTE and similar organizations urgently expound upon the assumed and unquestioned worthiness of educational technology and do not invite students, teacher, parents, or other stakeholders to reflect on decisions about embracing computer technology. It is viewed with unbridled enthusiasm in an overwhelmingly positive light. Technological fundamentalism has no place in cogent conceptions of technology literacy; mature conceptions of technology literacy explicitly decry technological fundamentalism.

Vital components of authentic technology literacy include a robust conception of technology, resistance to technological determinism, and rejection of technological instrumentalism and technological fundamentalism. These elements are critical for technologically literate citizens to live examined lives where we control technology rather than it controlling us without our awareness, to participate fully in democracy, and to contribute to achieving ecological sustainability.

DEMOCRACY AND TECHNOLOGY LITERACY

In a democracy, the core purpose of technology literacy must be to prepare future generations to actively shape the country's technological future in a morally responsible way (Cordes & Miller, 1999). Children in K-12 schools today will face an array of complex scientific questions that will evoke a multitude of thorny social conundrums. They are inheriting a world abound with technologies that were considered the realm of science fiction just a few decades ago. How should we determine what reproductive technologies are developed and implemented and which are not? Should genetically engineered food crops be labeled in the grocery store? Should we permit cloning of humans? How will climate change affect food production? What are some potential social consequences of introducing advanced robotics and nanotechnologies into the human body? What is gained and what is lost when educational experiences are replaced by online interactions? How do online exchanges impact our sense of moral reciprocity towards each other? This sampling of questions only begins to scratch the surface of the profound ecological and social dilemmas stemming from emergent technology today and in the future.

Another important dimension of the nexus between technology literacy and democracy lies in issues of power and control. How will we ensure democratic decision-making that serves the common good and not only the elite? A compelling conception of technology literacy will prepare citizens to aptly contribute to the important democratic debates pursuant to questions like these. One could point out that an ICT approach to technology literacy supports democracy by ensuring an informed citizenry capable of locating, organizing, scrutinizing, creating, and using information. While ICT literacy does indeed perform this function, it also fosters determinism and fundamentalism about computer technology while remaining silent on technology as a process that has shaped the unfolding of human civilization.

Technological determinism, technological instrumentalism and technological fundamentalism diminish our citizens' readiness to contribute to a healthy democracy. Unexamined acceptance of a deterministic view of technology is incongruent with a robust democratic system that oversees the restrained and responsible use of technology because it assumes human agency in controlling the expansion of technology is futile. Those who espouse deterministic views of technology can share the optimism of technological fundamentalists or can be pessimistic about technology in society (Nye, 2006), but because determinism is veiled from the idea that the infusion of technology into our lives is under human control, it can erode engagement in productive democratic debates on such topics. Conversely, instrumentalist perceptions contend human agency in controlling the expansion of technology is assured because technology is benign and under human control, leaving such citizens dangerously ignorant of the sometimes furtive ways that technology penetrates our lives. Those who hold technological fundamentalist views may very well engage in social debates seminal to robust democratic processes, but are more likely to approach debates with unbridled enthusiasm for technological advances, and thus regard possible choices about technology through a narrow lens. Such a perspective is more likely to contemplate which technologies to use and when to use them, rather than to question if certain technologies should be developed or used at all. Technological determinism, instrumentalism and fundamentalism are dangerous perspectives that feed complacency and blind acceptance of technological advances without analytical reflection, thus eroding our capacity for robust democratic debates.

Instead of regarding ourselves as passive consumers of technology that is ineluctably penetrating ever deeper into our lives, we need a shift towards actively evaluating new technologies by considering costs, benefits, and risks that are both anticipated and potentially unanticipated, as well as recognizing which social groups will gain and which will lose (AAAS, 1990). This active engagement in a deep conversation about technology needs to happen at all levels of society and in all realms. It cannot be ceded to engineers or the corporate world; it needs to happen in both K-12 and postsecondary classrooms, around kitchen tables, in teacher staff rooms and in coffee shops. To equip citizens to partake in these debates, our children need to be exposed to a potent conception of technology and need to be empowered to actively and openly deliberate the role of technology in society. Ultimately, the degree to which students are prepared to undertake reflective and responsible technological citizenship is vastly more important than teaching students to operate the latest generation of computers (Sclove, 1995). This vital aspect should be captured in any cogent perspective on technology literacy.

ECOLOGICAL SUSTAINABILITY AND TECHNOLOGY LITERACY

Another reason why a robust and insightful conception of technology literacy is vital is that human decisions about technology greatly affect our planet's potential for ecological sustainability. Consider this perspective from AAAS (1990, p. 107):

> Many parts of our world are designed – shaped and controlled, largely through the use of technology – in light of what we take our interests to be. We have brought the earth to a point where our future well-being will depend heavily on how we develop and use and restrict technology. In turn, that will depend heavily on how well we understand the workings of technology and social, cultural, economic, and ecological systems within which we live.

We need a conception of technology literacy that not only acknowledges and explores the powerful connections between technology and ecological sustainability, but also commits to pursuing ecological sustainability through restrained and responsible use of technology. Technology literacy standards steeped in technological fundamentalism bring the danger of a propensity towards a blind faith that technology will provide solutions to the most dangerous and extreme environmental problems in ways that will not require us to make painful choices or difficult changes. We need to reject the hubris inherent in technological fundamentalism and recognize that quick technological fixes and industrial applications of technology on massive scales stem from an arrogant and dangerous belief that humans know enough to outwit nature (Vitek & Jackson, 2008). Technologically literate citizens would understand that technology cannot extend the finite resources of our planet to infinite capacities and cannot solve all problems, but can be used by people to increase our prospects for ecological sustainability if managed in prudent, restrained, and responsible ways.

ITEA's conception of technology literacy (2003) is to be commended for including the natural environment as one of the reasons technology literacy is vitally important. This approach aims to recognize the full lifecycle of materials and products, that technology can be used in ways that are both beneficent and nefarious for the environment, and that sometimes using technology puts economic and environmental concerns in competition with each other. Exposure to these ideas may better prepare students to make sound decisions regarding technology and our planet's prospects for ecological sustainability. ITEA's documents delve into knotty and value-laden questions surrounding ecological sustainability, and congruent with a scientific approach, ITEA's work upholds a commitment to objectivity and balance.

Absolute objectivity, however, is a myth because every human being operates from his or her own worldview. Balance is often equated with neutrality, but even centrist views are anchored in particular values and assumptions; they are simply harder to recognize because they do not raise controversy within the mainstream. The ITEA standards attempt to be as sanitized from subjectivity and bias as possible, but absolute objectivity and absolute neutrality are impossible. For example, there is an anthropocentric worldview inherent in the ITEA documents (1996, 2003, 2007) – an assumption that humans are the most important beings in the universe and that the world exists primarily for instrumental human uses. Anthropocentrism can be contrasted with ecocentrism, which is grounded in an assumption that all of nature has intrinsic value and human are an integral part of the ecosphere rather than masters of it. Our prospects for ecological sustainability would be greatly improved by tapping into the ecological resilience inherent in participating in biodiversity rather than destroying it. It would be better for technology curriculum materials to openly acknowledge that while objectivity and balance are desirable and worthy of pursuit, they are ultimately impossible to attain in an absolute sense. Students need to be equipped to identify frames of reference and tacit assumptions imbedded in curriculum as well as their own values and beliefs, and to recognize which may be congruent with ecological sustainability and which may not.

Another connection between technology and our prospects for environmental sustainability resides in children's interactions with one particular form of technology; digital technology is altering children's sense of place in the natural world. In particular, children are becoming less connected to local, natural environments as their time and energy for play are directed more and more towards screen time and electronic gadgets (Louv, 2005). Connecting to one's local natural environment is key to developing a sense of place (Sobel, 1996, 2005) and ultimately to developing a bond with the natural world that will cultivate a drive to protect it (Sobel, 1996). Stephen Jay Gould (1991, p. 14) wrote, "We cannot win this battle to save species and environments without forging an emotional bond between ourselves and nature as well – for we will not fight to save what we do not love." Even though computer technology permits children to be exposed to information about natural environments all over the world through virtual interactions, spending excessive amounts of time in front of a computer threatens the development of children's local sense of place, which is seminal to fostering a sense of responsibility towards the planet. It sad that most children in the United States probably know more about the tropical rain forest than the local ecosystem they inhabit. In this regard, the ICT literacy model is problematic.

Another way that children's interactions with computers lessens our prospects for ecological sustainability is by favoring certain ways of thinking and types of knowledge that are incongruent with ecological sustainability. Bowers (2000, 2006) argues that computers represent the next phase of the Industrial Revolution and as such are cultural carriers of values and attitudes that support unsustainable consumerist behaviors. Orr (2002) cautions that "fast" knowledge leads to a hurried pursuit of technological progress regardless of our ignorance about unexpected, adverse, and irreversible consequences. Fast knowledge is supplanting "slow" knowledge – intergenerational wisdom acquired through cultural maturation and calibrated for preservation of ecological harmony. Both Bowers (2000) and Orr (2002) warn of the dangers of losing intimate knowledge of the land and ways of thinking that were built up slowly over hundreds of generations and exist in localized social and ecological contexts.

Ecological sustainability will require active participation of an informed citizenry in debates about the benefits, challenges, and dangers stemming from the uses (and misuses) of technology in our personal lives, our communities, and our society as a whole. A truly technologically literate citizenry would gain an ever-increasing sophistication of understanding about the reticular and potent relationship between technology and our prospects for ecological sustainability.

ROBUST CONCEPTION OF TECHNOLOGY LITERACY

Qualities of a robust conception of technology literacy include a foundation in a sophisticated conception of technology, resistance to technological determinism, aversion to technological instrumentalism and fundamentalism, and commitment to the purposes of democracy and ecological sustainability. Further, technologically literate citizens must also possess a potent analytical lens for understanding the nature of technology, its intended and unintended consequences, and its impacts on individuals and society. Sophisticated conceptions would resonate with the following concise definition of technology literacy:

> [Technology literacy is] the mature capacity to participate creatively, critically, and responsibly in making technological choices that serve democracy, ecological sustainability and a just society. (Alliance for Childhood, 2004, p. 4)

The Alliance for Childhood (2004; Cordes & Miller, 1999) has forged a compelling conception of technology literacy that recognizes the necessity of ecological sustainability, highlights the profound value of children's natural play as well as face to face human relationships, and raises important questions about the role of computers in the education of children.

Another cogent vision for technology literacy has been set forth by ITEA (1996, 2003, 2007) and is encapsulated in these quotations:

> Technological literacy is the ability to use, manage, assess and understand technology...Technology is the modification of the natural environment in order to satisfy perceived human needs and wants. (ITEA, 2007, p. 242)

> Technology literacy is much more than just knowledge about computers and their application. It involves a vision where each citizen has a degree of knowledge about the nature, behavior, power, and consequences of technology from a broad perspective. (ITEA, 1996, p.1)

> A technologically literate person understands, in increasingly sophisticated ways that evolve over time, what technology is, how it is created, and how it shapes society, and in turn is shaped by society. (ITEA, 2007, p.9)

This approach explicitly eschews technological determinism, instrumentalism, and fundamentalism. It aims to meet individual needs of citizens and consumers; societal needs for democracy, economy, and shared responsibility; and environmental needs so the Earth can continue to support human life (ITEA, 1996). It is based on a structure for the study of technology that elucidates the connections between processes, contexts, and knowledge. And while ITEA's approach to technology literacy falls short with regards to ecological sustainability, an egregious shortcoming to be sure, it offers a much richer model of technology literacy than the current dominant conception.

CONCLUSION

Today, the nation's operating conception of technology literacy resides in an information and communications technology approach, one that is rooted in a shallow conception of technology and resonates with technological determinism and fundamentalism regarding computer and other digital technology. Vigorously directed by government and the private sector, schools are committing billions of dollars to computers, other digital devices, services, and training (Oppenheimer, 2003; Cuban, 2001) and thus it appears we are making a sincere effort to fulfill our obligation to foster technology literacy among today's children. This approach falls egregiously short, however, as a conception of technology literacy that will prepare a citizenry who will understand the nature of technology and will be poised to make responsible decisions regarding technology on personal, societal, and ecological levels. Robust technology literacy is grounded in a sophisticated conception of technology, one that recognizes technology as a potent process by which we are modifying the natural world to meet human needs and wants. An important quality of an insightful and veritable approach to technology literacy is a careful analysis of technological determinism and an open rejection of technological instrumentalism and fundamentalism, essential for resonance with the needs of democracy and ecological sustainability.

There is an urgent need to open up a dialog about the differences between educational technology and technology education to increase awareness about the frightening dearth of robust technology education across the country. It is time for all stakeholders to critically examine the ideology that underpins an ICT approach to technology literacy, to reassess our usage of computer technologies in schools, and make commensurate changes to our allocation of educational resources. Our nation needs to undertake an ambitious, committed and immediate effort to implement a cogent form of technology education because our democracy and our ecological future depend on a technologically literate citizenry. Our children deserve no less.

NOTE

1 David Orr commits the sixth chapter of his book, *The Nature of Design: Ecology, Culture, and Human Intention,* to explaining technological fundamentalism.

REFERENCES

Alliance for Childhood. (2004). *Tech tonic: Towards a new literacy of technology*. College Park, MD: Author.

American Association for the Advancement of Science. (1990). *Science for all Americans: Project 2061.* New York: Oxford University Press.

Bowers, C.A. (2000). *Let them eat data: How computers affect education, cultural diversity, and the prospects of ecological sustainability.* Athens: The University of Georgia Press.
Bowers, C.A. (2006). *Revitalization of the commons: Cultural and educational sites of resistance and affirmation.* Lanham, MD: Rowman and Littlefield.
Carr, N. (2010). *The shallows: What the Internet is doing to our brains.* New York: W.W. Norton & Company.
Cordes, C. & Miller E. (1999). *Fool's gold: A critical look at computers in childhood.* College Park, MD: Alliance for Childhood.
Cuban, L. (2001). *Oversold and underused: Computer in the classroom.* Cambridge, MA: Harvard University Press.
Education World. (2008). *Technology standards by state.* Retrieved from http://www.education-world.com/standards/state/toc/index.shtml#technology
Gould, S. J. (1991). Unenchanted evening, *Natural History, 100*(9), 4–14.
International ICT Literacy Panel. (2007). *Digital transformation: A framework for ICT literacy* [Panel Report]. Available from http://www.ets.org/ictliteracy/digital1.html
International Society for Technology in Education. (1998). *National educational technology standards for students.* Eugene, OR: Author.
International Society for Technology in Education. (2003) *Use of NETS by state.* Retrieved from http://www.iste/org/Content/NavigationMenu/NETS/ForStudents/1998Standards/NETS_for_Students_1998.htm
International Society for Technology in Education. (2007). *National educational technology standards for students: The next generation.* (2nd ed.). Eugene, OR: Author.
International Society for Technology in Education. (2011). *NETS home page.* Retrieved from http://www.iste.org/standards/global-reach.aspx
International Society for Technology in Education. (2008). *Your learning journey.* Retrieved from http://www.iste.org/Content/NavigationMenu/EducatorResources/YourLearningJourney/EnvisioningtheFutureofEducationandTechnology/Envisioning_the_Future_Home_Page.htm
International Technology Education Association. (1996). *Technology for all Americans: A rationale and structure for the study of technology.* Reston, VA: Author
International Technology Education Association. (2003). *Advancing excellence in technological literacy: Student assessment, professional development, and program standards.* Reston, VA: Author.
International Technology Education Association. (2007). *Standards for technological literacy: Content for the study of technology.* Reston, VA: Author.
Januszewski, A., & Molenda, M. Eds. (2008) *Educational technology: A definition with commentary.* New York: Lawrence Erlbaum.
Louv, R. (2005). *Last child in the woods.* Chapel Hill, NC: Algonquin Books of Chapel Hill.
Niederhauser, D.S., Lindstrom, D.L., Strobel, J. (2007) Addressing the NETS*S in K-12 classrooms: Implications for teacher education. *Journal of Technology and Teacher Education, 15*(4), 483–512.
Nye, D. E. (2006). *Technology matters: Questions to live with.* Cambridge, MA: The MIT Press.
Oppenheimer, T. (2003). *The flickering mind: The false promise of technology in the classroom and how learning can be saved.* New York: Random House.
Orr, D. (2002). *The nature of design: Ecology, culture, and human intention.* New York: Oxford University Press.
Postman, N. (1982). *The disappearance of childhood.* New York: Delacorte Press.
Postman, N. (1995). *The end of education: Redefining the value of school.* New York: Alfred A. Knopf.
Sclove, R. (1995). *Democracy and technology.* New York: Guilford Press.
Sobel, D. (1996). *Beyond ecophobia: Reclaiming the heart of nature education.* Nature Literacy Series, Number 1. Great Barrington, MA: The Orion Society and The Myrin Institute.
Sobel, D. (2005). *Place-based education: Connecting classrooms and communities.* Great Barrington, MA: The Orion Society.
UNESCO. (2008). *Information and communication technology competency standards for teachers.* Retrieved from http://cst.unesco-ci.org/sites/projects/cst/default.aspx
United Nations Department of Economic and Social Affairs. (2005). *World youth report 2005: Young*

people today and in 2015. (United Nations Publication No. 05 IV.6 05 44737 – 3000). United Nations Publications.

Universal Service Administrative Company. (2008). *Schools and library E-Rate program.* Retrieved from http://www.universalservice.org/sl/applicants/

U.S. Department of Education. (2001). *Enhancing education through technology act of 2001.* Retrieved from http://www.ed.gov/policy/elsec/leg/esea02/pg34.html

U.S. Department of Education. (2004). *National education technology plan: Toward A new golden age In American education – How the Internet, the law and today's students are revolutionizing expectations.* Retrieved from http://www.ed.gov/about/offices/list/os/technology/plan/2004/index.html

U.S. Department of Education. (2007). *Enhancing education through technology (Ed-Tech) state program.* Retrieved from http://www.ed.gov/programs/edtech/index.html

Vitek, B., & Jackson, W. (Eds.). (2008) *The virtues of ignorance: Complexity, sustainability, and the Limits of Knowledge.* Lexington, KY: The University Press of Kentucky.

Teresa J. Shume
School of Teaching and Learning,
Minnesota State University Moorhead

CHAPTER 6

DAVID H. JONASSEN

TRANSFORMING LEARNING WITH TECHNOLOGY

Beyond Modernism and Post-Modernism, or Whoever Controls the Technology Creates the Reality

INTRODUCTION

Throughout the industrial age, technology has promised to improve the lives of those who used it. Modernists believe that technology can produce faster, better, and more efficiently. During the 20th century, education has embraced technology. Technology has promised smarter, happier, better educated, and more fulfilled learners. Technology has always been zealously promoted as a modern solution for the problems of education—lack of productivity, inefficiency, and lack of focus. During the twentieth century, each new technology emerged as the panacea for education's socio-cultural problems. Unfortunately, each new technology has failed to deliver on its promise. Why? In this paper, I argue that modern and post-modern conceptions of technology impede the emergence of personal identities and learning, and we must redefine the relationships between learners and technology in order to transcend post-modernists' cynical fears and to truly empower learning

To support these two positions, I will contrast modern and post-modern views of technology and suggest a newer, transformative view of technology in education in an effort to explain why technology has failed and to provide a vision for how it could work.

VIEWS OF TECHNOLOGY

What is the future of educational technology? At best, I can provide a number of visions that have been promulgated during the latter part of the 20th century. Because each of us constructs our own understanding of reality based on our real experiences and mediated experiences, these views can provide options for reflecting our own view. Since we also socially co-construct meaning, they can also provide a focus for conversation as we predict and hopefully co-determine our own future with technology.

Michael P. Clough, Joanne K. Olson and Dale S. Niederhauser (Eds.), The Nature of Technology: Implications for Learning and Teaching, 101–110.

Modern Views of Technology

During this century, technology has become the voice of society. In education, its role has been to transmit knowledge, culture, and meaning.

Triumph of technology: A technophilic view. For the past century and a half, technology has represented the vehicle to the future. From mechanical advantage to information edge, learning how to harness the power of various technologies has provided the pipeline to future success. The use of technology for social and material advancement has been a major goal of the past two centuries. For many, technology is the modern salvation to societal problems. In the nineteenth century, technology supplanted our physical work. In the twentieth, it has tamed time by transporting us faster. Moving into the 21st century, technology will fulfill our knowledge needs by thinking for us. Technologies are becoming more intelligent, with fuzzy logic controllers adjusting the performance of washing machines to accommodate the dirtiness of our clothes. We are increasingly represented by intelligent agents in our interactions with the world. They of course interact with other intelligent agents on our behalf rather than with us. HAL, the ascendant computer in Arthur C. Clark's and Stanley Kubrick's *2001: A Space Odyssey*, is waiting for us around the corner. In this technophilic view of technology as the answer to nearly every material and social problem, we began the abdication of personal identity, responsibility, and authority that matured in post-modern conceptions (discussed later).

Technophilia began in education with the highly specialized vocational education and home economics courses of the early twentieth century that sought to fulfill current vocational needs. However, the technophilic view of technology was best instantiated in the 1980s by an intense focusing on computer literacy. Computer literacy assumed that computers provided completely new symbol systems or formalisms to mediate knowledge sharing. It also assumed that computers were more than a symbol system; they should be the object of instruction. Computers should be studied as technological phenomena. Unfortunately, what too many students learned on the way to becoming computer literate was how to memorize the parts of a computer based on the "strong belief that vocabulary implies knowledge" (Bork, 1985, p. 34). We zealously believed that having students embrace and understand this silicon-mediated reality was essential.

Technology as teaching medium: Educational view. Since their inception, modern educational technologies have been conceived most frequently as instructional communicators, mediated teachers, and knowledge conveyers. Information is encoded visually or verbally in the symbol systems afforded by various technologies. During the "instructional" process, students perceive the messages encoded in the technology. A generation of research and teaching averred that information-based messages that are more effectively designed and encoded will naturally produce better communication, so result in greater learning gains (Fleming & Levie, 1978;

Salomon, 1979). Generations of instructional television and computer-based instruction operationalized learning in terms of presentation of information on screens to students, whose understanding or memory would occasionally be assessed by making pre-scripted responses to the technology. This view of "communication as transmission centers on the ancient practice of transmitting messages over long distances in order to exert control" (Pea, 1994). Technologies (books, teachers, slates, pictures) have been designed and intended for centuries to more efficiently transmit socially acceptable beliefs and values, that is, to exert intellectual authority over learners, thereby influencing individual constructions of reality.

This transmissive role for technology was intended to ameliorate the job of teaching to predictably control the learning of their students. Unconsciously, the role of teachers were also usurped. Technology as transmitter is based on an Aristotelian world view which relies on two essential components of reality–objectivity and causality–both integral components of western consciousness (Jonassen, 1983). Educational communications rely on objectivity to define the physical world (determine reality) which is transmitted to learners so that they can acquire the same objective reality. Educational communication also assumes that we can isolate cause-effect relationships so that we can be sure that our instructional interventions will affect learning predictably. To the degree that these beliefs are deemed true, technologies will work as information transmitters. And if we assume that information assimilation is a meaningful form of learning, learning technologies do enhance learning. However, from a critical perspective, this view of technology is both naive and ungrounded.

Post-Modern Views of Educational Technology

Post-modern conceptions of technology augur the further erosion of personal identity, responsibility, and authority because they are most concerned with power relationships. Their view is that technology represents a focus for power, a lever to lull society into believing again in democracy, when, in reality, technology, like any other value-laden tool, benefits some (those in power) more than others.

Technology as power: Controlling the masses. The eminent French post-modern philosopher Michel Foucault believed that thought can be instantiated in buildings. In *Discipline and Punish,* Foucault described the Panopticon (Figure 1), a circular building with a tower at its center with windows into each cell surrounding the tower. All that was needed was to put a supervisor in the middle and populate each cell with prisoners, workers, or school children enabling the supervisor to see and control everything. Each individual "is the object of information, never a subject in communication" (Foucault, 1977, p. 200).

The Panopticon is about power. The residents of the Panopticon are constantly reminded that they can be watched but will never know it when they are. In addition to surveillance, the Panopticon can function as a laboratory to try out experiments on its inhabitants. The Panopticon is "a generalizable model of functioning; a way of defining power relations in terms of the everyday life of men."

Modern learning technologies, especially networked computers permit, and have been used, to gather information on the ways we conduct our daily lives. Amazon.com does that constantly. Thus, they are Panopticons, but perhaps more insidious because they operate behind the scenes and we are typically not aware that it is happening.

In a post-modern critique of some of my research (provided in a blind review process), unknown reviewers argued that technologies such as hypertext and hypermedia, like prisons, observe, discipline, regulate and control their users. When learners interact with technology, they are told what to think, observed for what they do, and then evaluated for their understanding of the lessons.

The Internet has the potential to become the ultimate electronic Panopticon. It not only collects information about the habits of every user, it also deposits cookies onto their computers to observe them when they are not connected.

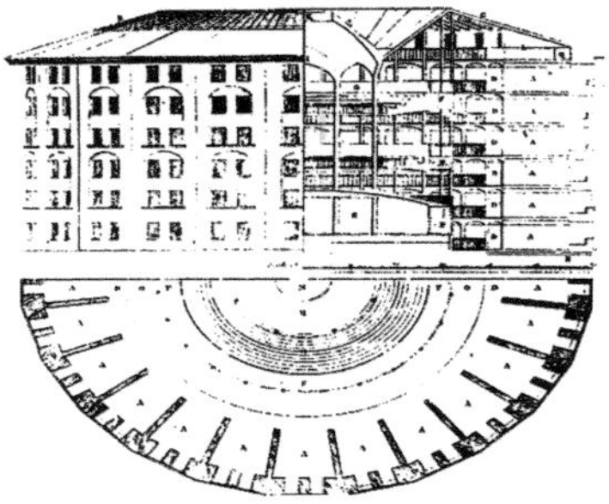

Technoglobalism: The commoditization of education. Another post-modern view of technology is provided by David Noble (1998; 2003), a Canadian professor of political science. He is one of the most outspoken critics of modern universities that seek to extend their influence through distance education technologies. Why are universities investing so heavily in distance technology infrastructure? Noble argues that in addition to the fear of getting left behind, there are the modern pressures of progress. However, technology is a Trojan horse that is not intended so much to bring education to the masses anytime, anywhere as it is the commercialization and commoditization of higher education. The acts of "transforming courses into courseware, the activity of instruction itself into commercially viable proprietary products that can be owned and bought and sold in the market." He argues that it stands to reason that publishers and hardware and software manufacturers, who have the most to gain, are the most ardent supporters of this movement. As universities become larger and more entrepreneurial in order to survive and grow (their modern imperative), this transformation of higher education is being initiated and implemented from the top down without any input from faculty or students. Its goal is economic advantage: technoglobalism. Its result, post-modernists argue, is the further erosion of personal identity, responsibility, and authority.

Online instruction has become pandemic in K-12 and higher education. While the putative goal of online instruction is to make it more accessible, the real goal of this commoditization is to control or eliminate the need for professors, Noble claims. Once courses go online, administrators gain greater direct control over faculty performance and course content (a higher education Panopticon). Once the knowledge is owned by the institution, the professors become redundant, and so they are outsourced, according to Langdon Winner (1997), so universities reduce their direct labor and plant maintenance costs, making them more efficient (another modern imperative). Technology is the new medium for administrative scrutiny, supervision, regimentation, discipline and even censorship, Noble argues. The result will be a reduction of faculty autonomy, independence and control over their work. Once the professors' ideas and identities become commodities, there will be no need for the real professors. Just like in other skilled industries, faculty activity is being restructured by technology in order to discipline, deskill and displace the labor force, and place as much control as possible into the hands of the administration.

What will be lost if education continues to devalue face-to-face instruction in favor of on-line courses. Anyone who has taught online realizes that mediated interpersonal interactions are substantively different from real-time interpersonal interactions. When knowledge and wisdom are objectified by online representation, how will learners view it? Bauerlein (2008) has concluded that youth are using themselves as experts and devaluing and distancing themselves from the adult world, and that despite claims that they are "digital natives," special and wonderful, they're actually dumber than ever.

Paranoid or prescient? Noble (2003) and Winner (1996) prophesy fundamental changes in education, as universities compete in the global information market and new educational companies bid against school districts for the privilege of educating their children. As society increasingly accepts the minimalization of education as information transmission and credentialing, we have less need for publicly held, publicly supported educational systems. When that happens, we accede power to corporations (including universities) who use technology as the medium for acquiring and using power.

Reflections on a Post-Modern world. What have we learned in our postmodern world? In the 20th century, we have witnessed the meltdown of the nuclear family. Parents have abdicated their responsibilities for caregiving and transmission of cultural values, relying instead on technology (especially commercial television, cellular phones, and now the Internet) to educate their offspring. What have children learned from these technologies? Obsessions with sex and violence; personal wealth is the only goal worth pursuing; peers are the only arbiters of reality (certainly not parents or teachers); education is unfulfilling and worthless; and knowledge is a commodity that can be charged on a credit card when you need it, or simply

Googled. The values that are conveyed by commercial technologies are violent, lurid, avaricious, petty, vapid, vulgar, and dumb. Are these the symptoms of social changes or the means for centralizing corporate power? Who is responsible for the degeneration of social values? The fabric of society is being rewoven with a substantively different warp, but who is controlling the loom?

Are there solutions? Can we reform society? What roles, if any, can technologies play in any solutions? Can we wrest control of the technologies? Can technologies foster and support positive change?

In the next section, I describe a different vision for how to use technologies to empower learners and to transform the relationship of learners and technologies. Transformative technologies can foster meaning making and strong identity formation among students, parents, and teachers who believe that learning is the construction, expression, and negotiation of personal beliefs, conceptions, and identities rather than the inculcation of doctrine. When no one has the right idea, but some have better ideas, and the best ideas emerge from social co-construction of reality, transformative technologies can help to learners articulate meaningful beliefs, rather than the self-serving perspectives promulgated by the Internet and mass media.

TRANSFORMATIVE VIEWS OF EDUCATIONAL TECHNOLOGY

Pea (1994) argued that in order to transcend the transmissive view of education, it is necessary to provide new ways of thinking, knowing, and acting in education. How can that happen? Rather than transmitting information more efficiently and (hopefully) effectively, and rather than controlling the thoughts and behavior of learners, encouraging learners to reflect on and represent what they know and believe and to use technology to support and amplify those activities is necessary. Why? Because whoever controls the technology creates the (perceived) reality.

Technology as Intellectual Partner

Students do not learn *from* technology (or teachers, for that matter). Rather, students learn from thinking in meaningful ways. Thinking naturally results from meaningful activity, such as representing what students know, rather than memorizing what teachers and technologies tell them. When learners use technologies to represent what they know, they are learning *with* technologies rather than *from* technologies. In this way, learners enter into an intellectual partnership *with* the technology. When students work *with* computers, for instance, they enhance the capabilities of the computer, and the computer in turn enhances their thinking and learning. The result of this intellectual partnership is that the whole of learning becomes greater than the potential of learner and computer alone. Learners use technologies as intellectual partners in order to.....

- articulate what they know (i.e. representing their knowledge
- reflect on what they have learned and how they came to know it
- support the internal negotiation of meaning making
- construct personal representations of meaning, and
- support intentional, mindful thinking.

Learning *with* technologies transforms the role of learners from receiver (classic, communications conception of learners) to producer, creator, and sender. That is, rather than requiring learners to react to transmitted messages, Jonassen (2000, 2006) has argued that learners use the technologies to construct representations of what they know using technologies, such as:

- semantic organization tools (databases, semantic networks) for organizing what they know;
- dynamic modeling tools (expert systems, spreadsheets, systems modeling tools) for building simulations and representing mental models;
- microworlds for exploring and experimenting with phenomena;
- synchronous and asynchronous conferencing environments for socially co-constructing meaning
- knowledge construction environments (hypermedia, multimedia, web publishing);
- information interpretation tools (visualization tools, information search engines) for better understanding information they encounter; and
- video for visualizing the range of ideas that students generate and providing feedback on learning performance.

These technologies function as cognitive tools for helping learners to elaborate on what they are thinking (Jonassen, 2000; 2006). I have always been more interested in what students know than what teachers know and are inclined to distribute. Providing learners with multiple knowledge representation tools enables them to understand ideas in different ways. When learners are designers who teach the computer, they are conceptually more engaged. The key to meaning making is ownership of the ideas that are created. When technologies are used as the tools for organizing, creating, and expressing those ideas, learners are learning with the technologies and necessarily engaged in meaningful learning. Cognitive tools represent a constructivist aplication of technology. When students develop databases, for instance, they are constructing their own conceptualization of the organization of a content domain.

Mediating the Social Co-Construction of Reality

Modern, transmissive views of technology have always assumed the objectivity of knowledge. Knowledge can be transmitted to individual learners who acquire the same objective reality. The process of learning is like filling up your automobile. The higher the octane of the fuel and the bigger the tank, the better the learning. This view has nearly always conceived of learning as an individual, acquisitive process. Teachers and technologies tell students what they know; students acquire what they know.

Contemporary conceptions of learning in discourse communities, communities of practice, learning communities, and knowledge-building communities, challenge this individual, acquisitive conception of learning as filling up students' knowledge tanks. Rather, they see learning "as a social phenomenon constituted in the experienced, lived-in world, through legitimate peripheral participation in ongoing social practice" (Lave, 1991, p. 64). When a goal is really important people collaborate to socially co-construct shared meaning and negotiate shared responsibilities. Knowledge is socially mediated rather than dictated. Although these processes naturally occur in non-formal situations, they are seldom allowed in formal educational processes. They are often mistaken for cheating. However, if we are serious about using technologies to transform existing practice, then we need to focus on how to use technologies in ways that support social negotiation and co-construction of knowledge.

New computer networks that facilitate immediate access to the world's information and nearly instantaneous communication with anyone anywhere have provided a level of global connectivity that was inconceivable a mere decade ago. This connectivity is redefining culture. Rather than being constrained by simultaneous location, communication is being redefined by need and interest. Computer networks have evolved to support discourse communities through different forms of computer conferences. Facebook connects millions of users who daily converse about their lives, their dreams, and their interests. The number of active and interactive discourse communities has expanded exponentially in the past five years. While such global access may be emancipating, because of the absence of peer review, it may also lead to narrow mindness, biases, hatred, etc. Clearly, unfettered communication is a two-edge sword.

In education, networked technologies have fostered the development of knowledge building communities. The goal of knowledge-building communities is to support students to "actively and strategically pursue learning as a goal," that is, intentional learning (Scardamalia, Bereiter, & Lamon, 1994, p. 201). Using Knowledge Forum and under the guidance of an effective teacher, students produce their own knowledge databases in their own knowledge-building community of students. Thus student knowledge can be "objectified, represented in an overt form so that it [can] be evaluated, examined for gaps and inadequacies, added to, revised, and reformulated" (p. 201). Through KIDLINK, the Global Schoolhouse, Learning Circles, and many other educational telecommunications projects students are forming global learning communities where participants conduct research (reading, studying, viewing, consulting experts), share information in the pursuit of a meaning, and reflect on the knowledge that they have constructed and the processes used to construct. Telecommunications have created keypals, global classrooms, electronic mentoring, information exchanges, electronic publishing, electronic field trips, pooled data analysis, parallel problem solving, collaborative electronic writing, serial creations, and social action projects. Students are escaping the boundaries of their classroom and community to play in an ever expanding information field. Telecommunications are redefining the concept of classroom and the culture of learning. But wait. In what ways are the concept of classroom and culture of learning creating students who lack social skills, are uncomfortable in face-to-face settings, are unable to follow lengthy logical arguments, seek entertainment rather than wisdom, fall for political sound-bytes rather than positions on important issues, and have shortened attention spans? What will they lose by escaping the social interactions of the classroom? Are these environments more nurturing than distributed ones. These are the questions that will be answered by this generation.

These communications technologies are capable of transforming the culture of education as well. By empowering students to negotiate their own beliefs and ideas, the balance of power has shifted from the educators to the students. I am not arguing here for a complete abdication of teacher responsibility. That would be destructive. Rather, I am arguing for a shift in the balance of power, new supportive guidance roles for teachers, and more amplifying roles for technology. For instance, the author pointed out earlier how the technologies can destroy face-to-face higher education, and the same case can be made for secondary school education – many "futurists" have stated as much. The key here is pointing out how the technologies themselves have a bias. Understanding these is the first step in wisely using them in education.

CONCLUDING THOUGHTS

In this chapter, I have reviewed modern, post-modern, and transformative conceptions of educational technology. I do not presume that these are the only views of technology that can inform our deliberations, but they do provide a rich set of options.

We live in a post-modern world. Values that were endemic to the modern world, progress and efficiency, have dissolved in a cultural cynicism. This is especially prevalent among our youth, who constantly question why they have to do anything that is not immediately self-aggrandizing. Modern solutions, using technology to transmit cultural values more efficiently will no longer affect today's youth. They are cynical about the goals of the institutions that seek to control their lives. They are post-modern, after all. From a Huxlean perspective, we have tacitly allowed this to occur.

However, educators who seek to transform education, to reorganize its foundational goals and values, can emancipate learners from the obligation to regurgitate that which has no relevance to them, to empower them to reflect on and represent what is important to them. Technology can support that goal. When used as tools for personal and social reflection, articulation, and creation, technology can help to transform learning and learners—to help them to become independent, self-regulated, life-long seekers and constructors of knowledge.

The future of technology in education will depend on who controls the technology and what their goals, values, beliefs, needs, and purposes are. Educators must reflect on their own beliefs and answer questions, such as:

- Who should control when, where, and how technologies in schools and universities should be used: corporations/institutions, teachers/professors, or students/ learners? What is the proper role of each stakeholder? Is collaboration possible?
- Whose knowledge and ideas are more important for learning: corporate and institutional agendas, teachers/professors who may know best, or learners who define their own purposes for learning, convey their own beliefs, and create their own reality? Are students willing to assume the responsibility that accompanies freedom? Shall we ignore the classics in favor of comics? Do children immersed in virtual worlds understand reality enough to deal with it? Are all discourses equally valid? I think not.
- What is the true mission of education—to transmit knowledge and values, to exert power over students, to expose students to multiple perspectives, or to empower learners to reflect, construct, and express their own knowledge and beliefs? Whose values and beliefs are more viable? How can these various ends be affected collaboratively?

REFERENCES

Bauerlein, M. (2008). *The dumbest generation: How the digital age stupefies young Americans and jeopardizes their future.* London: Penguin Books.

Bork, A. M. (1985). *Personal computers for education.* New York: Harper & Row.

Fleming, M. & Levie, W.H. (1978). *Instructional message design: Principles from the behavioral and cognitive sciences.* Englewood Cliffs, NJ: Educational Technology Publications.

Foucault, M. (1977). *Discipline and punish: The birth of the prison.* New York: Pantheon.

Jonassen, D.H. (1983). *The tenuous relationship between research and policy making: Lessons from the new Physics.* Media Management Journal, (Winter), *2*, 32–33.

Jonassen D.H. (2000). *Computers as Mindtools for schools: Engaging critical thinking.* Columbus, OH: Merrill/Prentice-Hall.

Jonassen, D.H. (2006). *Modeling with Technology: Mindtools for Conceptual Change.* Columbus, OH: Merrill/Prentice-Hall.

Lave, J. (1991). Situating learning in communities of practice. In L.B Resnick, J. M. Levine, & S. D. Teasley (Eds.), *Perspectives on socially shared cognition.* Washington, DC: American Psychological Association.

Noble, D. (1998). Digital Diploma Mills: The Automation of Higher Education. *Educom Review, 33*(3).

Noble, D.Noble, D. (2003). Digital diploma mills: The automation of higher education. In B. Johnson, P. Kavanagh, & K. Mattson (Eds.), Steal this university: *The rise of the corporate university and the academic labor movement* (pp. 33–47). New York: Routledge.

Pea, R. (1994). Seeing what we build together: Distributed multimedia learning environments for transformative communications. *Journal of the Learning Sciences, 3*(3), 285–299.

Salomon, G. (1979). Interaction of media, cognition, and learning. San Francisco: Jossey-Bass.

Scardamalia, Bereiter, & Lamon (1994). The CSILE Project: Trying to bring the classroom into World 3. In K. McGilly (Ed.), Classroom lessons: *Integrating cognitive theory and classroom practice* (pp. 201–228). Cambridge MA: MIT Press.

Winner, L. (1996). Who will we be in cyberspace? *Information Society, 12*(1), 63–72.

Winner, L. (1997). The Handwriting on the Wall: Resisting Technoglobalism's Assault on Education. In M. Moll (Ed.), Tech High: *Globalization and the Future of Canadian Education.* Ottawa: Fernwood Publishers.

David H. Jonassen
University of Missouri

SECTION II

TECHNOLOGY'S FAUSTIAN BARGAIN FOR EDUCATION

CHAPTER 7

JOCELYN WISHART

VISUALLY DOMINANT, DYNAMIC AND YET DECEPTIVE

The Nature of Simulation Technology as Displayed in Secondary School Science Teaching

INTRODUCTION

When Rogers and Finlayson (2003) reviewed science teachers' self reported use of Information and Communications Technology (ICT) in lessons, they learned that simulations were the most popular category of software. Over 95% of teachers reported that using simulations enabled them to achieve their teaching objectives and their reports referred to simulations stimulating thought and clarifying ideas as well as being an efficient use of time and motivating for students. Yet using simulations effectively in teaching is not as straightforward a task as it first seems.

Owen (2002) points out that two groups of simulations exist, those that are praised for their potential to extend the scope of science school experiences allowing students to access phenomena that would be too dangerous or expensive, and those that are purposely designed to be more explanatory so students are engaged in abstractions that would otherwise be too difficult or unavailable. For example:

> "In real life one cannot see the particles emitted from a radioactive source that are detected by a Geiger-Counter. A simulation of a ripple tank can allow the teacher to discuss observations with students in a controlled and predictable way." [Owen, 2002, p1]

However, educators need to look beyond the visually exciting and attractive graphics and animation to the way simulations are deployed within the science lesson. The UK Office for Standards in Education (Ofsted, 2001, p. 12) complained that too many teachers "select software packages for their visual appeal rather than their relevance to lessons". For example, they noted a primary science lesson being dominated by students' passive viewing of a simulation of materials dissolving.

Effectively using computer simulations is not as straightforward a task as it first seems. For instance, Baggott la Velle, McFarlane and Brawn (2003) describe the

Michael P. Clough, Joanne K. Olson and Dale S. Niederhauser (Eds.), The Nature of Technology: Implications for Learning and Teaching, 113–128.

complex and interrelated processes of subject, pedagogical, pupil, technological, curricular and contextual knowledge transformations that a science teacher must make in order to teach successfully through simulation software. Wellington (2000) cites several concerns inherent in science simulations. For example, computer simulations are idealized versions of reality built upon invisible, unquestionable, and often simplified models of a scientific process that give the students the impression that every variable is easily controlled.

However, if what may be gained and lost in using computer simulations is made evident in planning science lessons, then the potential positive outcomes of using simulation software may be realized while mitigating what may be lost. Osborne and Hennessy (2003) consider this potential to lie in six key areas. The first is in increasing the scope of reference and experience for the student through exploiting the power of visual representations to develop understanding – particularly of abstract phenomena like electricity flow. The second is in supporting exploration and experimentation through providing an immediate link between an activity and its results increasing the likelihood that students will relate the visual representation of relationships to the activity itself. This not only provides immediate opportunities for study but can also encourage students to pose exploratory ("what if...") questions and to pursue these by conducting follow up activities. This leads immediately to a third area of potential — structuring and supporting active engagement in learning. Genuinely interactive software requires active learner contribution and engagement (i.e. an element of reflection on choices and their effects), and this may include prediction, trial and evaluation (Rogers, 2004). Fourth, interactive computer simulations have potential to expedite and enhance work production, releasing students from laborious processes such as setting up equipment and recording results. For example, students often have great difficulty in successfully constructing electrical circuits which are known to give rise to problems in identifying and locating faults. Osborne and Hennessy (2003) describe simulations as yielding less 'messy' data and illustrating phenomena without the 'noise' of unwanted variables and human error in measurement. A fifth related area of potential is the way simulations enable students to focus on overarching issues by increasing the salience of underlying features of situations and abstract concepts such as current and voltage in electrical circuits; helping students to access ideas more quickly and easily, to formulate new ideas and transfer them between contexts. Finally, related to all of the above are the well documented motivational effects of using ICT, which seems to be intrinsically more interesting and exciting to pupils than using other resources (Denning, 1997, Cox 1997). In particular, improved motivation and engagement can be seen in students when using tools such as simulations and games which permit active engagement and offer pupils a degree of control over their own learning (Wishart, 1990).

Many of the researchers cited in this paper (Rogers and Finlayson, 2003; Baggott la Velle, McFarlane and Brawn, 2003; Wellington, 2000; Newton and Rogers, 2003) agree that for students to directly benefit from the multi-faceted potential of simulation software to enhance teaching and learning, the science teacher must carefully consider their planned use of this technology.

Newton and Rogers (2003) describe simulations and other ICT tools as adding value to science lessons through: (a) the intrinsic properties of the software such as the speed with which it can process large amounts of data and the way it can display or animate changes, and (b) potential student learning benefits that derive from the mode of application of the software such as clearer understanding and thinking. These are broad claims, and students may well not understand what it is that the technology is making easier for them, thus clouding the targeted science concepts. Again, if potential learning benefits are to be realized, issues such as these must be addressed beforehand when deciding whether and how to use a simulation. In this planning the teacher needs to bear in mind the possible Faustian bargains, for instance, that students may develop one set of ideas they attach to the simulations and another set of ideas they attach to the real world or that students may gain knowledge of a single scientific process at the expense of a wider understanding of the complex nature of science as found in the real world.

The study reported here investigated five experienced science teachers' planning regarding incorporating simulation software in their teaching and what they saw as the intrinsic properties of simulations that enhance learning.

METHOD

The five study participants took part in the Interactive Education Project (http://www.interactiveeducation.ac.uk/) run by the Graduate School of Education at the University of Bristol (Sutherland, Robertson and John, 2004). More than fifty teachers, including six science teachers, partnered with teacher educators and the researchers to create new ICT lesson designs for teaching. The project was predicated on two assumptions: (1) teachers are central to learning in school, despite prior ICT pedagogical research underemphasizing their role (Sutherland & Balacheff, 1999); and (2) ICT should be incorporated into a designed learning situation as appropriate, with attention directed to the whole classroom context including classroom talk, work on paper and other technologies that are usually available to a teacher.

During the three year project the teachers worked in partnership with both teacher educators and the researchers and were interviewed and observed throughout the design, implementation and review of these new lessons. For the purposes of the project, the lesson designs were termed Subject Design Initiatives (SDIs) and were informed iteratively by theory, research-based evidence, teachers' craft knowledge and the research team's expertise. Reflecting Rogers and Finlayson's (2003) findings, each science teacher chose to involve simulation software in their SDI.

Baggott la Velle *et al.* (2004) have previously reported on the study participants' views toward computer simulations in science teaching prior to implementing the SDIs. All the teachers were interested in the role of ICT in schooling, were supportive of its use in science, and had realistic expectations about its potential to enhance teaching and learning. They did, however, express the view that computer simulations presented an impoverished version of practical work and science inquiry. They did not appear to acknowledge the potential of simulation to mentally engage students in learning, or to build content knowledge and facilitate understanding through dynamic, visual representation. Baggott la Velle *et al.* (2004) speculated that the subject culture in the UK, where the National Curriculum separates science inquiry from science content knowledge, contributed towards this perception.

Once the SDIs were under way each lesson was video recorded in its entirety and followed up by a semi-structured post-project interview lasting 60 to 105 minutes. The questions that framed these interviews addressed:

- teachers' perceptions of the successes, problems and challenges of working with simulation software;
- changes, occurring during the project duration, in teachers' approach to incorporating simulation software in their practice and how this related to student learning;
- teachers' views about using simulation software in teaching, and how they changed during the project; and
- teachers' views about other processes they experienced during the project (e.g. working in SDI teams).

Five of the original six science teachers were available for the post-project interview. These reflective interviews were transcribed in full and their content analysed qualitatively in order to identify any salient or reoccurring themes. The results of this thematic content analysis appear in the next section.

FINDINGS

In all cases, simulations were planned to be used in hands-on mode by the students. The teachers, in particular Teacher A, considered that offering students a degree of control over learning activities, such as in choosing the parameters for each run of a simulation, can provide challenge, motivation and engagement. In answering the interviewer's questions Teacher A compared the use of the Crocodile Physics electricity simulator (see http://www.crocodile-clips.com/crocodile/physics/ for a recent version) to making physical circuits using cells, switches, wires and bulbs with grade-7 (age 11–12 years) students.

He was delighted to find his lesson free from the usual barriers to students learning about the way electricity flows in circuits caused by problems with rusty connectors, broken wires, cells that quickly lose their charge and 'blown' bulbs.

> "no matter how hard and how efficiently the kids work, they sometimes are dogged by just basic resistance problems in the circuit which are nothing to do with them."

As part of his SDI he had measured improved learning amongst the students using the simulation:

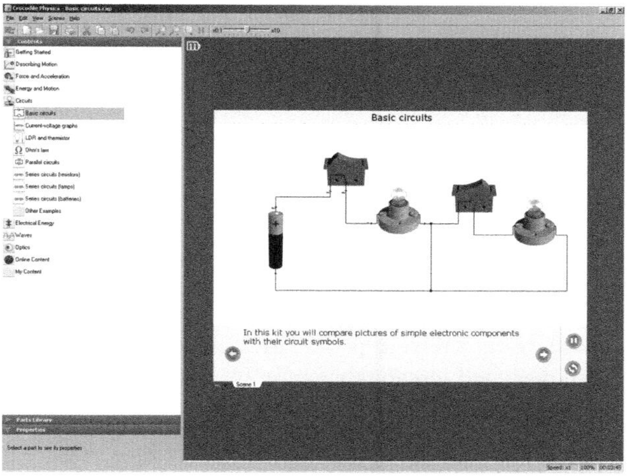

Figure 1. Simulation of a simple electrical circuit using Crocodile Physics.

> "We did some tests before and after to see what they came to us knowing about basic circuit work and conduction and various things, and we did a test afterwards. Most improved considerably in their understanding of what was going on.....I think possibly it was even better than I expected because I was a bit dubious about whether they were going to be able to cope with the concepts at that age."

Teacher A noted that students could safely explore a greater range of options — including creating short circuits and 'blowing' bulbs — that would have been problematic in working with physical equipment. Because students could play around with the circuit components just to see what would happen, things they might be reprimanded for in class, teacher A noted students "felt that they were in control of their learning. They had this idea that they were able to use it for themselves. And that, to a certain extent, was the case. I allowed them to have a little play on occasions, particularly as extension work." This view resonates with Wishart's (1990) earlier observations of students using both educational computer games and simulations where the opportunity to choose the path through a simulation game was found to be closely related to the amount learned from it.

The discussions in the teacher interviews tended to corroborate the six key learning opportunities of simulations identified earlier in the introduction to this chapter. For example Teacher D (who chose to use a Web-based interactive simulation of a voltaic cell in her SDI) notes the problem of ensuring students observe the relevant changes in studying how a real voltaic cell works:

> "For example with voltaic cells, sometimes when you put two different metals together and you're trying to create a voltage the difference is so small that

you can't actually pick up that anything is happening at all. Or sometimes the reaction is over so quickly, that I thought if they could actually ... if they could get their hands on the interactive software and do it over and over again they could actually stop and start and use different combinations of metals and see what was going on in principle [...] that would be more of a sort of useful experience for them, rather than me sort of saying 'Oh it does happen really, it does, you have to believe me on this.'"

Teacher D used simulations, such as that appearing in Figure 2 (see http://www.chem.iastate.edu/group/Greenbowe/sections/projectfolder/animationsindex.htm), with grade-9 students (age 13–14 years). The teachers linked the student learning they perceived occurring with the power of animated visual representation, and their lessons were designed to capitalise on this. For example Teacher D provided the following justification for why she thought the voltaic cell simulation helped students learn:

"For me I think it's visual, isn't it? I mean like for example seeing the electrons actually moving, as opposed to thinking of electrons moving. [...] I think that has more to do with learning. [...] It's all right saying the electrons [or ions] are there, but it's another thing to actually see them doing what they should be doing, and sort of having the effect that they should be having."

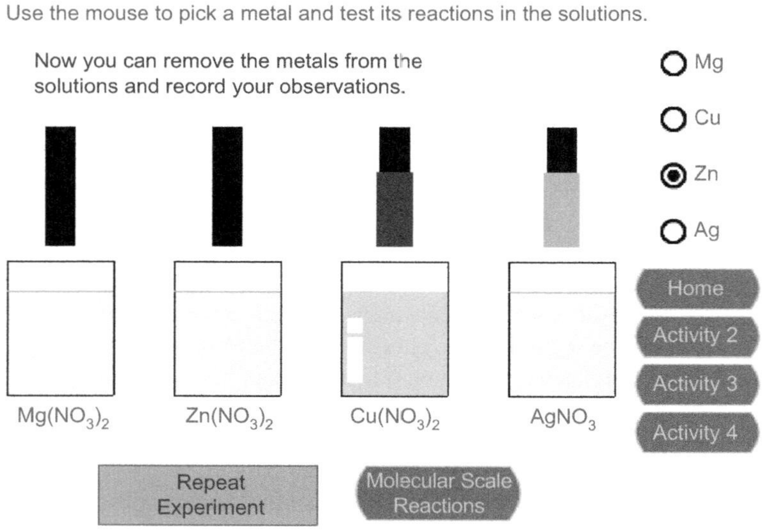

Figure 2. Simulation of reactions of metals from the Chemical Education Research Group at Iowa State University.

Simulation was also employed in order to release students from laborious (and often confusing) manual processes. Teacher A found Crocodile Physics useful because it allowed the students to interact directly with the concepts being modelled without the interference to their thinking that too often arises from the poor connections found in electrical circuits constructed in school.

Teacher J also noted a shift in his interaction with grade-10 (age 15 years) students when he used simulation software (see Figure 3) with a standard school laboratory practical investigating the effect of temperature on enzyme activity. He reported that conducting the experiment with authentic materials requires him to devote significant time to checking problems with laboratory equipment which often supersede the teacher's 'intellectual input'. Once the computer simulation was running he spent less time helping children to understand what the task was and more time "discussing the learning points that the simulation was there to demonstrate". However, he did raise a concern about the predictability of the dataset that was programmed into the simulation, and was pleased that a number of his more able students realized this limitation of computer models:

> "Of Set 1 probably about 4 or 5 [students] came up and said 'This is no good as coursework because we can't vary anything. We're varying things but we're all coming up with the same results.' Brilliant. They'd actually seen the top end limitation of the computer simulation… And in a way that's more useful to understand – that computers are limited."

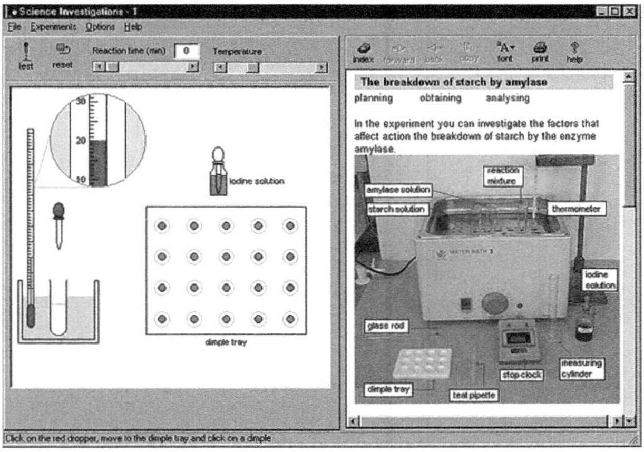

Figure 3. Simulation of an investigation of the effect of temperature upon enzymes from Focus Educational Software.

Teacher J's experience highlighted the importance of the relationship between the students' ability and the way that the simulation software influenced their learning.

On reflection, he realized that he needed different teaching strategies for using the simulation with his two student groups. With the less able group the power of the simulation to increase their scope of experience through visual representation was paramount. It also allowed them to repeat experiments as often as they needed in order to perceive patterns in the data, which could not be done practically in the time available. He noted: "You can generate far more data, and see the whole curve rather than four points on it". However, with the more able group, he needed to go beyond the more salient points and relationships so nicely presented in the simulated experiment to plan a lesson that addressed the premises on which the model underlying the simulation was based.

In the above examples, the teachers were replacing a hands-on based lesson with a computer based simulation. Teacher B, however, used both with one group. Through doing so, he also spotted the learning opportunities offered through unpacking the simulation and considering the underlying model. He was teaching electrical circuits to a group of able grade ten students. After introducing the content with the usual accompanying hands-on activities he then used a CD-ROM based simulation called 'Furry Elephants' (see http://www.furryelephant.com/ for a recent version) (Figure 4) with the group in order to clarify the underlying theory of electricity as a flow of energy carrying, charged particles.

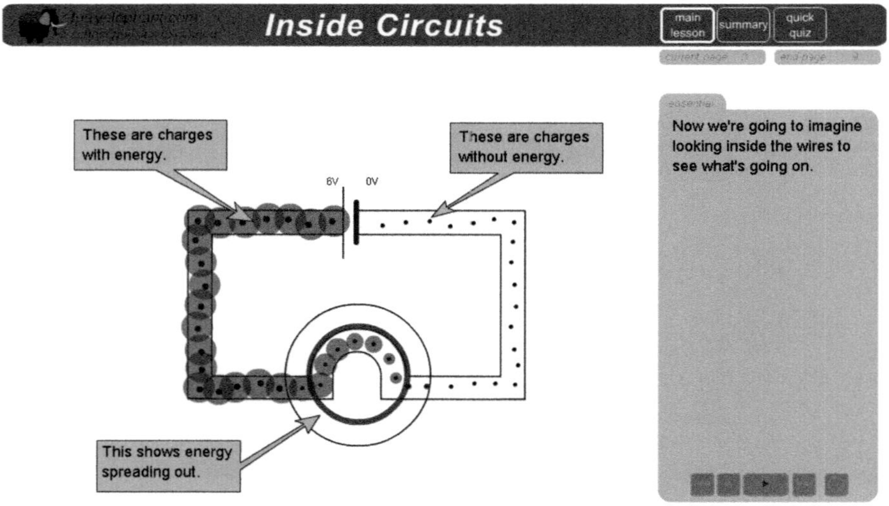

Figure 4. Animation of charge within an electric circuit from Furry Elephant.

However, some of the more able learners, prior to using the computer simulation that illustrated energy being dissipated at specific components, had already developed the accurate idea that energy is dissipated throughout a circuit. Teacher B learned from this experience to review the models used in many simulations carefully in developing a second SDI for older students. This second SDI focused on grade-12 (age 16–17 years) pre-university physics students using the Internet to research and review examples of photoelectric effect simulations.

Therefore, both of Teacher B's SDIs included discussion regarding how the students' understanding of the topic was actually in conflict with what was being represented in the simulation. Teacher B was particularly pleased with the students' responses to the second SDI. He concluded that this was an effective way of using the simulation resources on the Web to reinforce learning, because it could circumvent the problems of incorrect science in the simplified models used:

> "[Students] have to be critical. They're being more active and proactive in their learning, rather then just reacting to what they're seeing in front of them and automatically grabbing it off the Web because it looks pretty. They're being critical. And teaching critical thinking has got to be a good way forward."

Teacher B hit upon a method of planning for the use of simulations in science that might somewhat allay Wellington's (2000) legitimate and important concerns about them being idealized and simplified models of reality. By immediately acknowledging that the models on the Internet were not all perfect and asking able students to review them in the light of how effective they were at illustrating the photoelectric effect, he encouraged students to reflect on and review their own understanding, thus reinforcing and consolidating the targeted concepts. Teacher B reported that:

> "Following the SDI the year-12 students displayed a greater confidence in their use of scientific terminology associated with the photoelectric effect and had gained a deeper understanding of the concepts underpinning this."

He considered that as an activity:

> "It worked very well and I think the fact that they were able to do it in their own time was a better utilisation of time and didn't have the associated problems with computers [in school]."

Teacher B also asked the students to compare the explanations provided by him, the textbook, and the Web. In this way they were encouraged to explore and check the scientific explanations from these various sources. In effect they were engaged in meaning-making activities that prompted them to contrast their own ideas developed from their earlier practical work with the scientific models presented in the computer simulation. Here the teacher promoted cognitive change by employing a strategy advocated by Doise and Mugny (1984) whereby group-generated conflict stimulates the joint construction of a more advanced concept. This strategy has also proved effective for teaching science concepts to younger secondary school children (Howe *et al.*, 1991).

Teacher T had intended to use Multimedia Science School simulation software to investigate terminal velocity (Figure 5) with grade-9 (age 13–14 years). His most able students were experiencing difficulty grasping the relationship between terminal velocity and the concept of balanced forces. For him, the significant advantage of this simulation was that it allowed a slower 'investigation' of terminal velocity than possible with standard school laboratory equipment. However, the experience did not go according to plan as the school had problems running the software over its network.

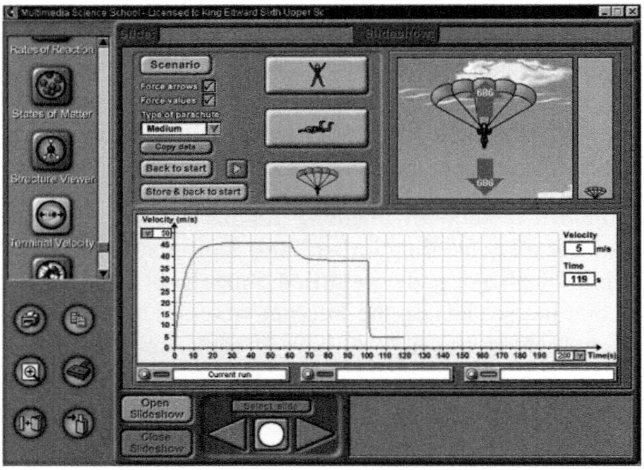

Figure 5. Simulation of terminal velocity from Multimedia Science School.

In responding to the interviewer's questions based on his general experiences with simulation software, Teacher T noted that he prefers to use simulations for their ability to extend pupils' thinking and consolidate their understanding rather than to replace practical work.

> "The concept – do an experiment and then use the simulation to help kids think about what the … say, particles [are doing] would be the ideal example of this, wouldn't it?"

> "we clean up the experiment, get rid of all the messy worldliness of it, and then use that cleaning up process."

This and other commonalities in the way science teachers perceived the use of simulation software to support teaching and learning in science are discussed in the next section.

DISCUSSION

When reflecting on the intrinsic properties of simulation that affect teaching and learning in secondary school science the teachers in this study fully supported the range and variety of potentials reported by Hennessy and Osborne (2003). The five teachers were clearly seizing on several salient aspects of simulation software and *attributing student learning to them*. In Orwellian terms simulations are viewed by science teachers as 'doubleplusgood', especially for their visual animations with their power of multimodal explanation and making the invisible visible, a feature that is viewed as key to their pedagogical value.

Teachers A and T both emphasized the improved motivation and engagement of pupils when using simulations. Teacher A described his pupils as 'switching into overdrive' when using the computers. However, Teacher B noted that such engagement is less likely if the teacher operates the simulation themselves using a data projector to display the computer screen to the whole class. The latter is likely to lead to the scenario observed and criticized by Ofsted (2001) where teachers, impressed by the visual power and clarity of explanation in simulations, unthinkingly if not unknowingly, operate the software themselves and display the results to a physically and mentally passive audience.

Simulation software also has the potential to turn a misplaced connection (e.g. an overloaded circuit or an insufficient model) into a learning point. Teacher A noted that his students actively 'blew' their simulated light bulbs with glee and learned from the process. Teachers B and J both used student identified problems with their simulations to develop students' knowledge and understanding. In one case the visual representation was at fault and in the other, the underpinning model was unexpectedly limited.

Teacher J and Teacher B found that, with forethought and planning, most simulations can be used to provide challenge for a wide range of groups, improving motivation and engagement of the pupils, clarifying and reinforcing their learning. Whilst they both chose to follow class practical or research tasks with simulation based activities to clarify and reinforce the underlying scientific principles, their pedagogical rationale for this order was not tested in this study. An interesting question worth pursuing in future work is whether students must have an initial conceptual understanding of the simulated phenomenon in order to get the most out of a simulation? If so, how is this initial conceptual understanding developed, and what is the role of tactile experience with authentic phenomena in developing that understanding? Further research is needed to determine the learning efficacy of simulations alone, hands-on experiences alone, and a combination of the two (tactile experience previous to simulation and vice-versa). The University of Bristol ChemLabs team is already reporting success (Harrison et al, 2009) in having students 'practice' laboratory experiments in simulation before the actual physical experience.

Teachers A and J highlighted the importance of simulations in freeing students from the laborious processes that so often interfere with their understanding of the underlying science concepts. They reported that using simulation software allowed them, as teachers, to spend less time than they would otherwise have done sorting out issues to do with the practical tasks and more time focusing on the key learning outcomes for the lesson.

However, in the name of instructional efficiency, might using simulations actually learn less about the real but messy nature of natural phenomena? Whilst, the two teachers both considered that this property of simulations led to student learning of the relevant concepts (as indeed Teacher A found in his post SDI tests), we don't know that what students learned was superior to what they may learned from hands-on practical work. We may be accepting a Faustian bargain whereby we gain the efficiency of sanitized and easily assimilated learning experience while sacrificing the practice in problem-solving and authentic learning that is associated with messy real world problems. We can't really pedagogically approve the use of simulation solely for the efficiency gains made by skipping the more problematic practical tasks.

In mitigating the messy and complex reality of natural phenomena, simulations offer a clear advantage of focussing both teachers and students on the salient concepts that can otherwise be masked by problems associated with lab set-ups, materials, and inattention so characteristic of young secondary school children. Nevertheless this is accompanied by a substantial disadvantage signalled by Wellington's (2000) concern that, through simulations, students will learn only sanitized and idealized science. Owen (2002) points out that "science is messy - simulations tend to be tidy". This view is reflected in Teacher T's comment that simulations "get rid of all the messy worldliness", and the valuable role of simulations after authentic laboratory work for "cleaning up" the science. The simulation technology's bias is therefore a clean and tidy outcome; this is the inbuilt expectation both of the technology and of the teachers who use it.

Taking the points made above a step further, the nature of simulation technology is to downplay and perhaps dismiss direct experience. This bias is inherent in: (a) the way the technology is so attractive to potential users; (b) the reduction in complexity and time that are part of effective hands-on inquiry based activities; (c) the sanitized outcomes and (d) administrators who would love to see simulations displace authentic costly laboratory experiences.

Teachers also need to be mindful of simplifications and even errors appearing in some simulations, both in the graphical representations and the underpinning model. A multimedia simulation is designed first and foremost for visual clarity and effect. This is of particular concern where simulations portray unobservable entities in models used by scientists to explain their observations. For instance, simulations often portray such entities (e.g. electrons) as spherical solids, but that is not how scientists view them. Science teachers must be aware of this predisposition of simulations, how it may influence students' learning, and carefully plan when, where and how to use such simulations so that they add to rather than detract from authentic science concepts.

Another bias arises in the way children approach simulations in light of their experiences with computer games, where strategic trial and error succeeds without any deep consideration of the reality represented by the simulation. Whilst Teacher A reported that allowing "a little play on occasions" led unexpectedly to higher levels of thinking among more able students regarding electricity, current and charge, he also noted that for many others in the class "they didn't think about what they were doing" and constructive exploration did not result. Sins *et al.* (2005) too found that students don't always approach a simulation task constructively. In an in-depth investigation of 11th grade students modelling how friction affects an ice skater, they identified students working through trial and error until they got the desired answer and not considering the reasoning behind it.

As Owen (2002) pointed out, a clear need remains to develop simulations that reflect the real difficulties that scientists encounter, and provide the possibility of addressing these difficulties. A more complete simulation environment might provide a richer vision of science and how science works as a contested and contestable activity. Though, in light of the above point, scaffolds to focus students' attention on relevant prior knowledge would need to be included to ensure the students worked constructively with the underpinning model. Results would need to be unpacked from 'noisy' data and referenced to the work of other scientists, and proper attention given to explaining and defending the outcomes and methods to a variety of audiences. Only then would students be in a position to truly understand the nature of scientists' work.

CONCLUSIONS AND RECOMMENDATIONS

These science teachers clearly moved on from their original perspective that computer simulations are an impoverished version of practical work. Their reflections conveyed that they perceive the following to be key ways that simulations impact teaching and learning in science lessons:

- increasing the scope of student experience;
- supporting exploration and experimentation;
- structuring activity and supporting active learning;
- releasing students from laborious processes; and
- enabling focus on key issues and enhancing motivation to engage with learning (Osborne and Hennessy, 2003).

However, only two of the five teachers showed awareness of the associated Faustian bargain of exchanging the "messy worldliness" of real science for a sanitized version that clearly follows simplified models of reality. Thus, as Wellington (2000) feared and Ofsted (2001) reported, science teachers are in danger of teaching only a sanitized version of science through uncritically embracing this technology.

Not surprisingly, teachers can indeed use simulations to enhance teaching and learning. However, this necessitates awareness of the inherent biases of simulation software so that appropriate care is taken when planning for their use in science classrooms. Effective science teaching and learning demands that planning for simulations goes beyond simply booking the computer suite and ensuring the simulation software runs over the school network. The visual dynamism of simulations, their imagery and animation, and the opportunity to make the invisible visible are open seductions that obscure the Faustian bargain. Digging beneath the overt and alluring promises of simulation software is crucial for making the most out of what may be gained from this technology, and recognizing and effectively addressing what is lost. For instance, the validity of science simulations, their underpinning models and representations, ought to be examined for their possible impact on students' understanding of concepts and authentic science. Teachers must consider how they will help students map the simulation onto the real phenomenon. For instance, what role should hands-on practical work play, and should it occur before, during, and/or after a simulation experience?

In this study two of the five teachers engaged their students in discussions regarding the limitations of the models underpinning the simulations. However, they did not address the limitations or inherent biases of the simulation technology itself. Teachers must consider in what ways simulations increase and decrease the scope of student experience and learning. This is imperative because perhaps the most alluring bias of simulations is the ease of abandoning the complex and messy use of real objects in favour of the clarity of an explanation made through moving visual images on a screen.

ACKNOWLEDGEMENTS:

The author wishes to acknowledge the teachers involved for their invaluable contribution to this work, Ros Sutherland, the director of the Interactive Education Project and Richard Brawn, Linda la Velle, and Angela McFarlane, the other members of the Science team, for their support with the Interactive Education Project data collection.

REFERENCES

Baggott la Velle, L., McFarlane, A.M. & Brawn, R. (2003). Knowledge Transformation through ICT in Science Education: A case study in teacher-driven curriculum development. *British Journal of Educational Technology, 34*(2) 183–200

Baggott la Velle, L., McFarlane, A., John, P.D., & Brawn, R. (2004). According to the Promises: the subculture of school science, teachers' pedagogic identity and the challenge of ICT. *Education, Communication and Information, 4*(1) pp. 109–129

Barton, R (1998). IT in practical work: assessing and increasing the value-added. In J. Wellington (Ed.). *Practical Work in School Science*. London, Routledge.

Cox, M. (1997). *The effects of Information Technology on Students' Motivation.* NCET and London: King's College.

Denning, T. (1997). *IT and pupil motivation: a collaborative study of staff and pupil attitudes and experiences.* Keele University/Coventry: NCET.

Doise, W. & Mugny, G. (1984). *The social development of the intellect.* Oxford, Pergamon Press.
Harrison, T.G., Shallcross, D.E., Heslop, W.J., Eastman, J.R, & Baldwin, A.J. (2009) Transferring best practice from undergraduate practical teaching to secondary schools: The dynamic laboratory manual. *Acta Didactica Napocensia, 2*(1), ISSN 2065–1430.
Howe, C., Tolmie, A. & Anderson, A. (1991). Information technology and group work in physics, *Journal of Computer Assisted Learning, 7*(2), pp. 143–153.
Newton, L. R. & Rogers, L. (2001). *Teaching Science with ICT.* London: Continuum.
Ofsted (2001). ICT in Schools: *The Impact of Government Initiatives, An Interim Report April 2001.* London: Office for Standards in Education.
Owen, M. (2002) *ICT and the Future of Science Education. Report of a Science Year Seminar.* Available at http://www.nestafuturelab.org/viewpoint/learn06.htm [Accessed 6.9.04]
Osborne, J. & Hennessy, S. (2003). *Literature Review in Science Education and the Role of ICT: Promise, Problems and Future Directions.* A Report for NESTA Futurelab. Available at http://www.nestafuturelab.org/research/reviews/se01.htm [Accessed 6.9.04]
Rogers, L. (2004). Integrating ICT into the curriculum. In R. Barton (Ed.). *Teaching Secondary Science with ICT.* Open University Press.
Rogers, L. & Finlayson, H. (2003). Does ICT in science really work in the classroom? Part 1, The individual teacher experience. *School Science Review, 84*(309), 105–112.
Sins, P.H.M. Savelsbergha, E.R. and van Joolingen, W.R (2005). The Difficult Process of Scientific Modelling: An analysis of novices' reasoning during computer-based modelling. *International Journal of Science Education, 27*(14), 1695–1721.
Sutherland, R., Robertson, S. & John, P. (2004). InterActive Education: Teaching and Learning in the Information Age. *Journal of Computer Assisted Learning Special Issue, 20*(6), 410–412
Sutherland R. & Balacheff N. (1999). Didactical complexity of computational environments for the learning of Mathematics. *International Journal of Computers for Mathematical* Learning *4*, 1–26.
Wellington, J. (2000). *Teaching and Learning Secondary Science - Contemporary Issues and Practical Approaches.* London: Routledge.
Wishart, J. M. (1990). Cognitive Factors relating to User Involvement with Computers and their effects upon learning from an Educational Computer Game. *Computers and Education, 15*(1–3), 145–150.

Jocelyn Wishart
Graduate School of Education,
University of Bristol, UK

CHAPTER 8

JOANNE K. OLSON, MICHAEL P. CLOUGH, &
KIMBERLY A. PENNING

SOCIAL NETWORKING TECHNOLOGY AND SOCIETAL EXPECTATIONS FOR TEACHERS AS ROLE MODELS

Education without values, as useful as it is, seems rather to make man a more clever devil.

C. S. Lewis, *English essayist & novelist (1898–1963)*

INTRODUCTION AND BACKGROUND

Honorable character and virtuous behavior have always been associated with the noble title "teacher." This is reflected in the high expectations society has for teachers that go well beyond effective teaching of subject matter content. Teachers work with students, often impressionable children, who are undeniably influenced by the behavior of adults—particularly those in close proximity to them. Thus, teachers are also expected to teach and model high character and moral standards. At the very least, teachers' questionable conduct and particular behaviors they may exhibit that are reserved for adults are not to be made public. Any teacher who advertises such behaviors will likely suffer career-ending consequences.

However, today's prospective teachers have access to technologies that blur the distinction between public and private space (Bugeja, 2005). The ease of making and sharing videos, taking and distributing unlimited digital images instantly, talking on the telephone in places outside the home and phone booths, and posting information that can be viewed by anyone with an internet connection has certainly expanded the number of ways people communicate and the very nature of communication. As communication technologies have shifted dramatically, a corresponding decrease in social skills have been noted. For example, Japanese theatre companies installed scrambling devices due to complaints about audience members receiving and answering cellular telephone calls during live performances (Poupee, 2002). Others have lamented individuals who loudly carry on phone conversations in places where such conversations had previously been limited to discreet private talk, such as restaurants and elevators (Bugeja, 2005). A common theme to these concerns is the

Michael P. Clough, Joanne K. Olson and Dale S. Niederhauser (Eds.), The Nature of Technology: Implications for Learning and Teaching, 129–162.

conflict between traditional social norms and new technologies that attack norms of politeness. Bugeja writes of the negative consequences of this "connected" world in which we find ourselves. He argues that "the greater the convenience, the greater the interpersonal consequences and ethical concerns." The result is a *loss* of community, at the very same time the technology is touted as creating "connections."

The ethical concerns raised by a technologically-connected world are not always readily apparent, but they are indeed very real. For instance, people have always sought social connections and ways to express themselves, and today's college students, including preservice teachers, are no different. However, the internet combined with the increased isolation of the American college student (Nathan, 2005) has resulted in growing use of social networking websites as a way to make friends and communicate with them. Prospective teachers uploading content to the web that is intended for friends and peers may find that same material being accessed by students, parents, administrators, and strangers. This content may portray behaviors and a personal character deemed unacceptable for teachers of children. This study investigates how elementary education majors at a Midwestern university portray themselves on social networking sites.

SOCIAL NETWORKING ON THE INTERNET

Facebook is one of a growing number of social networking websites that allow people to post information about themselves and others. Started by a Harvard student who wanted an easier way to network with classmates, Facebook has blossomed into a robust online community that included more than 12 million people in 2006, 300 million people in 2009, and grew to 835 million monthly active users in 2012 (Carvin, 2006b; Facebook, 2009; Internet World Stats, 2012). Because these tools started as a way for college students to network, the majority of the users were originally of college age. However, Facebook is available to the general public and anyone in the world can have access to the network—all that is required is an e-mail address. Users of Facebook now include millions of individuals from multiple age groups.

The popularity of social networking has exploded, and problems have followed. Several published studies and reports address the dramatic increase in cyber-bullying occurring on social networking sites. When children have ready access to instantly post messages to the internet, not surprising are the creation of voting sites for the "Ugliest Girl on Campus" and hate messages to fellow students. Warren Blumenfeld addresses this issue in more detail in chapter 14 of this book. Problems regarding postings on social networks are not limited to school affairs. Corporate secrets, opinions about bosses, and other workplace issues have been posted by employees' children, with parents often unaware of what images or statements their children are posting (Finnigan, 2007). Post-secondary faculty who use Facebook to communicate with their students can be surprised at what they find readily available online. Pablo Malavenda, an administrator at Purdue University, came across Facebook pages indicating that students were selling cocaine on campus. The students were expelled, and photos from Facebook were admissible in court as evidence. In retaliation, the students started a Facebook group called "We hate Pablo," and posted his home address and instructions to hurt and eliminate him (Carvin, 2006b). These and many more examples illustrate some of the very serious criminal and ethical issues that are raised by the use of Facebook. Administrators of Facebook do not edit the content and rely on users to report inappropriate conduct. Illegal and questionable behavior can be posted without review or reprimand. For students, this may imply that what is placed on the internet is simply personal expression and an exercise of freedom of speech.

Teacher education students also take part in online networking and make regrettable decisions. In spring 2007, a pre-service teacher at Millersville University was denied her degree in Elementary Education (and subsequent state licensure) because of a photo that she placed on her MySpace account. The photo depicted her in a pirate hat, drinking from a plastic cup, with a caption of "Drunken Pirate" (Carvin, 2006a). She unsuccessfully sued the university, and her lawsuit made national headlines. The judge ruled that the "Plaintiff learned at the practicum's outset that she was required to 'maintain the same professional standards expected of the [school] teaching employees' and that those professional standards were violated when she posted inappropriate material on the internet that was subsequently viewed by her cooperating teacher and her students" (Diamond, 2008). Her case is not the only one involving universities, online postings, ethics, and students involved in specific programs that have standards for appropriate behavior. Athletes at the University of Iowa were removed from the football team and eventually arrested when university officials noticed three players' Facebook photos showing the players holding large amounts of cash. Their Facebook photos appeared on the evening news on television stations across the state, and in *The Des Moines Register* newspaper (Peterson & Barton, 2007).

The personal information that individuals freely choose to place about themselves on the web has attracted the interest of universities wanting to learn more about students, employers screening potential candidates, and others seeking to know more about those in their purview. An estimated twenty percent of companies are using the internet to search online profiles before they interview candidates (CBS, 2006). Parents and children also search these sites. A high school art teacher in Austin, Texas was fired for photos posted on the website Flickr.com. These photos depicted the teacher in the shower, lifting weights, getting dressed, in bed and doing other routine activities (Associated Press, 2006). Students who saw the pictures showed them to another teacher in the school, who notified school officials. The school district fired the teacher because the photos were considered inappropriate and violated the "higher moral standard" expected of public school teachers (CBS, 2006). Bill Shaw, professor of law and ethics in business, commented on this case by stating that "School teachers are supposed to be mature enough not to titillate their students," and "A teacher is more or less expected to be a guide or ... demonstrably mature." Teachers' personal behavior as well as their judgment regarding what to make public online or through other avenues are clearly grounds for disciplinary action, and they illustrate the moral and ethical standards that teachers, prospective or practicing, are expected to convey.

PUBLIC AND INSTITUTIONAL EXPECTATIONS OF ELEMENTARY TEACHERS

Since the foundation of public schools, educating children in character and ethics has been part of the school's charge (Lortie, 2002). As employees serving the public and entrusted with children, teachers have been expected to not only teach appropriate behavior and ethics to children, but to also be models of upstanding character. In the early years of public education, "teachers could be dismissed easily not only for incompetence but for a variety of infractions against morality stringently defined" (p. 8). According to Murray (2007), historically a community granted an individual permission to become a teacher based on at least one of the following: "(a) an assessment of the prospective teacher's character, values and beliefs, usually by a member of the clergy; (b) an assessment of the prospective teacher's knowledge in select domains, usually by a common or standardized test in a teaching subject; and/or (c) an assessment by a faculty with regard to the prospective teacher's course of professional study, usually with a major emphasis on pedagogy and teaching skills" (p. 381). Today's standards have changed somewhat, but the expectation that teachers be models of high character continues. An article in *NEA Today* describes at least eight cases of teachers who were reprimanded or dismissed from their jobs after inappropriate behaviors were posted on personal webpages (Simpson, 2008). Simpson, a member of the NEA Office of General Counsel warns that "school employees can be disciplined for off-duty conduct if the school district can show that the conduct had an adverse impact on the school or the teacher's ability to teach. And it wouldn't be too difficult to make that showing if the teacher's blog includes sexually explicit or other inappropriate conduct and is widely viewed by students" (p. 17).

In addition to the expectation for teachers to practice appropriate behavior, formal programs to teach good character to children are becoming more commonly used in public schools. A popular program is "Character Counts"—a program endorsed by thousands of schools, the National Education Association, and public groups such as Little League, the YMCA, 4-H, Boys and Girls Clubs of America, to name a few. The goal of the program is straightforward: teach and reinforce character, because:

> The ethical values that define good character are pretty basic. We can all agree what they are. The trick is to express them using a consistent language so that messages about ethics and character resonate across the community, from the home to the classroom to the playground to the workplace. (Josephson Institute, 2009)

This program includes lessons that focus on the teaching of trustworthiness, respect, responsibility, fairness, caring, and citizenship. The program developers acknowledge, "It is always primarily a parent's job to teach a child how to behave and make wise choices, but other institutions and adults working with young people play critical support roles." Teachers play a pervasive role in children's lives from preschool through high school. Thus, the character, behavior and judgment of teachers are crucial to the well-being of children and society.

This emphasis on character and values in the teaching profession is apparent in the inclusion of "dispositions" in the preparation and evaluation of prospective teachers in the United States. In 2001, the National Council for Accreditation of Teacher Education (NCATE, 2001) included three focal points of standards for teacher preparation programs, including "knowledge, skills, and dispositions" (Damon, 2007). While the first two terms refer to knowledge and skills related to content and pedagogy, dispositions are a bit more difficult to define. NCATE (2001) provides the following definition:

> Dispositions. The values, commitments, and professional ethics that influence behaviors toward students, families, colleagues, and communities and affect student learning, motivation, and development as well as the educator's own professional growth. Dispositions are guided by beliefs and attitudes related to values such as caring, fairness, honesty, responsibility, and social justice. For example, they might include a belief that all students can learn, a vision of high and challenging standards, or a commitment to a safe and supportive learning environment. (p.30)

NCATE further elaborates on the expectations teacher preparation programs should have of prospective teachers:

> Candidates work with students, families, colleagues and communities in ways that reflect the professional dispositions expected of professional educators as delineated in professional, state, and institutional standards. Candidates demonstrate classroom behaviors that create caring and supportive learning environments and encourage self-directed learning by all students. Candidates recognize when their own professional dispositions may need to be adjusted and are able to develop plans to do so. (NCATE, 2008, p. 20)

Teacher preparation programs in the United States, whether NCATE-accredited or not, have means by which teacher education candidates are evaluated and recommended to the state's teacher licensure agency. Teacher preparation programs can remove a teacher candidate from the program for violations of appropriate conduct. The institution involved in this study has written guidelines to assist teacher candidates in knowing dispositions expectations. The document lists expectations in the following areas:

- *Caring* (includes empathy, compassion, rapport, respect, passion, and cultural competence)
- *Communication* (includes presence, responsiveness, attentiveness, authenticity, collaborativeness, and voice)
- *Creative* (includes flexibility, inventiveness, resourcefulness)
- *Critical* (includes reflectiveness, initiative, open-mindedness, efficacy, and humility)
- *Professional Requirements* (includes professionalism, personal and professional ethics and integrity, work ethic/responsibility, and confidentiality)

Included under "professionalism" is the expectation that the candidate "endeavors to meet the standards expected of a teacher such as appropriateness of dress, grooming, demeanor, punctuality, tact, discretion, courtesy, etc." A copy of the full document appears in Appendix A.

Concerns have been raised about the assessment of dispositions. Burant, Chubbuck and Whipp (2007) assert that "dispositions" is not a single construct—that conceptualizing and identifying dispositions falls into three distinct domains. "These are listed as beliefs, personality traits and inference from behaviors. Each approach offers a perception on dispositions, but each one is also considerably flawed" (p. 400). Personality traits are defined as a disposition that produces consistent patterns of behavior in individuals (Mullin, 2003). In 1963, Gage argued that the personality characteristics of teachers is one of three central variables of teaching, and Getzel and Jackson (1963) viewed the personality of the teacher as the most important variable in the classroom. However, Washburne and Heil (1960) found that the personality of the student interacts with the personality of the teacher, making it relatively impossible to identify personality traits of the ideal teacher (in Burant *et al.*, 2007).

Efforts are underway to develop assessments of dispositions (Bonnstetter & Pedersen, 2005; Wasicsko, Callahan & Wirtz, 2004). This task is particularly challenging because, as Burant *et al.* (2007) note,

> There are currently flaws in the way we think about and assess teacher dispositions. Belief statements are best understood as acquired knowledge, not dispositions. Personality traits are too static, and teaching context is too fluid and complex to conceive of dispositions as a reduction of ideal personality traits for teachers. (p. 405)

Thus, Burant, Chubbuck and Whipp (2007) argue that we should avoid the term "disposition" when assessing teachers and instead use two separate variables; moral sensibility and code of ethics. Moral sensibility addresses the way in which a teacher thinks and acts. Code of ethics refers to a specific foundation of ethics connected to teaching. This would include those qualities that are directly related to, and give clear descriptions of, the behaviors desired in the teaching profession.

Despite the challenges in the development of an instrument to measure dispositions, institutions and accreditation agencies remain committed to using dispositions in the evaluation process of prospective teachers. "Dispositions" are difficult to measure, but the teaching profession has a general sense of when such expectations are violated or considered inappropriate. Data from the institution that participated in this study suggests this is the case when problems arise with their student teachers—almost all students who were asked to leave their student teaching placement (either by the cooperating teacher, the university supervisor, or a school administrator) are removed for demonstration of a less-than-professional disposition. For example, these include the perpetually late or unprepared student teacher, the student teacher who was included in the faculty photo and decided to put "bunny ears" with her fingers behind the principal's head, violations of student confidentiality, use of inappropriate language with students, excessive absences, and lying. And the cases of teacher dismissal noted earlier for poor judgment regarding online postings illustrate that the assessment of dispositions goes beyond what occurs in school buildings. The dispositions exhibited online by preservice elementary teachers and their prevalence are the subject of the study reported here.

STUDY PURPOSE AND DESIGN

The purpose of this study is to determine how college students majoring in elementary education at a large, public Midwestern university portray themselves in the publicly-accessible domain of Facebook during and after their teacher education program, and to compare these portrayals to the dispositions expected of K-12 prospective teachers by that institution. This study is useful beyond the institution examined here. Teacher preparation programs more broadly are impacted by the increased use of social networking sites by prospective teachers, and these students are likely facing similar access to and use of technology as their peers in other institutions. Teacher educators cannot assume that their students have the same kind of college experience as previous generations. Today's college students are under tremendous pressure to communicate electronically (Nathan, 2005) and their use of such communication tools enables others to have access to that information, with potentially negative consequences. Students in a teacher preparation program may be unaware of such consequences, and when faculty in preparation programs understand the nature of their students' self-portrayals, they can better determine how to prepare these students for a profession that demands high moral and ethical standards.

Research Questions

This study was designed to answer the following questions:

1. What percent of the elementary education majors at a large, public Midwestern university have a profile on Facebook?
2. What percent of the elementary education majors at a large, public Midwestern university have a fully-accessible profile on Facebook?
3. What percent of the elementary education majors' profiles contain material that is considered inappropriate?
4. What percent of the elementary education majors' profiles contain material considered marginally inappropriate?
5. What kinds of inappropriate images/messages/references are portrayed on elementary education majors' profiles?
6. What changes, if any, exist in elementary education majors' Facebook profile presence and security settings after two years—when they are in student teaching or graduated?
7. What changes, if any, exist in the content of fully-accessible profiles of elementary education majors?

Study Context and Methodology

The study was conducted by examining publicly-accessible profiles on Facebook posted by students enrolled in the elementary education major at a large, public Midwestern university. Participants include students who declared "Elementary Education" as a major, even if they are freshmen and have not yet been admitted into the teacher education program—a process that begins during the sophomore year. The study was submitted to the Institutional Review Board and was declared exempt from human subjects consent due to the public nature of the data under federal guidelines, 45 CFR 46. To protect students, however, we have chosen not to disclose full names or other personal identifiers. Initial data (research questions 1–5) were collected in early summer, so all but a few students in the study had completed a full year of the major, including a freshman orientation course that addresses standards for teacher education (including dispositions and field placements), and a foundations of education course.

All students in the major were entered into the "search" feature of Facebook and their profiles printed between 5/22/07 and 6/5/07, and again in 6/2009. Facebook profiles are updated regularly (sometimes hourly) by students. The purpose of printing the profile was to ensure that the researchers examined a single point in time. Important to note is that only publicly-accessible information and profiles were examined. We had no access to profiles for which students had restricted access. Thus, the information we saw is the same information that could be seen by children in schools, administrators, parents, and other members of the public.

Students in the major are predominately white females, aged 18–24. A small number of nontraditional, male, and minority students are part of the study population. The major had an enrollment of 471 students at the time of data collection. Of these students, 85.7% are female (n=404), and 14.2% are male (n=67). The major attracts a large number of transfer students, and this is reflected in the lower number of students at the freshman/sophomore levels; 15.3% of this population are freshmen (n=72), 19.7% are sophomores (n=93), 25.3% are juniors (n=119), and 39.7% are seniors (n=187). The larger number of seniors is likely due to students who add an additional semester to their coursework to complete a reading endorsement, thus maintaining their "senior" status for an additional semester.

This study is a naturalistic inquiry conducted in a similar manner as the dorm room door analysis reported by Nathan (2005) who, as an anthropologist, enrolled as a freshman at her own university. She used grounded theory and a constant-comparative method to develop categories of photos and text content that captured students' self-portrayals in a descriptive manner. We realize that readers may disagree with some aspects of our coding process, or the conclusions we draw from these data. However, consistent with methodological perspectives in qualitative research, the objective is to make the methodology as transparent as possible so that "given the same theoretical perspective of the original researcher and following the same general rules for data gathering and analysis, plus a similar set of conditions, another investigator should be able to come up with the same theoretical explanation about the given phenomenon" (Strauss & Corbin, 1990, p. 251).

To reduce bias in coding, the first and third authors coded over one-third of the same data independently. Inter-coder agreement in each category met or exceeded 93%. In addition, the behaviors that were coded as inappropriate are consistent with those reported in the literature and at the university as causing teachers or student teachers to be removed from their positions. Second, we have chosen to include our position on the findings in the Implications section. While some may disagree with our position, we feel that the findings warrant the opening of such important conversations in the field of teacher education and that the final section should serve this role.

Coding procedures: Profile sections. From students' printed Facebook profiles, we examined six main areas that correspond to spaces on a Facebook profile (See Appendix B). Each section of the profile was read by two researchers and a three category scale (*appropriate*, *marginal*, or *inappropriate*) was used to assess each section. The content of each section was independently reviewed to determine if it was (1) offensive to the researchers and potentially to parents and administrators (*inappropriate*), (2) possibly offensive to those stakeholders but probably acceptable to others (*marginal*), or (3) not likely to offend parents and administrators (*appropriate*). These decisions were guided by the following documents:

- the university's dispositions document (See Appendix A),
- the NCATE dispositions definition,
- student behaviors that have resulted in school requests to remove a prospective teacher from a school placement. A report of these behaviors is developed annually by the university's field placement office and is made available to program faculty.

Sections of the profile were coded after reviewing the entire contents of that section. If a single inappropriate item was found in that section, the section was coded as *inappropriate*. This was also the case with a *marginal* rating. A section was coded *marginal* if the most questionable content on it was *marginal*. To be coded *appropriate*, the entire section needed to be *appropriate*. Efforts were made to code in the prospective teacher's favor when difficult judgments had to be made about a particular section. For example, in one section of a webpage coded as *marginal*, a student writes, "lol damn sunday night is going to be awesome!!!! i can't wait! did you want to come up to the rugby house or do you want me to come to dm?? i am up for anything!!!" This may refer to a party on Sunday night, and being up for "anything" might suggest some rather inappropriate behavior, however, because the posting does not explicitly state the inappropriate behavior, it was given a coding of *marginal* as it is suggestive, but less explicit than an *inappropriate* posting. Each section of the page was coded independently by two researchers and every effort was made to evaluate the section based only on its exact content.

Examples from profiles are provided below for each category to illustrate the nature of the content that was coded at each level. All content has been left unedited for grammar or spelling, with the exception of the use of substitute symbols (e.g. $%#^) in place of vulgarity.

Inappropriate:

- Photos: A photo of 3 college men holding another upside down over a keg of beer with a tube emerging from the keg and being held in the upside down student's mouth. The caption of the photo reads "Beerfest 07 Event #5"
- Information: After multiple statements about enjoying "exploding" and "imploding," a student lists some of his favorite quotes as "Doggy style is a pain in the a$#" and "'God says he can get me out of this, but you're f*#ked' –Irish guy on Braveheart" and "'Is it dead?' -Rocko on Boondock Saints after splattering a cat when his pistol misfired." He lists his current job as a caregiver at a community preschool center. Under "job description" he states, "I get payed to watch your kids, be afraid, be very afraid."
- Groups: A student who claims membership in groups that include "'Dixie Chicks' suck my left nut" and "In Heaven There Is No Beer, Thats Why We Drink It Here"
- Main Photo: A main photo depicting the 19-year-old student drinking from a bottle of whiskey. (Age 21 is the legal drinking age where the study took place.)
- The Wall: Postings from friends that include, "Hahah!! i was looking at pictures today and i completely forgot about how last year when we came to veisha and how i kept making you drink all of those beers, and then i made you go into the porta potty to do it, haha oh man what was i thinking, you should definitely come up this weekend or i will see you at home on saturday!!"

Marginal:

- Photo: An image of a student in a very revealing bikini lying on a dock next to a lake.
- Groups: A student whose groups include: "Jews for Hillary," "Hillary Clinton: One Million Strong!" "Hillary Clinton," "A love for wine," "Students for Hillary Clinton," "Jewish girls are HOT!!!!" "I FREAKING LOVE Hillary Clinton (a.k.a. Vote Hillary 2008)," "Hillary Clinton Supporters," "Hillary Clinton for President and Equality," "Hillary Clinton 4 Prez! It always takes a Clinton to clean up after a Bush!" "Hillary '08" and "Hillary Clinton Rocks!!!"
- Information: An otherwise appropriate description of herself, but her interests include: "golfing, soccer, swimming, ... painting, drawing, *no clothes*, running, watching TV...." (italics added)
- The Wall: A friend writes, "actually i was thinking that you could tickle me...and make it awkward, and then call me archiepoo!!! i would love that...NOT!!! lol damn sunday night is going to be awesome!!!! i can't wait! did you want to come up to the rugby house or do you want me to come to dm?? i am up for anything!!!"
- Main Photo: An image of a student kissing another student.

Appropriate:

- Photos: A group of images of the student riding a horse, watching a football game with friends, and camping—fishing, roasting marshmallows, and struggling to put up a tent.
- Groups: A student whose groups include "I live in Iowa and I am NOT a farmer," "Teach Bailey Lewis to make the waterdrop sound," and "When I was your age, Pluto was a planet."
- Information: A student with a list of favorite quotes that include, "Jealousy does not show how much you love someone, jealousy shows how insecure you are." Favorite things: "basketball, friends, hangin out, decorating my apartment, flavored water, working with children, smiling, watchin TV"
- Main Photo: An image of the student smiling in front of a restaurant, standing between two friends.
- The Wall: "Hi Sarah! Nice to see you on facebook. How are you doing? No job offers, but I haven't been looking. Ryan is trying to start a business so we're staying in Iowa for one more year (it's cheaper!) before moving back to Illinois. Once we move back I'll start looking."

Coding Procedures: Category Development of Marginal/Inappropriate Content

Once each area was given an overall rating of *appropriate, marginal*, or *inappropriate*, those areas given a *marginal* or *inappropriate* rating were examined to determine the nature of the content that was deemed questionable. These decisions were also guided by the institution's dispositions document and the NCATE definition of dispositions. The categories were data-derived, and include:

1. *Partying*: Depictions of groups of people engaged in the college party scene. These included images of reckless behavior at bars or dorms, wild sexually-oriented dancing, etc. This category did not include more serene get-togethers such as scrapbooking parties or students smiling at a football game. What defined an image as partying included a sense of reckless behavior with a somewhat large group of people. What defined written descriptions of partying was the clear reference to "partying" or "clubbing" in the text.
2. *Sex/sexual content*: Depictions of individuals engaged in sexual contact with same sex or opposite sex, and references to sexual encounters.
3. *Inappropriate clothing/too much skin*: Images of nudity, revealing shirts, skirts that are too short for a student to sit down without revealing private areas, underwear, and other appearances that would not be acceptable in most classrooms.
4. *Political views*: Bold statements about political references. "Students for Hillary" comments were not coded as marginal or inappropriate, but statements such as "Monica Lewinsky had more president in her than George Bush ever will" had more obvious potential to offend K-12 education stakeholders.

5. *Vulgarity*: Profanity, anything children would get in trouble for saying in a public school. Also includes text message abbreviations that involve profanity, such as WTF (What the F#@^) or MILF (Mother I'd like to F#$%).
6. *Alcohol*: Given the "Drunken Pirate" court case, we determined how many students portrayed themselves drinking alcohol.
   - Underage Drinking: A subset of category 6, we coded instances where students were under 21 and showed themselves drinking alcohol.
7. *Criminal References/Behavior*: A broad category that included any serious reference to violence (such as personal intent to commit a violent act against another), gang references, racist comments, hate messages, or criminal activity.
8. *Religious Views*: This category included strong comments about religious views that made a deliberate statement of proselytizing in the context of teaching. Based on the establishment clause in the U.S. Constitution, the government is not allowed to favor one religion over another. A teacher who claims to be converting others or promoting his or her religion in the context of teaching may be seen as being in violation of the establishment clause. Each instance of a religious statement was coded marginal, and none were considered inappropriate. A religious statement coded marginal was: "I consider myself a missional teacher. My purpose in life is to be the hands of God and help expand his Kingdom." The statement was coded as marginal as some parents could view this as ill-advised behavior for a public school teacher and in direct violation of the establishment clause. However, the quotation is vague enough that the researchers felt a marginal rating was more descriptive than an inappropriate rating. This category does not include general information about personal church affiliations or quotations from scriptures.
9. Other: This category includes behaviors such as sitting on a toilet, passed out drunk in a bathroom, smoking at a Hookah bar, describing specific incidents from a student teaching placement, and other information that might raise concern among K-12 stakeholders.

FINDINGS

Research Question One: What percent of the elementary education majors at a large, public Midwestern university have a profile on Facebook?

Of the 471 students in the elementary education major, 76% (n=358) had a profile on Facebook at the time of data collection. This does not mean that the remaining 113 students are not posting material on the internet, because they may have a profile on another networking site. Freshmen in particular may have been more likely to retain their involvement with MySpace, a networking site that initially targeted high school students and was popular when data were first collected.

Research Question Two: What percent of the elementary education majors at a large, public Midwestern university have a fully-accessible profile on Facebook?

Of the 471 students in the elementary education major, 32% (n=153) have an active, fully-accessible profile on Facebook. Another 44% (n=205) have a Facebook webpage, but have restricted access to that page. This means that only "friends" are allowed to view the contents of the page.

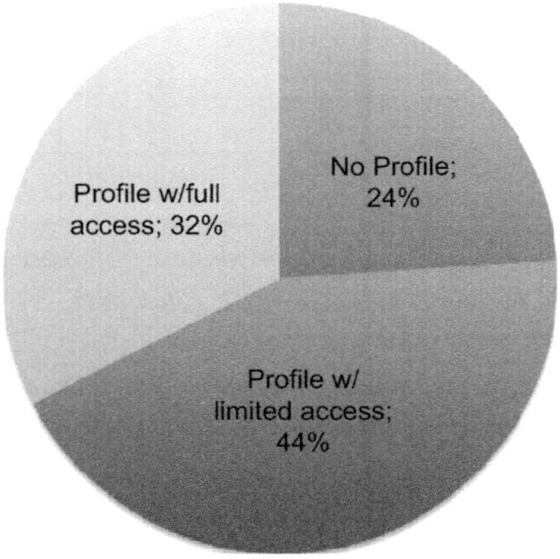

Figure 1. Participation in Facebook by Elementary Education majors.

Research Question Three: What percent of the Elementary Education majors' profiles contain material that is considered inappropriate?

Of the 153 fully-accessible profiles that were examined, 56% (n=86) contained inappropriate material. The location of inappropriate material varied, and is described under "Research Question Five."

Research Question Four: What percent of the Elementary Education majors' profiles contain material considered marginally inappropriate?

Of the 153 fully-accessible Facebook profiles included in this study, 22% (n=34) contained at least one category that was coded marginal, with no inappropriate material on the site. Important to note is that both marginal and inappropriate material may be offensive to a potential employer or parent. For example, using our coding scheme, the Millersville University student who was denied her education degree and licensure for the "Drunken Pirate" photo on her Facebook profile would have been coded "marginal." When the marginal and inappropriate categories are considered together, 78% (n=120) of the examined profiles contain material that could prevent an elementary education major from being allowed to work with children.

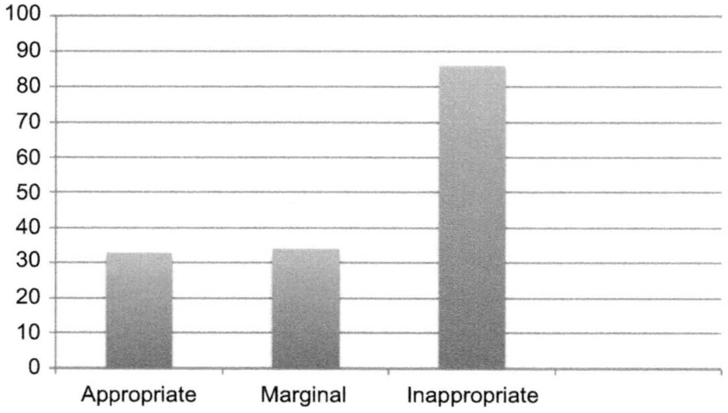

Figure 2. Number of elementary education majors' fully-accessible Facebook profiles categorized by appropriateness.

Research Question Five: What kinds of inappropriate images/messages/references are portrayed on elementary education majors' profiles?

The nature of the inappropriate material displayed by elementary education students varied by section on the profile (e.g., Groups or Photos). To assist the reader, frequencies of each type of inappropriate posting is provided, along with the percentages of profiles coded "appropriate," "marginal," or "inappropriate" *for that section*. Important to note is that only one inappropriate posting anywhere on the profile is sufficient for a student to be in jeopardy of facing disciplinary action by a school where he or she is working with children. Some areas of Facebook profiles contained a higher percentage of students posting inappropriate material than others. Therefore, a high percentage of appropriate material in one section does not mean that students' profiles are necessarily appropriate. The reader is reminded that 78% of the profiles examined contained at least one inappropriate or marginal item.

Section: Groups. The most common location of inappropriate material was in the "Groups" section. Of the profiles examined, 58% contained something marginal or inappropriate in the "Groups" section. Specifically, 41% (n=60) were coded as inappropriate, 17% (n=24) were coded as marginal, 30% (n=43) were appropriate, and 12% (n=18) students did not list groups.

In the "Groups" section, 69 students listed groups that contained vulgarity, 38 students had groups with references to alcohol (2 of them underage), 23 students had references to partying, 30 students had references to sex, 13 students had negative political references, 6 students had references to criminal activity (such as shoplifting and breaking parole), 2 students were categorized as belonging to groups that may be offensive with regard to religion, and 3 students belonging to groups in the "other" category (references to urinating, etc.).

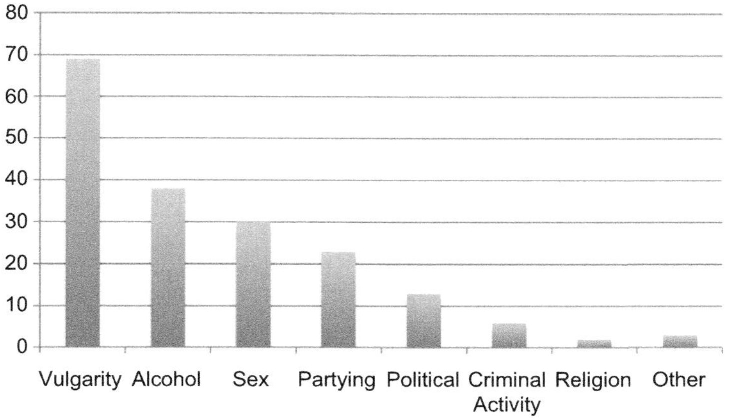

Figure 3. Categories and frequencies of inappropriate material in the "Groups" section.

Section: Photos. The second most frequent location of inappropriate and marginal material is the "Photos" section, with 52% of the examined profiles containing inappropriate or marginal material. Photos were not posted by 39 students. Of the 114 students who posted photos, 43% (n=49) were inappropriate, 9% (n=10) were marginal, and 48% (n=55) were appropriate.

This category was particularly insightful as images can portray what words cannot fully capture. The viewer can quickly see the appearance of students' living quarters, the view from the inside of local bars, the kinds of clothing students wear when they attend parties, games of beer pong, students doing a beer bong, and a kegstand. Students' spring break vacations were also commonly posted, and included local adventures as well as South Padre Island, Daytona Beach, South Africa, Panama, and New Orleans.

Alcohol use was the most frequent inappropriate event depicted in photos (48 students). While 32 of these 48 students were of legal drinking age, most of these pictures were far more than a glass of wine with dinner at a restaurant. Rows of students drinking out of whiskey or vodka bottles, students doing beer bongs or singlehandedly drinking a pitcher of beer while others cheer them on, students drinking shots while the caption reads "Shot Number Four!" and other such uses of alcohol were far more common than a legal adult sitting with a drink. The latter type of alcohol use was coded as "marginal," and this applied to only 3 students. Sixteen students posted photos of themselves drinking while underage. More than one of these students also posted a "countdown clock" to his or her 21st birthday, while including albums documenting their underage drinking.

Thirty students posted pictures of themselves partying. Many of these students had hundreds of photos from such parties. Photos from these parties showed entire walls covered with beer case packaging, students in very compromising poses, images of women puckering at the camera while others touched their breasts, students licking one another while sitting around a beer keg, students dancing on bar tables, and students engaging in multiple drinking "events" at a "Beerfest" in a fraternity house. One image showed an unconscious male student lying on a sofa, with his male friend squatting over him, exposing his naked buttocks and positioning them directly over the unconscious student's face. In most photos, the names of the students in the picture were labeled.

Nineteen students had photos that contained images depicting sexual content. More than holding hands or simply kissing one another (the latter coded as marginal), these images included women touching their tongues together, people on top of one another with captions that made joking references to rape, people in bed with one another, and excessively provocative poses.

Seven students included photos that showed excessive skin. In most instances, this included very revealing shirts that were accompanied by the student standing in such a position that the viewer could see down the student's shirt. In one instance, a student had her hands raised in a gangster-sign while leaning toward the camera wearing a gaping tank top, all while sitting on a toilet with her jeans down.

Six students showed some form of vulgarity in their photos, including direct use of middle finger hand gestures, t-shirts with profanity on them, and even a group of freshman girls standing in a line spelling a profane word with their bodies (the caption: "What's that spell?").

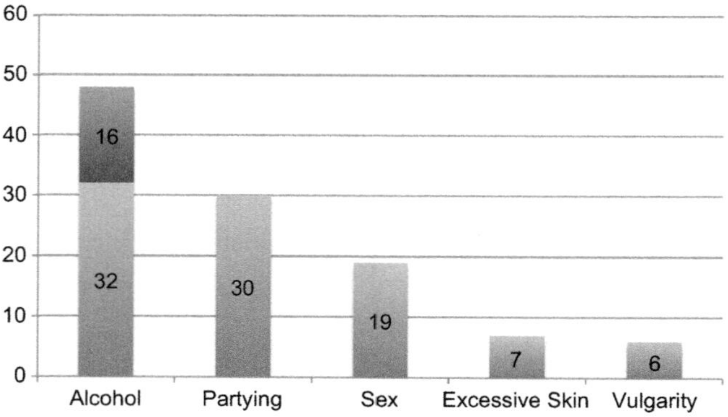

Figure 4. Categories and frequencies of inappropriate material in the "Photos" section.

"Alcohol" is divided into those of legal drinking age (n=32) and underage drinking (n=16).

Section: Information. The Information section is a rich description of students and their interests. Students wrote a variety of descriptions of themselves, quotations that inspire them, favorite music, and favorite movies. The largest area within the Information section tended to be the quotations space. Interestingly, most quotations were not from famous inspirational authors, poets, politicians, or sports heroes. Quotations overwhelmingly were funny or bizarre statements made by self or friends. For example,

"You're the whitest Native American I've ever seen." ~Jennifer P—

"Me: 'I can't sleep with socks on.'

Jenny: 'Me neither. My feet get too hot.'

Me: 'No, not because they're too hot. I just can't stand it!'

Jenny: 'So you're saying your feet are claustrophobic?'"

"You are seriously the fastest water drinker EVER!" – Me to Alissa

"I'm not sure what truck stop…Oh wow. I really want to spoon with you right now." -Alissa

Interestingly, four students mentioned in a section called "Favorite Books" that they do not like to read or do not read books. One student said, "Ha ha. What are books?" One student claimed, "Don't really have the time to read!" yet he updates his Facebook profile several times per day, and lists multiple television shows and movies as his regular pastimes.

In the "Information" section, 27% (n=37) of the examined profiles were inappropriate, 22% (n=30) were marginal, and 55% (n=82) were appropriate.

Across all areas of the "Information" section of the profile, the most common inappropriate category included references to sex (41 of the 149 students who posted information). Profanity (33 students), alcohol (28, with 4 underage), proselytizing religious views (12 students) and partying (11 students) were also common. Five students made reference to some criminal activity or hate message, and two referred to other behaviors (such as urinating or defecating).

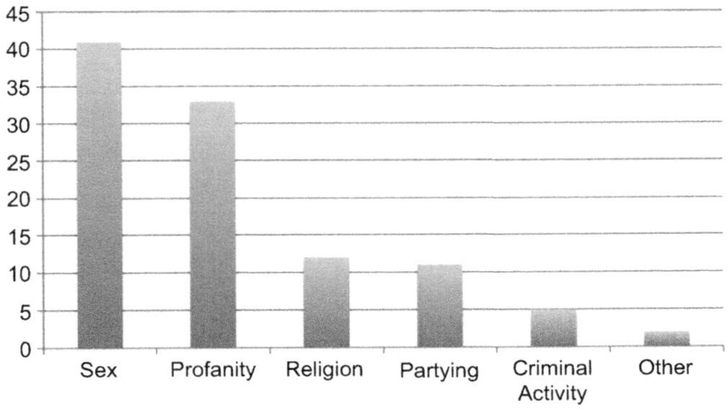

Figure 5. Categories and frequencies of inappropriate material in the "Information" section.

Section: The wall. Postings on the "Wall" are not posted directly by the student, but can be removed by him or her. One danger is that a student may try to keep an appropriate profile and a friend could post an inappropriate message that could be read by others before he or she has the opportunity to remove it. This situation was rarely the case, however. Students with inappropriate wall postings almost always had inappropriate material that they posted elsewhere on their site, and inappropriate wall messages had been on the sites for a substantial period of time. Of the 136 students who had an active "Wall" section, 18% (n=25) were inappropriate, 31% (n=41) were marginal, and 51% (n=70) were appropriate. Only 17 students had no wall available on which others could post.

Profanity was posted on the wall by friends of 37 students in this study. This was the most common inappropriate posting. References to alcohol (n=21), partying (n=15), sex (n=11), body parts (n=5), criminal activity/hate messages (n=4), and other references (n=3) appeared on students' walls. Interestingly, students made no references to politics or religion on the wall. Even students who identified themselves as very politically active did not have wall conversations related to politics, despite the fact that several high-profile presidential candidates had been campaigning on campus during the time of data collection.

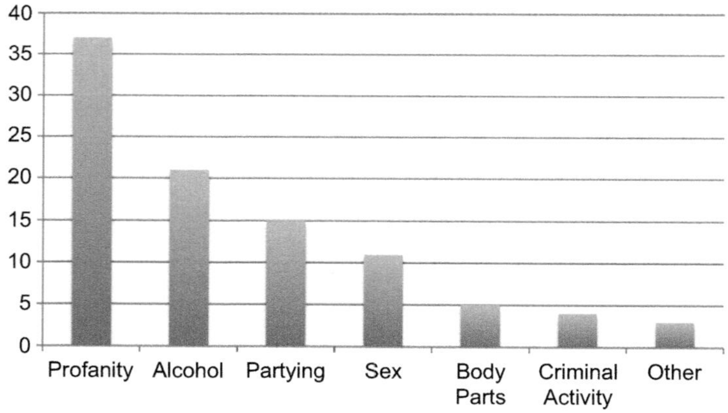

Figure 6. Categories and frequencies of inappropriate postings on the "Wall" section.

Section: Main photo. The main photo was appropriate for 85% (n=127) of students. Only 6% (n=9) of the students had an inappropriate main photo, and 9% (n=14) had a main photo considered marginal. Inappropriate or marginal ratings were given to photos that showed partying (6), alcohol (6), sex (5), excessive skin/inappropriate clothing (4), and underage drinking (3).

Section: Friends. Friends were more difficult to code since they were only displayed by their main photo on a student's profile in a small corner. Often these main photos were displayed in a very small format, so determining if the drink they were holding was a soda or beer was difficult, if not impossible. Given the limitations of these data, only 6 students had friends with identifiably inappropriate main photos, and 14 had friends with marginal main photos. The remaining 120 were considered appropriate. Consistent with the students' own main photos, their friends' inappropriate or marginal photos conveyed alcohol (7), partying (4), sex (5), and excessive skin/inappropriate clothing (4). A viewer could, however, go to these students' pages to determine if their use of alcohol was underage, but this was beyond the scope of the study.

Research Question Six: What changes, if any, exist in elementary education majors' Facebook profile presence and security settings after two years—when they are in student teaching or graduated?

Of the 471 students/former students studied, 41% (n=194) had no profile on Facebook in 2009. This increased from the 105 students (22%) who had no profile in the summer of 2007. This does not mean that these individuals do not have a presence on the internet, as they may be using another networking site. In addition, a student was entered as having "no profile" if he/she was not readily accessible. Some students have common names and several hundred people appear with the same name. Several of those individuals did not list a network or a photo that would easily identify them as the individuals we were seeking. As a result the number of students/alumni listed as "no profile" is likely to be somewhat overstated but is limited by the design of this study.

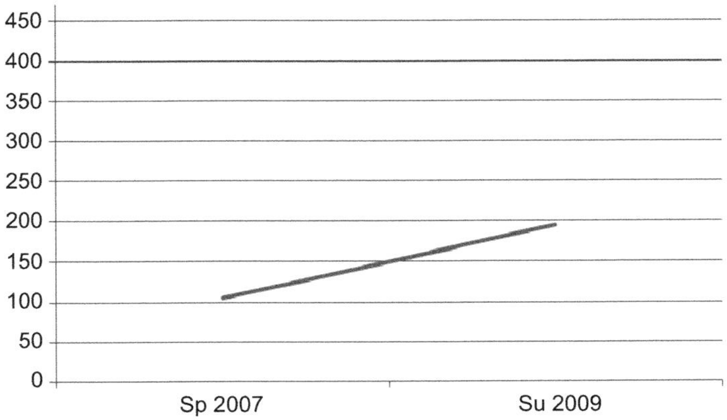

Figure 7. Elementary education graduates who do not have a Facebook profile.

In contrast, 217 (46%) of the students/alumni studied have a profile on Facebook that has limited access. This almost unchanged from 2007, when 205 students had profiles with restricted access (44%).

60 students (13%) still had a fully accessible profile on Facebook. This is down from 153 students (32%) in 2007. Trends over time are displayed in Figure 8.

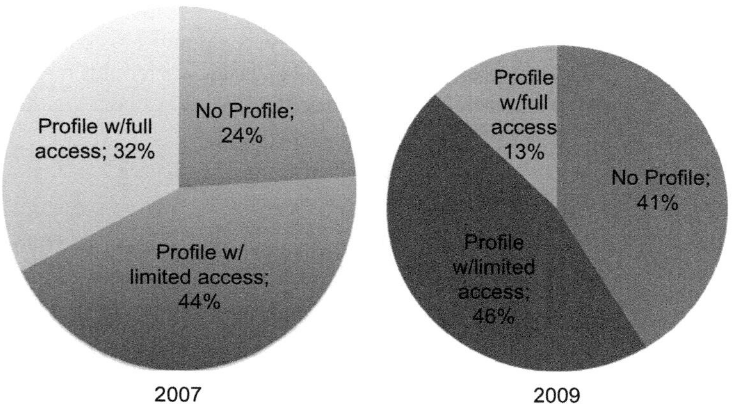

Figure 8. Elementary education majors' Facebook presence and security settings, 2007–2009.

Research Question Seven: What changes, if any, exist in the content of fully-accessible profiles of elementary education majors?

Of the 60 individuals who had fully accessible profiles on Facebook, 34 contain inappropriate material. 10 contain marginal material, and only 16 were appropriate.

The most common location for inappropriate material was in photos, on "the wall," and in personal information sections. Images of college parties from 2 years earlier are still on these sites, and one individual had more than 1,470 photos posted on her webpage. The distribution of open profiles was not skewed toward the college seniors, but distributed itself equally across students, recent alumni, and current teachers. One individual complained on her webpage that she hated her job as a food server, but that she can't find a teaching job; yet her webpage contained an image of her doing a beer bong from a funnel and tubing connected to a dildo with a caption of "go girl!" Another listed among his "favorite quotes" a conversation between a police officer and a suspect related to a pat-down search that involved unintentional contact with the suspect's penis. Just one line down on the webpage, he lists his profession as "I am a fifth grade teacher at [name removed] Elementary in [name of city removed]." Photos titled "When I'm drunk," "Not in jail," "Skittles Shots," "IDKwhatthiswasabout," "oneofmybestfriendsVodka," "crotch shot," "Beer lovers," "drunkmindstogether," and "excessive amountsofalcohol" with accompanying inappropriate images are more inappropriate than profile content that has resulted in media-reported teacher firings.

DISCUSSION

The findings of this study are sobering and an affront to the noble title "teacher." During the program, 78 percent of the elementary education majors' publicly-accessible profiles expressed content that is contradictory or potentially contradictory to the dispositions required of teachers by the institution where our study took place—dispositions developed as a reflection of what society expects of teachers (Simpson, 2008) and what schools reinforce to students in their field placements. Most of the 33 fully-appropriate profiles (22%) were rather neutral—students making arrangements to meet for lunch and displaying photos from a football game. While we did find ourselves smiling at the occasional picture of a student working with poverty-stricken children in Panama, rebuilding a house in New Orleans, or cradling a new baby niece, such positive personal messages were scarce and overwhelmed by activities unbefitting of those entrusted to teach children.

Behavior at odds with expected dispositions is the primary reason prospective teachers are removed from student teaching at the institution where this study took place. During the spring 2009 semester, sixteen preservice teachers received disciplinary action during student teaching (an all-time high) for behavior conflicting with expected dispositions. Failure to exhibit required dispositions, not teaching practices, accounted for most all the students who had to be removed from student teaching during the 2008–09 academic year.

In this study, the dispositions most frequently violated by students include expectations of "appropriateness of dress, tact, discretion, courtesy," and "adheres strongly to high moral principles and ethical standards…evidences integrity." A photo showing a prospective teacher being cheered as he drinks a pitcher of beer, and another photo titled "Kisses to redheaded sluts" that displays a prospective teacher downing shots with two other students, are not the behaviors parents and society want impressionable children to see from their teachers. Exhibiting and purposely posting such behaviors are the antithesis of tact, discretion, and integrity. Other photos titled, "Closet alcoholic," "Gotta love cheap wine," "Jenny wanted a 3-way," "Jenny doing her first kegstand right before the cops busted the party and Jenny went running," "Shake that ass," "OMG this is just wrong…hehehe Good old new years" and "90 beers will do it" are clear indications that these prospective teachers either do not understand or do not take seriously their position as role models for children.

Perhaps most surprising is that prospective teachers would *purposely choose to display themselves* in the public domain of the internet, placing peer pressure on others, succumbing to peer pressure, exhibiting intolerance of others, and engaging in unadvisable and illegal activities. Many schools where these preservice teachers are guests have put into place programs to promote character and decision-making that these prospective teachers' own chosen behaviors mock. Teachers are role models for children, whether or not they accept that responsibility. The issue is only what kind of role model teachers will be for children. While some readers may argue that this is a college phenomenon and that these individuals will "grow up" when they graduate, 44 students in the original group studied still displayed information after graduation that could prevent them from obtaining or keeping a job in teaching. Our program addressed these issues in a freshman orientation class, introduction to technology course, science methods, and all field placement experiences.

Some readers may argue that the concerns raised in this study are little more than a new version of generation-bashing and that the behaviors exhibited by prospective teachers in our study have always existed to some extent among college students. While the students' behavior noted in our study may be little different from past generations, this misses what *has* changed! What makes the current situation alarming is how new technology has altered the moral landscape, does influence thinking and behavior, and demands conscious awareness and heightened discretion. How technology impacts individuals and society is never simply a matter of how we use it, but how the very nature of the technology alters the way people think and act and ultimately shapes culture (Postman, 1985 & 1992). Facebook and other electronic social networks, like all technologies, have a bias. Electronic social networks favor and encourage personal expression through the immediate sharing of information, often pictorial, without review, but for everyone to see. Fighting against this bias demands purposeful attention to what electronic social networks favor, how they alter thinking, the consequences of sharing information, and prudent decision-making. While today's student behavior may have similarities with inappropriate behaviors of previous generations, electronic social networking promotes the rapid, unexamined and indiscriminate dissemination of those behaviors. Conscious effort is required to resist this bias of electronic social networking.

For example, not so long ago, pictures were captured by cameras that required film. The very cost of film and its development was a consideration that limited the number of pictures taken, and influenced to some degree the kind of pictures taken. Significant time would pass before the film was sent in for development. Profoundly inappropriate photos might have been flagged by the company performing the development, and this placed some limits on the kinds of photos that were originally taken or sent in for development. Time would also pass before a photo would be available for viewing and this made likely that the viewing of an inappropriate photo would take place in a different context that might provide a heightened perspective regarding the photo's appropriateness. Once available, the photo was likely first viewed alone and a decision was made whether to share it with others and, importantly, who those others might be. Particular pictures, because of their content, would be shared with some individuals, but not others.

In that moment where such pictures were shared, the reaction of viewers would send immediate feedback regarding the appropriateness of the content conveyed in the photo. With each reaction, information regarding the appropriateness of the photo's content would be provided, often subtly. Even though some viewers might chuckle at inappropriate content in a picture, their nonverbal reaction, voice-intonation, or advice to be careful about with whom to share the picture would send unmistakable messages about cultural expectations for behavior. The time that passed and the reaction that photos received along the way would all act as filters that shaped understanding of appropriate and inappropriate behavior and with whom else to share a photo or whether to share it at all.

But Facebook is faceless — individuals who post pictures often have received no feedback (e.g., outright concern, nonverbal reaction or voice intonation) that might convey messages regarding the appropriateness of a photo or written message. Photos can be immediately uploaded for all to see. The very nature of this technology is biased against the important filters — time and interpersonal cues — that help shape notions of appropriate and inappropriate behavior, or at least appropriate and inappropriate audiences for particular behaviors. For example, among the thousands of photos posted on Facebook by the elementary education majors, very few, if any, contained feedback or comments posted by others. This trend was also noticed by Cameron Marlow, a researcher employed by Facebook. Marlow found that what influenced the number of photos that people upload "wasn't based on how many of their friends showed approval for the photos by clicking that they liked them, or how many comments were left on each. Rather, it was based on how many photos your friends uploaded" (Marlow, in Ortutay, 2009). In a very real sense, the elementary education majors are using the technology in precisely the ways that the technology is designed to be used. So they are partially, but not fully, to blame here.

As Postman (1995) noted, technological change is not simply additive; it is ecological. New technology does not merely add a new option; it alters the landscape of our thinking associated with the media, and thinking more generally. Electronic social networks promote the rapid sharing of information, and in doing so those technologies alter thinking about what information to share, with whom to share it, and even what information was originally captured.

Of course, Facebook users may choose a limited-access profile that permits content to only be viewed by "friends." Ignoring that little on the web is truly secure, social networks confuse and trivialize the meaning of the very special relationship "friend." The true meaning of "friend" is profoundly different from "acquaintance," but social networks are an assault on that important difference. The word "friend" may be used loosely, but in sharing personal information about ourselves we all tacitly understand the difference between a friend and someone we simply know and see at school or work. This is why people have always first shared personal information with family, friends, or others who can be counted on. Even within these categories, we begin with certain individuals we particularly value. To varying levels, we trust such people with personal information, and look to them for feedback and a measure of protection. That is, these people are far more likely to provide feedback that we trust and value in making decisions. In referring to all who wish to be electronically connected as "friends," social networks like Facebook blur these important distinctions, and the important role they play in making appropriate personal and public decisions that help in shaping the wider culture.

Some readers may reflect on their past behaviors and be thankful that technology didn't capture their youthful dalliances. But technology *did* exist to capture and distribute inappropriate behavior. What *is* different is how new technology promotes the rapid and uncritical advertisement of what is being *purposely* captured. Without the social cues that have previously existed to moderate such behavior and, with time, create a sense of acceptable behavior, the ethical landscape changes. This moves the framework regarding decency from a shameful behavior that one might later regret to a behavior that is not recognized as shameful and is advertised for all to see, potentially including vulnerable and impressionable children. The former social framework acknowledges the moral/ethical issue and at the very least attempts to keep it private or limited to a smaller group of adults. The latter doesn't acknowledge the moral/ethical implications of the behavior. So readers who focus on how the technology captures youthful dalliances would be quite mistaken regarding what is going on here.

For instance, after observing inappropriate images of her own elementary science methods students placed on Facebook, the first author raised concerns to the students. Rather than removing the profiles or changing the content, or even being apologetic about their behavior, the students became angry that a professor had been viewing their profiles. They met at the café after class to complain about the professor. Some restricted access to their sites as a result, but no evidence exists that students questioned the content on their profiles.

IMPLICATIONS

> The function of education, therefore, is to teach one to think intensively and to think critically. But education which stops with efficiency may prove the greatest menace to society. The most dangerous criminal may be the man gifted with reason but no morals...We must remember that intelligence is not enough. Intelligence plus character — that is the goal of true education. The complete education gives one not only power of concentration, but worthy objectives upon which to concentrate....
>
> Martin Luther King, Jr. (1947)

The bias of electronic social networks toward the almost immediate sharing of information bypasses previous social filters that would help shape more acceptable decision-making, attitudes and behavior. Without that feedback, students' choice to purposely and widely advertise inappropriate behavior, and their sometimes puzzled response to disapproval is perhaps not so surprising. Bauerlein (2008) has noted how this younger generation uses technology to remain trapped in youth concerns, building a "camp in the desert" where they isolate themselves to exchange their photos, music, videos, status updates, and other entertainment—largely unaware of the greater adult world around them.

Clearly, students must be taught that the very public and potentially permanent display of inappropriate behavior may come back to haunt them in unanticipated ways. For instance, anything posted on the internet regarding a weekend of partying is potentially visible and downloadable by children, parents, and employers. New technology encourages immediate and thoughtless sharing of personal information in ways that have not previously existed. Moreover, inappropriate behavior may also be caught electronically by someone else who may, without permission, just as easily and quickly distribute it. At the very least, students should be aware of the potentially very damaging consequences of electronic social networks.

For teacher educators, the implications run deeper. We must explicitly and forcefully teach our students that their behavior in and out of the classroom *does* matter. We shirk our own responsibility if we ignore that parents, schools and society hold prospective and experienced teachers to high expectations of conduct. The study reported here makes clear that students are not merely making poor judgments regarding what to post online. Their bragging about inappropriate, offensive and sometimes illegal behavior is an indication that they do not see such actions as problematic. In the study reported here, the profiles of those in their student teaching semester were equally likely to contain inappropriate content as the profiles of their freshmen and sophomore peers. Fortunately, students tended to restrict access to or remove their Facebook profiles after graduation. However, nine percent of graduates, many of whom were teaching, had publicly-accessible material of questionable appropriateness posted online.

The view that prospective teachers are simply being college students and will likely outgrow their juvenile actions is an abdication of teacher educators' responsibility to prepare highly effective teachers. The prospective teachers in this study who posted inappropriate items on Facebook clearly did not consider that their conduct outside of their school-based experiences had anything to do with their role as a teacher. Simply making students aware that they must more carefully decide what to post online ignores the upright character that NCATE, many teacher education programs, parents, and the public expects of teachers. Teaching has never been solely or perhaps even primarily about effectively conveying a discipline's content knowledge. Socialization, values, and creating a public with an admirable character are undeniably part of the education establishment's responsibility. As Sizer and Sizer (1999) note, the students *are* watching!

Teacher education programs should explicitly *teach* dispositions and their importance, not simply assess dispositions and punish students who do not exhibit them. This effort must be ubiquitous in teacher education programs, not simply an esoteric course that students can easily dismiss as an anomaly in an otherwise unprincipled program. Students need consistent messages regarding how to be tactful, courteous, professional, and ethical, and the high standards of behavior that society places on teachers. For instance, teacher educators need to professionally confront situations where prospective teachers are bragging about inappropriate or illegal behavior that may result in their being removed from the program or a teaching position. This has always been important, but takes on added significance now that technology isolates students from mature others, including adults, who are in a position to guide students toward a more appropriate and noble character.

Perhaps the least obvious implication of this study is that all students, including preservice teachers, need and deserve a far more robust technology education (Olson & Clough, 2001; Kruse, 2009). Meaningful technology education is far more than learning how to use technology. It includes an understanding of what technology is, how and why technology is developed, how society directs, reacts to, and is sometimes unwittingly changed by technology. Technological literacy includes an understanding of the nature of technology, and addresses questions like those raised by Postman (1995):

- For every advantage of technology, what is the corresponding disadvantage?
- How are the advantages and disadvantages of particular new technologies distributed unevenly?
- What is the underlying philosophy of particular technologies? For example, how do particular technologies change the way we think and act?
- What are the intellectual, emotional, sensory, social, and content biases of particular technologies?
- What goals are promoted and ignored by particular technologies?
- How does technology change the ways we view learning, teaching and schooling?
- How does the technology promote and inhibit thinking and learning?
- How does technology use us without our awareness?

The National Educational Technology Standards for Teachers (ISTE, 2008) recommend understanding the social, ethical, and human issues inherent in technology, but many educators appear not to understand or take seriously that technology is not neutral or simply a matter of how people choose to use it. Any serious understanding of technology recognizes that it has inherent biases and promotes certain types of behaviors while suppressing others.

Facebook certainly promotes a kind of interaction among individuals, but it also promotes a personal bulletin board where students are pressured to fit in—to appear "fun," "likable," "adventurous" — just as the dorm room doors did in Nathan's (2003) study. For today's students, however, the prevalence of digital photos and the instantaneous manner of making them publicly accessible has led to a type of "race to the bottom" — the more outrageous the better. The result is prospective elementary teachers gone wild—not cognizant of their role as a teacher and model for impressionable children—and unaware or unconcerned that society is watching and is not amused.

REFERENCES

Associated Press (2006). *Teacher's topless pics cause stir*. CBSNews.com Accessed 9/10/07 from http://www.cbsnews.com/stories/2006/06/17/national/main1726404.shtml.

Bauerlein, M. (2008). The dumbest generation: *How the digital age stupefies young Americans and jeopardizes our future*. New York: The Penguin Group.

Bonnstetter, R. J., & Pedersen, J. (2005). How many hats can a teacher wear? *Paper presented at the Northcentral Association for Science Teacher Education conference*. Eau Claire, WI.

Bugeja, M. (2005). Interpersonal Divide: *Search for community in a digital age*. New York: Oxford University Press.

Burant, T.J., Chubbuck, S.M., & Whipp, J.L. (2007). Reclaiming the moral in the dispositions debate. *Journal of Teacher Education*. 58.5 (Nov-Dec 2007): p397(15).

Carvin, A. (2006a). *Avast ye knee-jerk reactions, Matey! Learning.now*. May 9, 2007. http://www.pbs.org/teachers/learning.now

Carvin, A. (2006b). *Facing up to Facebook*. Learning.now. December 15, 2006. http://www.pbs.org/teachers/learning.now

CBS (2006). *Employers look at Facebook, too*. Accessed 6/20/06 from http://www.cbsnews.com/stories/2006/06/20/eveningnews/main1734920.shtml.

Damon, W. (2007). Dispositions and teacher assessment: *the need for a more rigorous definition. Journal of Teacher Education*. 58.5 (Nov-Dec 2007): p365(5).

Diamond, J. (2008). U.S. District Court for the Eastern District of Pennsylvania. Snyder v. Millersville University, No. 07–1660. Retrieved September 1, 2009 from http://voices.washingtonpost.com/securityfix/Decision2008.12.03.pdf

Facebook (2009). *Press Room: Statistics*. Retrieved October 13, 2009 from http://www.facebook.com/press/info.php?statistics

Finnigan, S. (2007). Wired shut? *Teens in their own high-tech world. Family Circle Magazine*. Retrieved September 18, 2008 from http://www.parents.com/kids/safety/internet/wired-shut-teens-in-their-own-high-tech-world/?page=1

Gage, N.L. (1963). Preface. In N.L. Gage (Ed.), *Handbook of research on teaching* (pp. v-ix). Chicago: American Education Research Association.

Getzels, J.W., & Jackson, P.W. (1963). *The teachers' personality and characteristics*. In N.L. Gage (Ed.), Handbook of research on teaching (pp. 506–582). Chicago: American Educational Research Association.

International Society for Technology in Education (2008). *National Educational Technology Standards for Teachers*. Retrieved October 10, 2009 from http://www.iste.org/Content/NavigationMenu/NETS/ForTeachers/2008Standards/NETS_for_Teachers_2008.htm

Internet World Stats (2012). *Usage and Population Statistics*. Retrieved September 3, 2012 from http://www.internetworldstats.com/facebook.htm

Interstate New Teacher Assessment and Support Consortium (INTASC) (1992). Model standards for beginning teacher licensure and development. *A resource for state dialogue*. Washington, DC: Council for Chief State School Officers.

Josephson Institute (2009). Character Counts: *Frequently Asked Questions*. Retrieved October 11, 2009 from http://charactercounts.org/overview/faq.htm

King, M.L. (1947). The purpose of education. *Morehouse College Student Paper*, January-February. Accessed 9/13/09 from http://www.ncat.edu/~univstud/The%20Purpose%20of%20Education.pdf

Kruse, J.W. (2009). Teaching and assessing philosophical issues of technology in an 8th grade science classroom. Proceedings of the 10th International History, Philosophy and Science Teaching Conference, University of Notre Dame, June 24–28.

Lortie, D. C. (2002). Schoolteacher: *A sociological study*. 2nd edition. University of Chicago Press.

Mullin, D. (2003). Developing a framework for assessing teacher candidate dispositions. January, 2003. *Presentation at AACTE National Conference*, New Orleans, LA.

Murray, F.B. (2007). Disposition: a superfluous construct in teacher education. *Journal of Teacher Education*, 58(5), 381–387.

Nathan, R. (2005). My freshman year: *What a professor learned by becoming a student.* New York: Cornell University Press.

National Council for Accreditation of Teacher Education (NCATE). (2008). *Professional standards for the accreditation of teacher preparation institutions.* http://www.ncate.org/documents/standards/NCATEStandards2008.pdf

National Council for Accreditation of Teacher Education (NCATE). (2001). *Standards for professional development schools.* http://www.ncate.org/documents/pdsStandards.pdf

Olson, J. K. & Clough, M. P. (2001). Technology's Tendency to Undermine Serious Study: A Cautionary Note. *The Clearing House, 75*, 8–13.

Ortutay, B. (2009). *'Don't post that!'—Networking etiquette emerges. Associated Press.* Retrieved September 12, 2009 from http://www.usatoday.com/tech/webguide/internetlife/2009-09-04-facebook-etiquette_N.htm

Peterson, R., & Barton, T. (2007, August 22). Barta reacts to web images of Hawkeye players. *The Des Moines Register (Des Moines*, IA), p. 1D.

Postman, N. (1995). The End of Education: *Redefining the Value of School.* New York: Vintage.

Postman, N. (1992). Technopoly: *The Surrender of Culture to Technology.* New York, Vintage.

Postman, N. (1985). Amusing Ourselves to Death: *Public Discourse in the Age of Show Business. Penguin*: New York.

Poupee, K. (2002). All circuits are busy. Science & Technology. *Le Monde, Paris France.* October 25, 2002.

Simpson, M. D. (2008, April). *The whole world (wide web) is watching. NEAToday*, p. 17.

Sizer, T.R. & Sizer, N.F. (1999). The Students are Watching: *Schools and the Moral Contract.* Boston, Beacon Pres.

Strauss, A., & Corbin, J. (1990). Basics of qualitative research: *Grounded theory procedures and techniques.* Newbury Park, CA: Sage Publications.

Wasicsko, M. M., Callahan, C., & Wirtz, P. (2004). *Integrating dispositions in the conceptual framework: Four a priori questions." KCA Journal 23, no. 1 (fall 2004)*, 24–28. Retrieved June 25, 2008 from http://www.kyca.org/online/file.php/1/KCA_Journal/KCA_Journal_2004.pdf

Washburne, C., & Heil, L.M. (1960). What characteristics of teachers affect students' growth? *The School Review*, 68(4), 420–428.

APPENDIX A

ASSESSMENT OF DISPOSITIONAL PROFESSIONAL QUALITIES IN TEACHER EDUCATION PROGRAM CANDIDATE DRAFT.

This form will be used to evaluate dispositions that education students display during class and field experiences, to document professional progress, and to identify areas where improvement is needed.

Not Applicable or Not Observed	Serious Concerns	Needs Improvement	Emerging	Acceptable
NA	0	1	2	3
	Behavior displayed is contrary to expectations for this disposition	Behavior is displayed occasionally	Behavior is displayed frequently	Behavior is displayed frequently and consistently

Caring: Candidates with this set of dispositions value and appreciate all aspects of other persons' well being—cognitive, emotional, and spiritual—thereby enhancing opportunities for learning needs of other education students and in working with professionals. The following list comprises many, but not all, of the qualities, tendencies, and/or behaviors which characterize a set of caring dispositions:					
Empathy: Inclination to identify with, and see things from the perspective of others	NA	0	1	2	3
1.2 **Compassion:** Sympathy, often with a desire to help relieve the suffering of others	NA	0	1	2	3
1.3 **Rapport:** Ability to develop appropriate relationships with peers and other stakeholders	NA	0	1	2	3
1.4 **Respect:** Shows appropriate regard for the needs, ideas, and experiences of others	NA	0	1	2	3

Caring: Candidates with this set of dispositions value and appreciate all aspects of other persons' well being—cognitive, emotional, and spiritual—thereby enhancing opportunities for learning needs of other education students and in working with professionals. The following list comprises many, but not all, of the qualities, tendencies, and/or behaviors which characterize a set of caring dispositions:					
1.5 **Passion:** Demonstrates interest, enthusiasm and optimism for the people, content, and context of the teaching/learning process	NA	0	1	2	3
1.6 **Cultural Competence:** appreciates and capitalizes upon diversity; is aware of and acts to reduce one's own biases; employs culturally sensitive pedagogy	NA	0	1	2	3
Comments:					
Communication: Candidates with this set of dispositions are sensitive to and skillful in the various aspects of human activity. They have effective interpersonal relationship skills and attitudes that foster collaborative enterprises useful in enhancing the teaching/learning process. The following list comprises many, but not all, of the qualities, tendencies, and/or behaviors which characterize a set of communicative dispositions:					
2.1 **Presence:** Keen with-it-ness and engagement in human interactions and others' needs	NA	0	1	2	3
2.2 **Responsiveness:** Attentive to others' needs; the ability and inclination to act as best meets the needs, subtle as well as obvious, of others and their circumstances	NA	0	1	2	3
2.3 **Attentiveness:** Concentrates on others' communication; takes others' communication into account	NA	0	1	2	3
2.4 **Authenticity:** Genuineness that fosters and enhances the teaching and learning process while exercising judgment about personal and professional boundaries	NA	0	1	2	3
2.5 **Collaborativeness:** Seeks means to involve & work with others in planning, problem solving, and learning	NA	0	1	2	3
2.6 **Voice:** Speaks out when the need arises	NA	0	1	2	3
Comments:					

Creative: Candidates with this set of dispositions display the capacity to envision and craft things in novel and meaningful ways to meet the needs of students. The following list comprises many, but not all, of the qualities, tendencies, and/or behaviors which characterize a set of creative dispositions:					
3.1 **Flexibility:** Adapts, adjusts, and modifies practices to meet the needs of students and peers; thinks on one's feet; is comfortable with change	NA	0	1	2	3
3.2 **Inventiveness:** Uses the needs and interests of students to approach curricular and strategic decisions; visualizes and implements novel ideas and practices	NA	0	1	2	3
3.3 **Resourcefulness:** Utilizes resources in effective ways; adapts practices to unforeseen challenges	NA	0	1	2	3
Comments:					
Critical: Candidates with this set of dispositions have the ability to examine closely, to critique, and to ask questions. They do not accept the status quo at face value but employ higher level thinking skills to evaluate, analyze, and synthesize. Self-evaluation and reflection characterize candidates with this set of dispositions. The following list comprises many, but not all, of the qualities, tendencies, and/or behaviors which characterize a set of critical dispositions:					
4.1 **Reflectiveness:** Takes time consistently to evaluate effectiveness of instruction & behavior in terms of the larger goals of education; nurtures reflectivity in students and peers; reflects on own growth and accountability	NA	0	1	2	3
4.2 **Initiative:** Exhibits a willingness to pursue solutions to problems or questions; gathers relevant data and persistently seeks to improve situations or areas of need	NA	0	1	2	3
4.3 **Open-mindedness:** Exhibits an ability to look at different sides of an issue; recognizes the possibility of error in one's own beliefs and practices; does not display or act upon prejudices against people or ideas	NA	0	1	2	3
4.4 **Efficacy:** Nurtures high expectations, demonstrates self-direction and confidence, and empowers students and peers	NA	0	1	2	3
4.5 **Humility:** Places the needs of the learner and/or learning task above one's own ego; reflects on own growth and accountability	NA	0	1	2	3
Comments:					

Contextual: Additional qualities defined by and consistent with your institutional mission/conceptual framework:					
5.1 Not applicable	NA	0	1	2	3
5.2 Not applicable	NA	0	1	2	3
Professional Requirements: These are qualities and practices that teacher candidates must exhibit in order to be recommended for licensure, some of which are explicit in the State Code of Ethics and Code of Responsibilities. The candidates will display all of the following qualities and/or behaviors that characterize this set of professional requirements. Also, because each of these is considered absolutely necessary, each one will be separately assessed:					
6.1 **Professionalism:** Endeavors to meet the standards expected of a teacher such as appropriateness of dress, grooming, demeanor, punctuality, tact, discretion, courtesy, etc.	NA	0	1	2	3
6.2 **Personal and Professional Ethics and Integrity:** Adheres strongly to high moral principles and ethical standards as expressed in the [State] Code; evidences integrity	NA	0	1	2	3
6.3 **Work Ethic/Responsibility:** Attends to school policy for teacher attendance; completes teaching-related tasks in a thorough and efficient manner	NA	0	1	2	3
6.4 **Confidentiality:** Complies with federal, state, and school policies relating to confidentiality	NA	0	1	2	3
Comments:					

APPENDIX B

ANALYZED AREAS OF FACEBOOK PROFILES

Main Photo: The *main photo* is an uploaded photograph that is visible to anyone who searches the site and finds the student. Even if the student restricts who may access the profile, the main photo, name of the student, and network is displayed. A network is the group to which the person's e-mail address is registered (e.g. the university).

Information: The *information* section is an area students create by filling in pre-existing categories. The student can select which categories are used, but cannot add new categories unless they are created by the site designers. Typical categories include: name, hometown, network, sex, interested in (e.g. men, women, friendship, etc.), relationship status, religious views, political views, major, college, high school, year of graduation, home address, campus address, instant messenger address, phone number, website address (if the student has another website elsewhere), interests, favorite music, favorite books, favorite movies, favorite quotes, current courses, employer, dates of employment, and job description.

Friends: Students can become "friends" with one another by submitting a "friend request." This sends an e-mail to the profile holder, who can choose to accept, reject, or ignore the request. Once accepted, the name and main photo of the "friend" will appear in the "Friends" section on the profile. "Friends" are arranged by network, with the student's network friends displayed prominently. A link is visible to "Friends in other Networks." Each network is listed with the number of friends the student has in that network. For example, a student in the "Iowa State" network may have 257 friends in this network, another 25 friends in the "Drake University" network, 30 friends at his or her former high school, and 1 or 2 friends at other universities or corporations.

Groups: Students may create their own interest groups for others to join. Groups may be global or limited to a network. These groups may or may not meet in person, and often they are ways to post information relevant to the group, or just to meet people with similar interests. Some groups are political ("Students for Obama in 08"), where information on upcoming speeches is made available, along with links to the main campaign site. Other groups are less formal ("Mmm...Beer") and appear to have no value other than to gather members. Some seem to be designed for humor ("Are you a model? Oh, wait, you're an idiot who got dressed up to go to class.") or to make a statement ("I hate walking behind smokers to class."). The number of groups students join varies widely; some students have no groups, and other students may have 50 or more groups listed on their profile.

The Wall: The *wall* is a section on the profile where other students can post messages to the profile holder. The messages are visible to all persons who visit the site, and seem to have no expiration date. When more wall messages are posted than can fit onto the screen, a link appears that reads "View all # wall posts," indicating the number of wall posts that can be viewed, and sending the reader to the entire listing if he or she selects that link. At the time of data collection, the responses created by the profile holder to the wall messages were not visible on that page. To view how a student responded to another student, the reader simply selects the name of the person who posted the wall message (listed next to the message, along with his or her main photo) and reads their wall.

Photos: Students may upload photos for others to view and may arrange them in albums. Students may place captions on photos, name albums, and name the individuals in the pictures. Scrolling over the photo often reveals the names of people in the photo. In addition, students may link photos to another student's profile. Viewing photos on a Facebook profile occurs in one of two sections: photos uploaded by the profile holder, and photos added by others. Either photo type is equally easy to view—simply by clicking on the small view of the photo, the larger version is displayed. The name of the person who uploaded the photo is shown, along with the name of the album in his/her profile where the photo is stored. Also interesting to note is that copying photos is remarkably easy. Opening the photo, and then selecting "copy photo" will save a file of the photo on the viewer's computer.

An earlier version of this chapter appeared in Olson, J. K., Clough, M. P., & Penning, K. A. (2009). Prospective elementary teachers gone wild? An analysis of Facebook self-portrayals and expected dispositions of prospective elementary teachers. Contemporary Issues in Technology and Teacher Education, 9(4). Retrieved from http://www.citejournal.org/vol9/iss4/general/article1.cfm. Sections appearing in the earlier version are copyright by AACE and reprinted with permission of AACE (http://www.aace.org).

Joanne K. Olson
School of Education,
Iowa State University

Michael P. Clough
School of Education,
Iowa State University

Kimberly A. Penning
Minnesota State University

CHAPTER 9

MICHAEL TODD EDWARDS, SUZANNE R. HARPER, & ROBERT M. KLEIN

AND NOW THIS . . . PROBLEM

Neil Postman, Technology, and the Secondary School Mathematics Curriculum

INTRODUCTION

As mathematics educators, much of our professional time is spent preparing candidates for future careers as classroom teachers. Although most of our students use technology routinely in activities outside the classroom—for instance, text-messaging friends between classes or posting videos to *YouTube*—many are surprisingly inexperienced users of mathematics-oriented teaching technology. For this reason, we spend considerable instructional time investigating the "how-to's" of teaching technology with future classroom teachers.

Prior to reading *Amusing Ourselves to Death* (Postman, 1985), we identified ourselves as critical users of mathematics teaching technologies. But, truth be told, we really weren't. As avid users of graphing calculators and dynamic geometry software (DGS), we spent considerable time sharing the many *virtues* of technology with colleagues and students but too often overlooked potential *drawbacks* that calculators and computer software presented to those teaching mathematics—particularly for novice users and their students. For us, technology and mathematics had merged into a single area of study; instructional choices regarding the use of technology had become invisible to us—with technology's embedded ideologies and biases hidden from our view (Postman, 1993). However, as we've become more aware of the work of those critical of technology, in particular the writings of Neil Postman and Marshall McLuhan (1964), we've begun to examine our use of technology with teacher candidates more carefully, weighing instructional benefits and costs associated with graphing calculators and dynamic geometry software prior to their use in our classes.

Postman and McLuhan have encouraged us to view mathematics teaching technologies as non-neutral. They necessarily change the information we present to our students. Graphing calculators and DGS, for instance, have a social history and a set of biases that enter the classroom even if we don't explicitly acknowledge

Michael P. Clough, Joanne K. Olson and Dale S. Niederhauser (Eds.), The Nature of Technology: Implications for Learning and Teaching, 163–188.

their existence. The tools "encode" their history by influencing what content is said to "exist" in school classrooms (ontology) as well as the manner in which that content is represented and learned (epistemology). In a global economy, decisions regarding the appropriate role of various teaching tools in school classrooms are not limited to teachers; governments and businesses also influence the implementation of technology. What Postman points out in *Amusing Ourselves to Death* and *Technopoly* is that "globalization" and "politics"—although seemingly far removed from graphing calculator use in school classrooms—are issues that profoundly impact *what* our children learn (and value), *how* they learn this information, and *who* is most likely to benefit as a result. As we shall see, Huxley's *Brave New World* foreshadows a reality we may well be lulled into if we fail to critically assess the rightful role of technology in school classrooms.

MATHEMATICS EDUCATION, TECHNOLOGY, AND DISINFORMATION

In the foreword of *Amusing Ourselves to Death*, Postman (1985) compares the dystopic visions of Orwell and Huxley, noting that Orwell feared those who would deprive us of information.

> Huxley feared those who would give us so much that we would be reduced to passivity and egoism. Orwell feared that the truth would be concealed from us. Huxley feared the truth would be drowned in a sea of irrelevance (p. vi).

Later in the same text, Postman maintains that "in America, Orwell's prophecies are of small relevance, but Huxley's are well under way toward being realized" (p. 156). In Postman's view, as communication increasingly shifts away from print-based media to visual forms such as television, video, and digital imagery, we become increasingly "narcoticized by technological diversions" (p. 111). Rather than living in the *information age*, as one might expect, we find ourselves struggling to make sense of our world in an *age of disinformation*—information that is "misplaced, irrelevant, fragmented or superficial" (p. 107).

Although policy organizations such as the National Council of Teachers of Mathematics (NCTM) recommend the use of calculators and other technological tools as "vital components of a high-quality mathematics education" (2008), critics contend that today's students have been *disinformed* through haphazard application of technologies that overwhelm (or "underwhelm") learners rather than focus adequate attention on mathematics content (Askey, 1997; Koblitz, 1996). Depending on its use, technology can either be a teaching liability or an instructional asset. Clearly, teachers need to understand "how technology affords and constrains student actions and thoughts, when and how use of technology can advance learning and critical thinking, and when it can hinder the mathematical development" in order to use the tools in a competent manner (AMTE, 2007).

Any discussion of affordances and constraints depends on a concomitant examination of the goals and assumptions of education more broadly. Panels and policy boards such as the National Mathematics Advisory Panel (NMAP, 2008) often have an impact, however indirect, on the educational priorities expected of our classroom teachers. NMAP is notable especially for its emphasis on issues of national security in addition to economic competitiveness as guiding priorities. Such reports tie investment in Science, Technology, Engineering, and Mathematics (collectively "STEM") to the prosperity and global competitiveness of the nation. Inasmuch as "technology" is not a traditional academic field in its own right, it would seem to be included as part of each of Science, Engineering and Mathematics. As such, mathematics teachers must consider carefully the extent to which decisions made about technology in the classroom may be influenced by broader discourses.

In this chapter, we examine effective and ineffective uses of technology through the analysis of two classroom teaching vignettes, constructed as amalgams of actual cases from secondary mathematics methods courses. In particular, we explore fundamental ways in which technology is misapplied in the teaching of school mathematics—namely through *information underload* and *information overload.* Information underload refers to misapplications that focus undue emphasis on technology itself, at the expense of mathematics content. Its converse, information overload—more specifically, *macroglut* and *microglut*—refer to misapplications that overwhelm learners with unnecessary content-oriented information. Macroglut, microglut and information underload, the enemies of understanding in a technological world, each undermine teachers' efforts to present mathematics content in coherent, connected ways. Through analyses of common misapplications of technology in school classrooms and discussions of more appropriate instructional alternatives, we connect Postman's work to the preparation of mathematics teachers.

INFORMATION UNDERLOAD

Jake's Sketch: A Classroom Vignette

We sat in the back of the classroom and watched as Jake, a teacher candidate enrolled in our secondary mathematics methods course, taught his lesson to a group of high school students. The students sat quietly as Jake projected geometric figures from his laptop onto a large screen in front of the classroom. Using a popular dynamic geometry software (DGS), Jake asked the students a series of questions related to his sketches. A figure similar to that depicted in Figure 1 was projected in front of the classroom.

"What are some things you see in the sketch?" Jake asked. Hands shot in the air and eyes widened as Jake dragged on a vertex of a right triangle. As Jake moved his computer mouse, the right triangle grew larger, and the numerical measures of the areas of various shapes on the screen changed along with the shapes themselves. All eyes gazed upon the colorful sketch at the front of the room. While the students watched the shapes and numbers change, one girl conjectured that the sum of the areas of the two smaller squares equalled the area of the largest square.

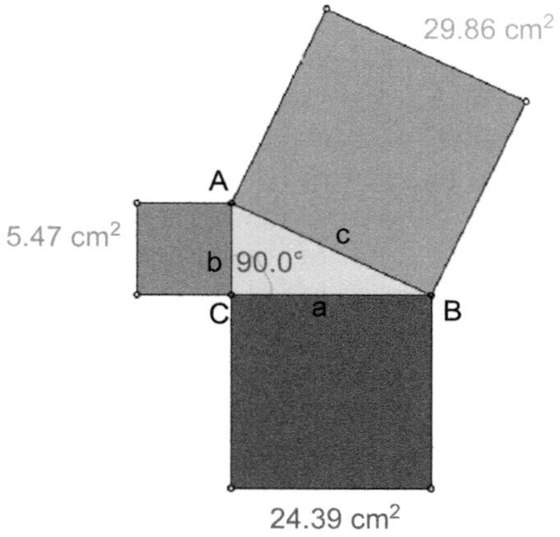

Figure 1. Jake's sketch generated with dynamic geometry software (DGS).

As Jake engaged the students in a discussion of this observation, it was apparent that Jake's cooperating teacher was impressed with his technological skills and his natural interactions with the students. We had to agree. Jake seemed quite successful at engaging the students in mathematical dialog.

"The areas of the two smaller squares are *a*-squared and *b*-squared, true?" Jake asked the class. Heads nodded, and Jake continued. "You all said that the sums of the areas of the two smaller squares equalled the area of the larger square. Can you write that as an equation?" Jake asked the class. The class grew quiet momentarily as students considered the question.

The silence was soon broken, however, by a curious onlooker. "*a*-squared plus *b*-squared equals…" the student hesitated. "*c*-squared?"

"The Pythagorean Theorem?" another student immediately questioned.

"Yes, yes. That is true," Jake agreed, encouraged by the apparent "aha" moment of the students. At this point, Jake asked if there were any questions about the sketch. When it was clear that there were not, Jake continued by writing a homework assignment on the board at the front of the classroom.

As university supervisors, we sat in the back of the classroom, whispering observations back and forth. As we talked, we became more concerned about the effectiveness of the lesson. "What do these kids actually *know* about the Pythagorean Theorem?" we asked. As we contemplated such questions, a buzzer sounded signalling the end of the class period.

"We'll continue our discussion of the Pythagorean Theorem tomorrow," Jake remarked over the din of closing books and zipping backpacks. As students departed, it was apparent that Jake was pleased with the lesson; however, nagging questions remained for us regarding Jake's use of technology. In our field notes, we wrote down two columns—"strengths" and "concerns"—to help sort out our observations. Table 1 summarizes many of our initial thoughts.

Table 1. Strengths and Concerns with Technological Aspects of Jake's Lesson

Technology Strengths	**Technology Concerns**
Technology fostered conversations about the Pythagorean Theorem with students.	Talk was teacher-directed with only a handful of students engaged in the conversation.
Technology supported student conjecturing and hypothesis-testing.	Student conjectures involved previously covered concepts.
The investigation reviewed (part of) an important result, namely that $a^2 + b^2 = c^2$.	The lesson focused exclusively on the result itself rather than its mathematical justification.

As we prepared our notes for Jake, we recalled Postman's comment that "our media-metaphors classify the world for us, sequence it, frame it, enlarge it, reduce it, color it, argue a case for what the world is like" (1985, p. 10). This certainly seemed true in the case of Jake's lesson. Whether he was aware of it or not, Jake communicated the following messages to his students as he explored the Pythagorean theorem with DGS:

1. *Numbers are more important than variable.* While variable was used in Jake's lesson to describe the general Pythagorean relationship, it wasn't a necessary component of the lesson.
2. *Mathematics is memorizing formulas.* Mathematics is primarily concerned with uncovering formulas and memorizing them rather than understanding *why* the formulas hold and *how* the formulas are revealed by and used to articulate the general relationship between quantities.
3. *Proof is inductive.* The existence of numerous cases in support of a hypothesis *proves* the hypothesis to be true (in the mathematical sense).

Although Jake did an excellent job executing many of the ideas we routinely explore in teaching methods courses—for instance, providing adequate wait time for learners and fostering classroom collaboration and discourse—it was clear that he missed an opportunity to foster students' deeper *mathematical understandings* of the topic. Jake's instruction was clearly an example of *information underload.* The lesson focused undue emphasis on technology itself—colors, motion, quickly changing numbers—at the expense of deep engagement with the Pythagorean theorem. And while the students were answering the teacher's questions, they weren't engaged in activity that required them to extend their understanding of mathematics content.

The discussion rehashed a formula that was already known by many students, but not well understood, namely "*a* squared plus *b* squared equals *c* squared." Although many students are capable of repeating this phrase (many undoubtedly in their sleep), too often they do so without a clear understanding of the derivation of the formula or its mathematical significance. As a mathematical mantra, "*a* squared plus *b* squared equals *c* squared" is an empty catch phrase and little more. As students repeat it, right triangles aren't mentioned—more importantly, neither are the *meanings* of variables *a*, *b*, and *c*.

A First Alternative to Jake's Lesson: Proof without Words

To engage students on a deeper mathematical level, either with technology or without, secondary mathematics teachers should provide students with significant opportunities to explore *unfamiliar* results in a manner that *leads to proof* (De Villiers, 1999). Jake might better appreciate such goals if we shared the following alternative—taken from the popular text *Proofs without Words* (Nelsen, 1997). In the activity, students are provided with the diagram shown in Figure 2 and asked to explain (argue, reason, deduce) how it illustrates the Pythagorean theorem.

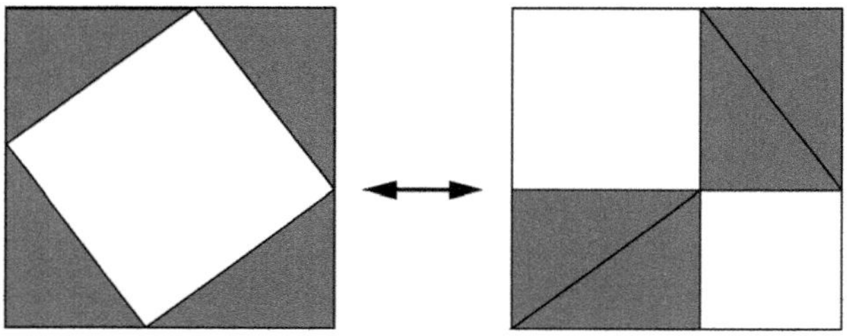

Figure 2. Alternative approach.

While Jake's software-based approach and the alternative share certain similarities (e.g. both appeal to students' visual intuition, both overtly explore the Pythagorean Theorem), careful inspection reveals significant differences between the activities.

First, unlike Jake's activity, the *Proofs without Words* task requires students to use variables meaningfully to explain the diagram. The fact that the illustration has no labels makes the act of labelling quantities with variables both intentional and meaningful. Noting that the eight shaded triangles in the diagram are congruent right triangles, students annotate the diagram as shown in Figure 3.

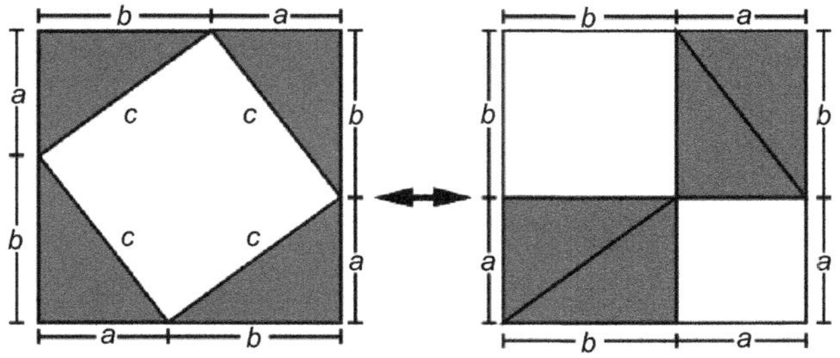

Figure 3. Diagram with side lengths of congruent right triangles annotated.

By expressing the area of each region in terms of its shaded parts, students generate the following algebraic expressions:

Area of Left Figure	Area of Left Figure
$= 4(\frac{1}{2}ab) + c^2$	$= 2(ab) + a^2 + b^2$
$= 2ab + c^2$	$= 2ab + a^2 + b^2$

Noting that the quadrilaterals on the left and right halves of the diagram are both squares with side lengths of $(a+b)$ units, students conclude that the squares are congruent. Hence, the area of the left figure equals the area of the right figure. This result may be represented algebraically as:

$$2ab + c^2 = 2ab + a^2 + b^2 \Rightarrow \quad (1)$$
$$c^2 = a^2 + b^2 \quad (2)$$

Since similar diagrams can be formed with eight copies of any right triangle, the argument holds *in general*. In this way, the activity remedies a second concern we had with Jake's lesson—namely, the alternative approach focuses on the "mathematical why" of the Pythagorean theorem (i.e. it offers students an opportunity to construct a *proof* of the Pythagorean Theorem). Admittedly, proof is just one avenue to "mathematical why" but it is a central focus of current curriculum efforts (NCTM, 2000) and an area much in need of improvement (Knuth, 2002).

Lastly, the argument is deductive in nature—that is, it is built upon statements that are known to be true. Unlike the inductive argument that Jake provided, using DGS to generate numerous true examples, the deductive argument associated with the *Proof without Words* is logically valid and establishes *mathematical truth.* Furthermore, the "truth" of the theorem isn't confirmed by the DGS numerical cases (externally) but through student reasoning in adherence to mathematico-logical principles (internal).

The *Proof without Words* avoids Milgram's "agentic shift" (quoted in Postman, 1993, p. 114) in which "humans transfer responsibility for an outcome from themselves to a more abstract agent" (Postman, 1993, p. 114). In this case, students are responsible for the reasoning and application of logic to justify the validity of the assertion as opposed to Jake's example where the DGS numerical examples "demonstrate" the theorem's validity.

Further Comparisons: Introducing the ACT Framework

A central theme in Postman's *Amusing Ourselves to Death* is that the medium through which ideas are communicated profoundly influences the content of those ideas.

> Each medium, like language itself, makes possible a unique mode of discourse by providing a new orientation for thought, for expression, for sensibility. Which, of course, is what McLuhan meant in saying the medium is the message (p. 10).

Postman reminds us that all technologies—indeed all *media*—are value laden. When considering appropriate uses of various technologies in secondary mathematics classrooms, educators must be mindful of the influence of technology on student discourse and perceptions of mathematics. With this in mind, we've developed a series of questions to help teachers compare instructional alternatives based on the needs of their students. For each instructional approach, we ask three questions (Postman, 1985) which we refer to as the "ACT framework":

Affect Concerns: What beliefs about mathematics are implicitly (or explicitly) communicated to students by the activity?

Communication: What modes of communication are encouraged (or discouraged)?

Trade-Offs: What is gained and what is lost when choosing one approach over the others?

In Table 2, we compare Jake's DGS lesson to the *Proof without Words* approach using the ACT framework.

The ACT framework provides teachers with a structured method for analyzing instructional options. In the case of Jake's lesson and the *Proof without Words* alternative, the framework favors the latter approach, although both options have particular strengths and weaknesses.

A Second Alternative: Enhancing the Task with the What-if-Not Strategy

The power of dynamic geometry software (DGS) lies not in its capacity to generate images on a screen for passive viewing; with DGS, students are active users. When

Table 2. Comparison of Jake's Lesson with Proof without Words alternative

	Jake's Lesson	Proof without Words
Affective Concerns	• The lesson reinforces common misconception that mathematics is a collection of formulas. • Students compare numerical areas rather than analyzing underlying math. • Students are encouraged to think inductively rather than deductively and that "seeing is believing."	• The lesson communicates that mathematics is concerned with underlying meaning and proof. • Students use variable to justify the Pythagorean theorem. • Students are encouraged to think deductively while considering proof as the basis of mathematical truth.
Communication	• Encourages low-level questions; visually appealing, however, variable is under-utilized in the approach (as is writing in general). • Since students already completed algebra coursework, variable should play a role in the communication. • Tasks are interactive but not mathematically engaging.	• High-level mathematical thinking is encouraged. • To be successful, students must represent geometric knowledge of area using algebraic symbols. • Furthermore, they must understand that two forms of area are equivalent and operate on these forms using algebraic manipulation. • Variable plays a central role.
Trade-Offs	• Using software, we are able to see numerous examples (e.g. animation) instantaneously. • Drawing more than one or two cases by hand might consume an entire class period. • Calculations of area are automated—this allows students to focus attention on relationships among areas rather than their calculation (which really isn't the focus of the lesson anyway).	• The activity provides students with more concrete, visual representations of algebraic structures. • The task is somewhat abstract. • Students are asked to develop algebraic representations of area on their own. • Because the figure isn't dynamic (only one figure is provided), students may fail to see that the geometric drawing suggests a general proof rather than a specific instance.

used strategically, tools such as *The Geometer's Sketchpad*, *Cabri*, and *GeoGebra* encourage students to construct mathematical proofs from hypotheses observed in sketches. Students engage in a computer-mediated "mathworld" that fosters exploration and pattern recognition that can lead to the formation of conjectures and then proof. Since the need to verify is more compelling when students explore *unfamiliar* mathematical situations, DGS is arguably most effective as students investigate *novel* tasks. As they search for new results to uncover and prove, students engage in work mirroring that of a research mathematician (de Villers, 1999).

The What-If-Not (WIN) strategy (Brown and Walter, 2004) provides mathematics teachers with a useful method for posing unfamiliar questions that can be used for exploration with DGS. Teachers identify crucial "features" or attributes of problems and ask, "WHAT IF that feature were NOT the case?" By identifying alternatives to the feature, teachers and students generate new but related problems or questions. Applying the WIN strategy to Jake's geometric sketch, we generate a related task that avoids the *information underload* of Jake's initial activity.

Recall that in Jake's sketch, squares were constructed on each side of a *right* triangle. The fact that the triangle contained a 90 degree angle was a key feature; it ensured that the sum of the areas of the two smaller squares necessarily equalled the area of the largest square. *On the other hand, what would happen if the triangle in the sketch didn't contain a right angle*? Figure 4 illustrates a revised version with the right angle feature removed.

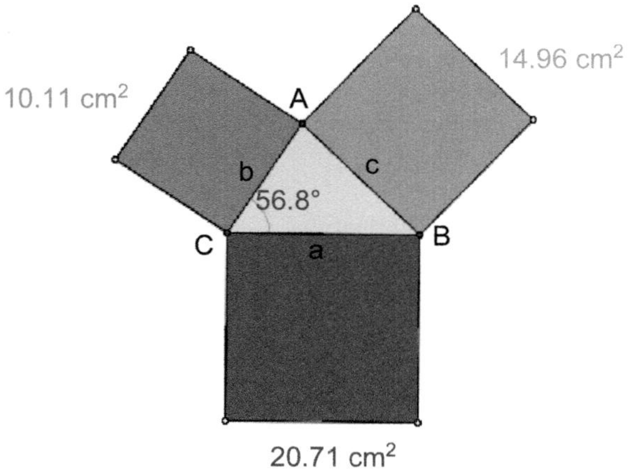

Figure 4. Revised sketch with right angle constraint removed.

In Jake's initial sketch, the measure of the marked angle, $\angle ACB$, was 90° rather than 56.8°. Using the modified sketch, students form new hypotheses. For instance, through dragging and careful observation, students uncover the following:

1. When the marked angle $\angle ACB$ is acute, the area of the square with side length c is *less than* the sum of the areas of the other squares (i.e.) $a^2 + b^2 > c^2$);
2. When the marked angle $\angle ACB$ is right, the area of the square with side length c *is equal to* the sum of the areas of the other squares (i.e. $a^2 + b^2 = c^2$);
3. When the marked angle $\angle ACB$ is obtuse, the area of the square with side length c is *greater than* the sum of the areas of the other squares (i.e. $a^2 + b^2 < c^2$).

Note that the second conjecture is, in fact, the Pythagorean relationship—a theorem already known to students. On the other hand, the first and third conjectures are *not familiar* to most school students. Below, we prove these unfamiliar conjectures with the aid of DGS. A proof is motivated by considering side AC as the radius of a circle centered at C (RUSMP, 2008). Constructing a dynamic sketch of the circle, students drag point A around the circle in a counter clockwise fashion such as illustrated in Figure 5.

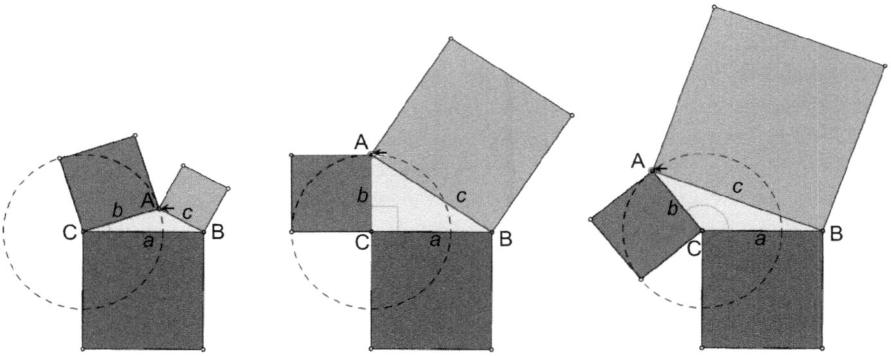

Figure 5. Dragging point A within a dynamic sketch provides insight regarding a proof of students' conjectures.

As students drag point A, they note that the lengths of two sides of the triangle remain fixed, namely CB (i.e. a) and AC (i.e. b). Only the length of AB changes (i.e. c). For convenience, we denote the length of AB when C is acute as c_1; when C is right as c_2; and when C is obtuse as c_3. This naming convention is illustrated in Figure 6.

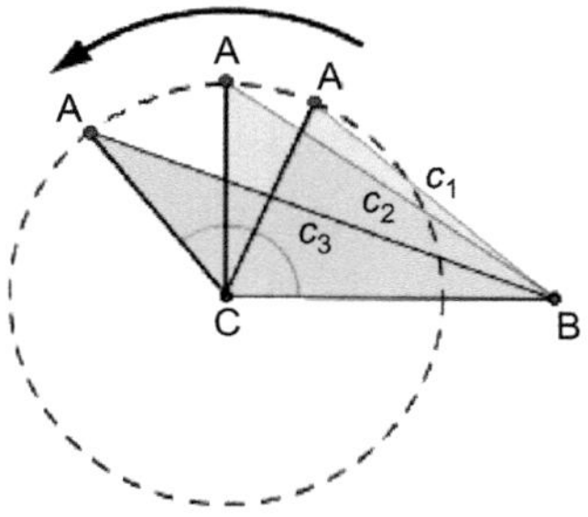

Figure 6. As A is dragged around the circle, lengths of sides BC and AC remain constant, however the length of AB changes. As angle C grows larger, segment AB becomes longer.

As A is dragged, the measures of AB and angle C increase together. Therefore $c_1 < c_2 < c_3$, hence $c_1^2 < c_2^2 < c_3^2$. Since c^2 denotes the length of AB when ΔACB is a right triangle, $c_2^2 = a^2 + b^2$ by the Pythagorean theorem. Thus we can substitute $a^2 + b^2$ for c^2 in the previous inequality. This yields $c_1^2 < a^2 + b^2 < c_3^2$, and we've proved the first and third conjectures.

Analyzing the WIN-enhanced Task with the ACT Framework

As was the case in Jake's initial lesson, the WIN-enhanced task has students use DGS software to explore conjectures related to the Pythagorean theorem. Yet Jake's activity and the WIN-enhanced task are markedly different. Therefore, it is helpful to compare the initial and revised task with the ACT framework.

As Table 3 suggests, students explore significant mathematics with DGS with the WIN-enhanced activity; however, the approach entails significantly more difficult symbolic reasoning that may be beyond the grasp of weaker students. Both the Proof without Words and WIN-enhanced alternatives are preferable to Jake's initial lesson because they require students to engage in more significant mathematics. In the latter case, DGS was used in a manner that supports (rather than undermines) deductive reasoning.

Summary of Jake's Vignette

In the above discussion, we used the ACT framework (based on questions from Postman (1985)) to uncover limitations in Jake's initial lesson. Jake's use of DGS provides an example of *information underload*, a teaching misapplication that focuses undue emphasis on technology itself, at the expense of mathematics content. While Jake's lesson generated discussion among students, his questions focused on previously known results—namely, the Pythagorean theorem. As a result, we explored two instructional alternatives constructed specifically to address concerns with Jake's lesson. The WIN strategy, in particular, provided us with a helpful

Table 3. Comparison of Jake's Lesson with WIN-enhanced alternative

	Jake's Lesson	WIN-enhanced Task
Affective Concerns	• The lesson reinforces the common misconception that mathematics is a collection of formulas. • Students compare numerical areas rather than analyzing underlying math. • Students are encouraged to think inductively rather than deductively.	• The task communicates that mathematics is concerned with underlying meaning and proof. • Student use of DGS encourages deductive thinking. • Students consider proof as the basis of mathematical truth.
Communication	• Encourages low-level questions; visually appealing; however, variable is under-utilized (as is writing). • Since students already completed algebra coursework, variable should play a role in the communication. • Tasks are interactive but not mathematically engaging.	• Students use variable, algebraic substitution, and previously known results to justify DGS conjectures. • High-level thinking is required. • Variable plays a central role, but notation is potentially difficult. • Students visualize the length of c changes as angle ACB changes—this is the foundation for our proof.
Trade-Offs	• Using software, we are able to see numerous examples instantaneously. • Drawing more than one or two cases by hand is time-consuming; Calculations of area are automated—this allows students to focus attention on relationships among areas rather than their calculation (which really isn't the focus of the lesson).	• Using the WIN approach, teachers pose a task that is clearly more difficult than Jake's original problem. • Representing the problem symbolically may present significant obstacles for struggling learners. In particular, use of subscripts may be a source of confusion.

method for modifying the lesson to generate more challenging tasks suitable for exploration with technology.

In our next vignette, we explore misapplications of technology that *overwhelm* learners with *too much* mathematics-oriented information. Whereas Jake's lesson failed to provide students with access to sufficiently engaging mathematics, the examples that follow illustrate uses of technology that provide students with a distracting overabundance of content. Postman (1992) described such a state of affairs as "information glut" (p. 72). For purposes of analysis, we subdivide glut into two fundamental types; namely, macroglut and microglut. We explore each term in more detail in the following section.

INFORMATION OVERLOAD

Macroglut and Microglut

When faced with macroglut, students and teachers find themselves struggling to make sense of vast stores of data *created by others*. Technology, for instance Internet browsing software, provides teachers and students with access to an overwhelming array of content and teaching-oriented resources. More often than not, however, these resources have conflicting instructional goals and intended audiences that include those other than young learners. The information provided by such sources is of questionable worth—providing incorrect or misleading information to students and teachers that leads to a *trivialization* of information (Postman, 1993). Macroglut involves information overload on a large scale with data from national and international sources (e.g. the Internet, national policy organizations, government agencies).

Microglut, on the other hand, involves information generated on a local level. For instance, students and teachers faced with microglut find themselves struggling to make sense of the vast amount of data of their own creation generated with the "aid" of mathematics-oriented tools. Too often, graphing calculators, spreadsheets and DGS overstimulate students with visual imagery, colors, and animation. The tools provide teachers and students with superfluous information or, worse yet, information that is misleading or incorrect.

Ironically, the remedy for both macroglut and microglut is a call for even more technology to "handle" the overabundance of information (Postman, 1993). This leads to a never-ending spiral of dependence on technology that takes students farther from understanding and meaning with each pass—an unfortunate situation exemplified by the following vignette.

Eileen's Linear Regression: A Classroom Vignette

Eileen, a graduate student enrolled in a master's-level certification program for secondary mathematics, asked us to observe the teaching of an Algebra I class at a local high school. Having earned a bachelor's degree in mathematics several years ago, she realized after graduation that she wanted to teach. Eileen's final semester in her master's program was spent student teaching in a rural school outside Cincinnati.

The lesson Eileen asked us to observe involved student exploration of linear functions, one of the cornerstones of first-year Algebra. In the activity, Eileen and her students analyzed data as they explored linear relationships in a context that connected school mathematics to the "real world."

As the bell rang, two students raced into the classroom before Eileen closed the door. "We have been studying equations of the form $y = mx + b$, where m represents what?" Eileen looked to a girl in the front row of the classroom.

"Slope," the girl responded.

"That is right, Jaya." "And Ryan, what does the b represent?" Eileen turned to face one of the boys who entered the classroom as the bell rang.

"Um… the x-intercept? No, wait. The y-intercept. It's one of the intercepts!" The boy looked around for help from his classmates.

"Yes, you are right. It is one of the intercepts—but which one?" Eileen sought clarification.

Another student spoke up, "It is the y-intercept, where it crosses the y-axis."

Eileen continued, "Great! Let's get out our pencils and calculators to take a look at today's problem." Students' reactions suggested that graphing calculators were a routine feature of their class. Eileen distributed the problem illustrated in Figure 7.

Name: ____________ Date: ____________ Bell: ________

Chirping Linear Functions

Joe noticed that crickets seemed to chirp faster in the Summer than in the Spring or Fall. His grandmother had told him that he could determine the temperature by listening to the crickets. Over the next season, he counted the chirps per minute of a cricket and recorded the temperature. His data is provided in the table below.

1. Input these values into lists of your graphing calculator. Be sure you have ten values in list L1 and ten values in L2.

Chirps per minute	67	75	83	91	99	119	134	140	149	164
Temperature (Fahrenheit)	54	55	58	58	60	67	69	70	74	77

2. Find a mathematical equation that Joe can use to estimate the temperature by listening to the crickets.

Figure 7. Eileen's handout of the Cricket Problem.

Although she considered herself a novice user of technology, Eileen was pleased to have access to graphing calculators during her student teaching. Graphing calculators were part of her experience as a high school student, and she frequently used them in master's coursework as well. She began the lesson by refocusing class attention on the task at hand. "What do I want you to do?" she queried.

"Put these numbers in our calculators," one student clarified.

"Okay, and then what?" Eileen prodded for further information.

"Find an equation," Jaya responded.

"Okay, let's put these numbers in our calculator then." Eileen projected her calculator on the screen at the front of the classroom so students could follow more easily on their calculators. "Enter in the chirps in the left-most column, and temperature into the second column, *L2*," Eileen directed. As the students tapped their calculator keys, the chirping data were recorded as two lists in the calculator's memory. The calculator screenshot in Figure 8 illustrates the end result of the data entry.

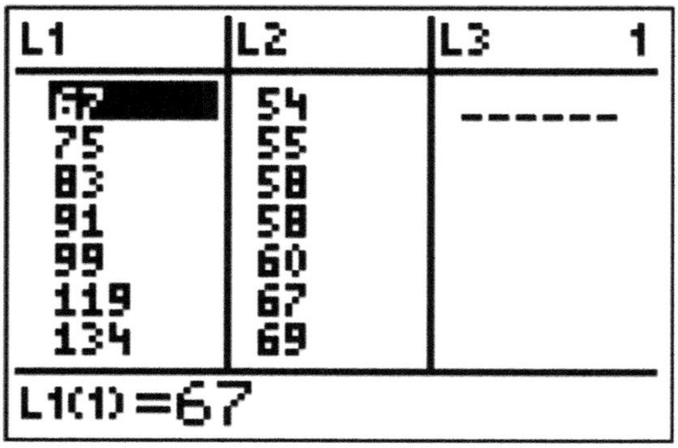

Figure 8. Eileen's calculator screen as projected in the front of the room.

Eileen continued, "So let's say we are on a camping trip, and we want to know the temperature but we don't have a thermometer with us. How can we find an estimate for the temperature outside using this data set?"

"Listen to the crickets, and count their chirps," Ryan suggested.

"How is that going to help us?" Eileen asked.

Ryan elaborated, "Well, we can find the temperature looking at the data we already have. If the crickets chirp 90 times, then we know the temperature is just below 58° since when the crickets chirped 91 times it was 58°." Ryan pointed to the list of data on the calculator screen, "The number of chirps can give us an estimate to look up in our table."

"Good. That is definitely one way we can find an estimate for the temperature. What if we find that the crickets are chirping only 45 times per minute? What temperature is it outside?" Eileen skilfully tried to elicit the need to find a linear regression line to make more accurate predictions. The class sat silently.

"Let's look at our data graphed and see if that will help us," Eileen suggested. As Eileen quickly graphed the data on the overhead calculator, the students mimicked her steps, generating a graph similar to that depicted in Figure 9.

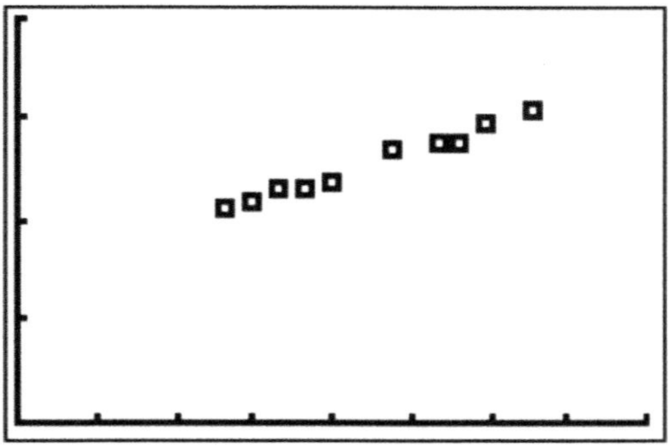

Figure 9. Eileen's calculator graph projected in the front of the room.

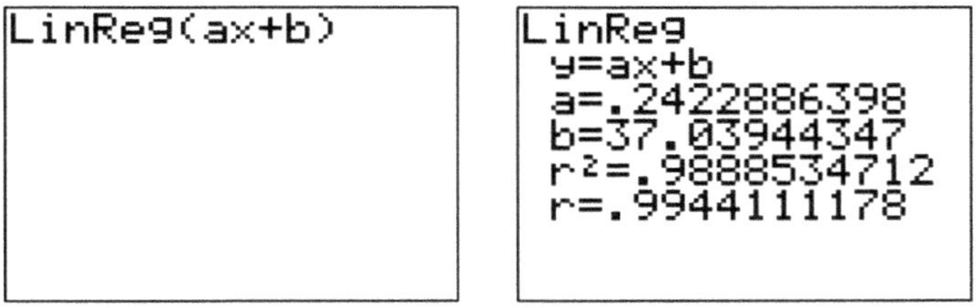

Figure 10. (Left) Calculator LinReg command; (Right) Resulting output.

In the absence of student suggestions, Eileen offered, "We can find an approximate line through the data points by asking the calculator to find the line of best fit." Eileen led her students to the statistical features of the calculator and obtained calculator screens similar to those highlighted in Figure 10.

As Figure 10 (right) suggests, the calculator-based regression generates a number of different outputs including the form of the regression equation, $y = ax + b$; the slope of the least-squares regression line, a; the corresponding y-intercept, b; the coefficient of determination, r^2 ; and the correlation coefficient, r.

Pointing to the resulting outputs, Eileen noted the following. "We can use this equation to find an estimate for the temperature when the crickets chirp 45 times per minute. How would we do that?" The room grew silent as nary a cricket (nor student) chirped in the classroom.

After an awkward silence, Jaya queried, "What are all of these numbers on the calculator? What equation are you talking about?"

Before Eileen could answer, the bell rang. Students began gathering their things together, creating the typical end-of-class rumble. Eileen tried to speak over the noise, "We will try to clear things up tomorrow—but in the meantime try to find an estimate for the temperature when the crickets chirp 45 times per minute BEFORE you come to class tomorrow."

As the students left the classroom, Eileen looked to us with a sigh of relief. "Well, I'm so glad I didn't run into any technology glitches while you were here. I was certain something was going to go wrong," she admitted. Eileen's use of technology raised a number of troublesome questions for us. Although the lesson enabled students to make mathematical connections to real-world phenomena, we were uncomfortable with the nature of student interactions with technology in the activity. For instance, in the lesson, the regression was presented as a mysterious feat of "magic" performed by the graphing calculator. The mathematics underlying the calculation of the fit line was entirely hidden from the students' view. To clarify our thoughts of the lesson, we wrote down two columns in our field notes—"strengths" and "concerns"—sharing our feedback with Eileen during a post-observation session. A sample of our remarks appears in Table 4.

Similar to Jake's lesson, Eileen enacted general pedagogical ideas well; however other options would have promoted deeper mathematical thinking and understanding among her students. Finding a line of best fit provides students with a practical application of linear equations. Moreover, the task also provides students with an excellent example of the use of mathematical models to make predictions with real-world data (Lesser, 1999). Although graphing calculators helped students generate best-fit lines quickly and accurately, the process of calculating coefficients was treated as a "black-box" process in Eileen's lesson. Buchberger (1990) suggests that when students first learn about linear regression, they should be engaged in activities that reveal, rather than hide or muddle, the underlying mathematical concepts.

> In the first stage... the mathematical theory (definitions, theorems, proofs) must be developed on which the (algorithmic) solution to the problems studied... is based. This is the stage where mathematical insight and new mathematical

Table 4. Technology strengths and concerns for Eileen's lesson

Technology Strengths	**Technology Concerns**
Students were able to analyze data and make connections to practical applications of linear models.	Talk was teacher-directed with only a few students engaged in mathematical conversation.
Students quickly calculate the parameters of a best-fit line using built-in statistical regression capabilities of the graphing calculator.	The calculator use was as a "black-box" providing students with little or no idea how the values for the slope, intercept and correlation coefficient were actually calculated.
Technology provided a visual representation of the data set.	The use of technology did not provide students with any further understanding of mathematics regarding the construction of linear models.

> techniques are acquired. It would be disastrous for the future of mathematics if the insights and techniques that can be taught and learned in this stage would be ignored because the area is "trivialized" (Buchberger, 1990, p.13).

Eileen's instruction provides a textbook example of *information overload* and *microglut*. Specifically, Eileen's students used calculators to generate more information than was necessary or appropriate given their current level of mathematical understanding. As a result, they struggled to make sense of a confusing cocktail of information displayed in an unexpected form (a list of parameters instead of an equation) with an unfamiliar set of data labels (for instance, school texts typically refer to the slope of a line as *m* rather than *a*).

In such a case, the microglut ensures that no pieces of information are elevated to greater importance than others, rendering all information as trivia. Moreover, the inattention to how the parameters came about (conceived of as either the algorithm used to calculate the parameters or the overall strategy underlying the least-squares method) promotes a sense of the calculator's workings as magical. This sense of magic, Postman warns, "directs our attention to the wrong place. And by doing so, evokes in us a sense of wonder rather than understanding" (1993, p. 94).

An Alternative to Eileen's Lesson

Most calculators hide from view the intermediate calculations of the least-squares parameters. "Rather, a calculator often acts as a 'black box' and merely provides students with the results of calculations" (Edwards, 2005, p. 415). One approach that casts light on the black box processes involves helping students to approximate lines of "best fit" by creating lines and adjusting parameters until visual inspection convinces them that they have found the "best fit."

Once Eileen's class performed a linear regression with the cricket data, their calculator screens were cluttered not only with the slope and intercept of the best-fit line but also with the correlation coefficient, r, and coefficient of determination, r^2, associated with the regression (see Figure 10). Students were overwhelmed and could not decipher the results of the statistical calculations meaningfully from the calculator display. Instead of using the linear regression feature of the calculator, we suggest using a different technology-oriented method for obtain a line of "best-fit" with novice algebra students.

While the best-known methods for computing the least square regression line by hand require calculus, alternative methods accessible to secondary level students exist. Unfortunately, these methods are arguably too advanced for typical first-year algebra students, requiring knowledge of residual mean squared deviations and scatter (Ehrenberg, 1983), matrix operations (Darlington, 1969), or extensive knowledge of quadratics and parabolic functions (Gordon and Gordon, 2004). For this reason, we suggest that students manually fit a line through the data points on their calculator screen. The steps involved in this process are illustrated in Figures 11 and 12.

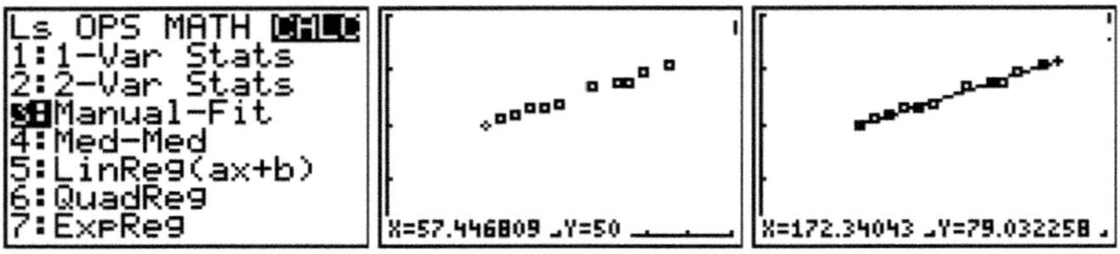

Figure 11. (Left) Students begin by selecting the Manual-Fit feature; (Middle) The user manually selects an endpoint of a line segment that fits the desired data closely by moving the calculator cursor to a desired location on the calculator screen; (Right) The user manually selects a second endpoint.

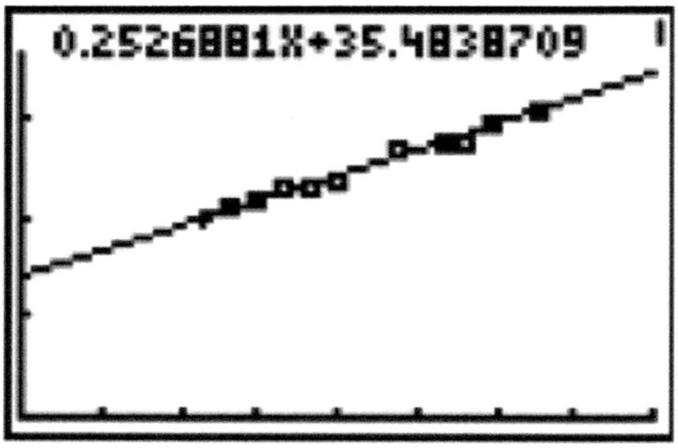

Figure 12. Once the user selects two endpoints of a "best fit" line segment, an algebraic expression of the line incident to the segment is displayed (in this case, 0.25x + 35.48).

The manual fit method allows the students to actively participate in the creation of a "fit" line, rather than merely observe the results of "black box" calculator computations. Having students estimate the line of best fit allows for further discussion in the classroom about appropriate values for slope and intercept, the meaningfulness of these parameters, and notions of "best" fit. The focus of the lesson then turns to comparing the students' understanding with the graphical and numerical representations and mathematically describing a possible relationship between the variables. The question of prediction encourages the students to use multiple representations of the data to make an estimate for the temperature given the number of cricket chirps. This investigation of the relationship between chirping crickets and temperature allows for rich mathematical discussion about linear relationships and promotes conceptual understanding, rather than meaningless "button-pushing" calculating the line of "best" fit. While least squares regression is certainly important, a more tangible first experience with fitting data to lines may form the basis for a better appreciation of the Least Squares algorithm later - *after* the student has an appreciation for the general goal.

Comparison of Approaches Using the ACT Framework

Mathematics activities using technology should address worthwhile mathematics concepts, procedures, and strategies, and should reflect the nature and spirit of mathematics (Garofalo, Drier, Harper, Timmerman & Shockey, 2000). While Eileen's lesson and the manual fit alternative both invite students to explore mathematical models using technology, use of least squares regression capabilities of the calculator are ill-advised for students who lack the background to fully understand the underlying mathematics. Because Eileen's students lacked such knowledge, the activity inadvertently depicted mathematics as "magic", forcing them to assume the role of "naïve empiricists" (Shoenfeld, 1985). In the case of linear regression, the graphing calculator performs computations that are inaccessible to first-year algebra students through any other means, thus encouraging a needless dependence on technology.

On the other hand, the manual fit approach automates procedures that students can complete with pencil and paper. By selecting two points and calculating the equation of the "fit" line passing through those points, students can replicate calculations that underlie manual fit. Unlike calculator-based linear regression, there is little mystery with such an approach. As such, the manual fit alternative promotes a view of mathematics as comprehensible, and reasonable. In Table 5, advantages of the alternative approach are highlighted within the ACT framework.

Summary of Eileen's Vignette

In the preceding discussion, we used the ACT framework (based on questions from Postman (1985) to uncover limitations in Eileen's initial lesson. Eileen's use of the graphing calculator provides an example of information overload. While Eileen's generated discussion among students, her questions focused on button pushing and using the calculator as a black box to obtain results - namely, the line of best fit. As a result, we explored an instructional alternative constructed specifically to address concerns with Eileen's lesson. In particular, we modified the lesson to generate more mathematical understanding of the line of best fit suitable for exploration with technology.

Table 5. Comparison of Eileen's Lesson with Manual-fit Lesson

	Eileen's Lesson	Manual Fit Lesson
Affective Concerns	• Reinforces the notion technology "does the mathematics" and provides the numerical answer to the problem. • Students compare numerical values in the table rather than analyzing underlying mathematics.	• The task communicates that mathematics is concerned with underlying meaning of "best fit." • Students use technology as a tool to develop deeper understanding of mathematics.
Communication	• Tasks are interactive but not mathematically engaging. • Technology provides feedback at a level that is inappropriate for novice algebra students. • Calculator actions are not transparent; • Calculator output marked by "microglut;" students don't know how to use information the calculator provides.	• Mathematical connections among a problem situation, its model as a function in symbolic form, and the graph of that function are explicitly made. • The manual fit line provides opportunities for students to engage in discussions involving the term "best-fit."
Trade-Offs	• The linear regression line provides students with a means to generate more accurate predictions. • Calculation of linear regression equation is inaccessible without technology; Technology renders inaccessible calculations accessible. • Calculations require little student thought beyond low-level data entry.	• Manual fit doesn't provide a true "fit" line in a strict statistical sense; the line is based on subjective input provided by students. • Mathematical connections among a problem situation, its model as a function in symbolic form, and the graph of that function are explicitly made using the manual fit approach.

IMPLICATIONS AND SUMMARY

The ACT framework is a semi-structured approach to making decisions about technology and its use in the classroom. Inasmuch as all decisions are political, involving negotiation of power structures, the ACT framework must be viewed as they are situated within the broader context of a globalized economy. Educational rhetoric, especially from corporate and political sources, frames the latest "crises" as resulting from failures of the educational system to adequately prepare students for the workforce. High-profile national reports (such as NMAP, 2008; NCEE, 1983) and international comparisons (such as TIMMS (IEA, 1995–2007); and PISA (OECD, 2003)) not to mention references in standards documents (NCTM, 2000; ISTE, 2007, 2008) similarly suggest that a central focus of educational systems should be to produced skilled laborers that can make the nation competitive against other nations in the global markets.

Mathematics and technology are half of the targeted STEM areas (measured alphabetically), and those school subjects are frequently the foci for responsiveness to global competitiveness and workforce readiness. If Technopoly is, as Postman (1993) describes it, "a state of culture... [in which the] culture seeks its authorization in technology, finds its satisfactions in technology, and takes its orders from technology" (p. 71), then Technopoly is well supported in American schools; mathematics and technology classes, especially. It is "what happens to society when the defenses against information glut have broken down"(1993, p. 72), as in the case of Eileen's class. Mathematics, science and technology, with their vast "formatting powers" (Skovsmose, 1994), are seen as areas key to competitiveness in the global economy. In the face of such pressures on discourse, it is not difficult to understand how practicing teachers may not have the resources needed to make effective decisions regarding technology use in the classroom. Some optimism is called for in that teacher preparation programs may hold the key to addressing the issue. Teacher candidates can, with guidance and tools like the ACT framework, take advantage of their position relative to the classroom and its discourses (close, but not too close) to engage in the "loving resistance" called for by Postman (1993).

The narratives of Jake and Eileen are instructive examples in that they demonstrate the degree to which mathematics teachers (pre-service teachers, and even teacher educators) may, in the absence of a practiced structure for critical reflection, embrace technology as an unquestionable good. Our experience suggests that they may willingly believe that they educate best by using technology whenever possible. We offer the alternative of actively knowing that the learning of mathematics content can be served best when decisions about technology use are coupled with a critical reflection such as that structured by the ACT framework highlighted above.

As in the story of Faust, there is no undoing of this bargain just as there is no undoing of globalization. Rethinking technology use in the mathematics classroom is an ongoing regimen, not an immunization. How then can this regimen of reflection be achieved given the constraints of time and the constant pressures of the "ed-tech" industry (directly or through reports, standards, and sycophantic research)? Some direction is suggested by Postman's description of the "Loving Resistance Fighter" who "maintains an epistemological and psychic distance from any technology, so that it always appears somewhat strange, never inevitable, never natural"(1993, p. 185). If educators aspire to a more cautious approach to technology, it must begin with our recognition as educators that at the table of educational goals, we have given the pre-eminent seat to efficiency. Nonetheless, education offers a way out, for, as Postman (1993) points out, in the United States "whenever we need a revolution, we get a new curriculum" (p. 185). The ACT and WIN strategies articulated here may serve as a starting point.

The lessons learned in the Jake and Eileen stories should not be taken as "anti-technology." In fact, they demonstrate a kind of discourse and strategy for re-imagining approaches to technology in mathematics instruction. In both cases, the suggested alternatives involve significant uses of technology, though structured with a keener understanding of the potential pitfalls involved in their use. They chronicle an awakening to the "Faustian bargain" of technology use. Superficially, the tale of Faust is the tale of a wager made with the fallen angel, Mephistopheles. A wager is made concerning Mephistopheles' ability to provide Faust with fulfilment on a level that Faust might never want to abandon. Despite numerous attempts by Mephistopheles, it is only when Faust is granted god-like powers to part the waters of the sea to create an Eden, that Mephistopheles produces in Faust a sense of pride in creation (nearly) sufficient to win the wager. Hence, the "Faustian bargain" involves more than a recognition of the simple tradeoffs of some good for some bad implicit in social appropriations of technology; it involves an underlying lust for the power that technology can bring and a power-induced blindness toward the negative aspects. As Hughes (2004) points out, "Goethe's *Faust* allows us to see the egotistical and controlling nether side of creativity" (p. 18). For Hughes, technology is a creativity embodied. For us, a set of teacher educators, Jake's and Eileen's use of technology resulted from their instruction in our methods courses, from our act of creation. Humility, it seems is called for at all levels of this narrative.

REFERENCES

Askey, R. (1997). What do we do about calculus? First, do no harm. *American Mathematical Monthly 104*(8). Washington, D.C.: Mathematical Association of America, 738–743.

Association of Mathematics Teacher Educators (AMTE) (2007). *AMTE Technology Position Statement.* Available:http://www.amte.net/sites/all/themes/amte/resources/AMTETechnologyPositionStatement.pdf

Brown, S. I., & Walter, M. I. (2004). *The art of problem posing* (3rd Ed.). Hillsdale, NJ: L. Erlbaum.

Buchberger, B. (1990). Should students learn integration rules? *Sigsam Bulletin, 24*, 10–17.

Darlington, R. (1969). Deriving least-squares weights without calculus. *The American Statistician 23*(5). American Statistical Association, 41–42.

De Villiers, M. (1999). *Rethinking proof with The Geometer's Sketchpad.* Emeryville, CA: Key Curriculum Press.

Edwards, M.T. (2005). Promoting understanding of linear equations with the median-slope algorithm. *Mathematics Teacher, 98*(6), 414–425.

Ehrenberg, A. (1983). Deriving the least squares regression equation. *The American Statistician 37*(3). American Statistical Association, 232–233.

Gordon, S. & Gordon, F. (2004). Deriving the regression equations without using calculus. *Mathematics and Computer Education 38*(1), 64–68.

Garofalo, J., Drier, H., Harper, S., Timmerman, M.A., & Shockey, T. (2000). Promoting appropriate uses of technology in mathematics teacher preparation. *Contemporary Issues in Technology and Teacher Education* [Online serial], 1(1). Available:www.citejournal.org/vol1/iss1/currentissues/mathematics/article1.htm

Hughes, T. (2004). *Human-built world: How to think about technology and culture.* Chicago, Ill: University of Chicago Press.

International Association for the Evaluation of Educational Achievement (IEA). (1995–2007). *Trends in International Mathematics and Science Study* (TIMMS). Available from the National Center for Educational Statistics (NCES).

International Society for Technology in Education (ISTE). (2007). *National Educational Technology Standards—Students* (NETS-S). Author.
International Society for Technology in Education (ISTE). (2008). *National Educational Technology Standards—Teachers* (NETS-T). Author.
Knuth, E. (2002). Secondary mathematics teachers' conceptions of proof. *Journal for Research in Mathematics Education 33*(5), 379–405.
Koblitz, N. (1996). The case against computers in K-13 math education (kindergarten through calculus). *Mathematical Intelligencer 8*(1), 9–17.
Lesser, L. (1999). Making the black box transparent. *Mathematics Teacher 92*(9), 780–784.
McLuhan, M. (1964). *Understanding Media: The Extensions of Man*, McGraw-Hill, New York, NY.
National Center for Fair and Open Testing (NCFOT) (April, 1998). *Science leader criticizes tests.* Available:www.fairtest.org/science-leader-criticizes-tests.
National Committee on Excellence in Education (NCEE). (1983). *A nation at risk: The imperative for education reform.* U.S. Department of Education.
National Council of Teachers of Mathematics. (2000). *Principles and standards for school mathematics.* Reston, VA: Author.
National Council of Teachers of Mathematics. (2008). *The role of technology in the teaching and learning of mathematics.* Available: www.nctm.org/uploadedFiles/About_NCTM/Position_Statements/Technology%20final.pdf
National Mathematics Advisory Panel (NMAP). (2008). *Foundations for success: The final report of the National Mathematics Advisory Panel.* U.S. Department of Education.
Nelsen, R. (1997). *Proofs without words: Exercises in visual thinking.* Washington, D.C.: MAA.
Organisation for Economic Co-operation and Development (2003). *Programme for International Student Assessment* (PISA). Author.
Postman, N. (1985). *Amusing ourselves to death: Public discourse in the age of show business.* New York: Penguin Press.
Postman, N. (1993). *Technopoly: The Surrender of Culture to Technology.* New York: Vintage Books.
Rice University School Mathematics Project (RUSMP) (2008). *Geometry Module.* Unit 6: Pythagoras. Houston, TX: Author. Available: rusmp.rice.edu/geometrymodule/PDFdocuments/Pythagoras.pdf
Skovsmose, O. (1994). *Towards a philosophy of critical mathematics education.* Dordrecht: Kluwer.

An earlier version of this chapter appeared in Harper, S.R., Klein, R. & Edwards, M.T. (2010). On using technology to confront information overload in a mathematics classroom. Journal of the Research Center for Educational Technology, 6(2), pp. 48–62. Sections appearing in the earlier version are copyright by the Journal of the Research Center for Educational Technology and reprinted with permission.

Michael Todd Edwards
Department of Teacher Education,
Miami University

Suzanne R. Harper
Department of Mathematics,
Miami University

Robert M. Klein
Department of Mathematics,
Ohio University

CHAPTER 10

JOANNE K. OLSON & MICHAEL P. CLOUGH

A CAUTIONARY NOTE

Technology's Tendency to Undermine Serious Study and Teaching

> One may assume . . . that what is called "computer literacy" does not involve raising questions about the cognitive biases and social effects of the computer, which I would venture are the most important questions to address about new technologies.
>
> Neil Postman

INTRODUCTION

In our increasingly polarized culture, taking a middle position on important issues is increasingly difficult. Few seem to appreciate individuals positioning themselves on a fence dividing two opposing camps and, interestingly, those on both sides of the fence often demand an allegiance to one or the other positions. "You can't sit on the fence," such moderates are told—despite their often being able to see further from that vantage point than those standing on the ground unable to see over the fence. So with some concern of being labeled "Luddites" by those who champion technology in the classroom and "Technophiles" by those opposing its intrusion, we take the position that while technology *could* assist teachers and students in effective schooling, in many ways it exacerbates the current problems.

DEEP AND ROBUST LEARNING REQUIRES SERIOUS STUDY

Developing deep, robust, and long-term understanding of science concepts is but one aim of the *National Science Education Standards (NSES)*. The desired state also includes an understanding of the nature of science and the attributes and skills that make for effective science inquiry (NRC, 1996). The task is daunting and reaching the desired state will not occur without a deep understanding of what learning means and the teacher's essential role in that process. Understanding how students learn—and why they sometimes don't—is the foundation of informed teaching.

Michael P. Clough, Joanne K. Olson and Dale S. Niederhauser (Eds.), The Nature of Technology: Implications for Learning and Teaching, 189–200.

Constructivist learning theory emphasizes that students use their pre-existing knowledge to grapple with and make sense of experience, and thus teaching should center on what the learner knows and their meaning-making. In the past decade, constructivism has received almost exclusive attention in education literature. During this time, a number of radical constructivist views have emerged that are now, for good reasons, being attacked (Matthews, 1997; Nola, 1997; Ogborn, 1997; Osborne, 1996; Phillips, 1995; and Phillips, 1997). Constructivism has come to mean so many different things to different people that it no longer conveys a specific meaning. But the foundations of constructivist learning theory are well supported, and if used wisely, help make sense out of the complexities associated with learning and teaching. First, learning is an *active* process and for learning to occur students must be mentally active—selectively taking in and attending to information, and connecting and comparing it to prior knowledge and additional incoming information in an attempt to make sense of what is being received. Second, because the incoming sensory input is primarily organized by the individual receiving the stimuli, the meaning intended by the incoming source is often not communicated in an intact manner. Third, knowledge that a student brings to current instruction may help or hinder the creation of meaning similar to that intended by that external source. Fourth, students' prior knowledge that is at odds with the intended learning is, at times, incredibly resistant to change. That is, in attempting to make sense of instruction, students interpret and sometimes modify incoming stimuli so that it fits (i.e., connects) to what they already believe. Fifth, as the number of links made to new learning increases, the likelihood of long-term and meaningful learning increases. These postulates help explain how learning occurs and they also raise concerns regarding whether particular technologies will promote or inhibit intended learning.

Clearly, effective science teaching is a highly interactive activity. Such teaching makes explicit students' relevant prior knowledge, engenders active mental struggling with that prior knowledge and new experiences, and encourages metacognition. Without this, students rarely create meaning similar to that of the scientific community. And this is also why most textbooks, audiovisual materials, multimedia, computers and other forms of technology do not promote and often hinder deep conceptual understanding—they do an extremely poor job of making apparent and playing off students' prior ideas, engendering deep reflection and promoting understanding of complex content. Worse yet, technology often undermines such serious study.

TECHNOLOGY'S TENDENCY TO UNDERMINE SERIOUS STUDY

Neil Postman (1995) wrote in *The End of Education* that all technological change is a Faustian bargain—that every advantage is tied to a corresponding disadvantage. To illustrate this point, we have chosen two unintended and rarely considered consequences of technology that appear to us to undermine the serious thinking and metacognition necessary for deep robust learning.

Education and Entertainment

All sorts of technologies fascinate students and have the potential to grab and maintain their attention in ways that interacting with a teacher, reading a book, seriously discussing ideas with other students, and thinking about their own thinking cannot. One only has to look at the hours children are willing to spend spellbound in front of a television or computer surfing the web (which by no coincidence is looking more and more like television) to see how rapidly changing sensory information plays to our biological bias attending us to changes in our immediate environment. Technology not only entertains, it also speeds up life by reducing the time we spend on dull time-consuming tasks. Calculators, graphing software, spelling and grammar-checkers, educational software, and hosts of other devices are invaluable tools to save us from mundane tasks. Not surprisingly, students, parents and even educators have enthusiastically and often uncritically endorsed technology in education. However, the wholly unforeseen Faustian bargain is the deep and pervasive attitude that now permeates attitudes toward learning—that it should largely be fun and entertaining, or at least not a struggle. Years ago, Neil Postman (1985) warned that whether particular technologies teach students their ABCs and how to count is of minor importance compared to what they teach students about learning and schooling. Using "Sesame Street" as merely one example, he writes:

> We now know that "Sesame Street" encourages children to love school only if school is like "Sesame Street." Which is to say, we now know that "Sesame Street" undermines what the traditional idea of schooling represents. Whereas a classroom is a place of social interaction, the space in front of a television is a private preserve. Whereas in a classroom, one may ask a teacher questions, one can ask nothing of a television screen. Whereas school is centered on the development of language, television demands attention to images. Whereas attending school is a legal requirement, watching television is an act of choice. Whereas in school, one fails to attend to the teacher at the risk of punishment, no penalties exist for failing to attend to the television screen. Whereas to behave oneself in school means to observe rules of public decorum, television watching requires no such observances, has no concept of public decorum. Whereas in a classroom, fun is never more than a means to an end, on television it is the end in itself. (p. 143)

The underlying attitude that education should be enjoyable and entertaining, or at the very least, not deliberate and measured, makes much of what we know about effective teaching appear stale and old-fashioned. Yet, engaging students' prior knowledge, grappling with new experiences, struggling to make sense of those new experiences, thinking about thinking, making new connections, and finding that prior connections no longer make sense are serious and difficult struggles requiring much effort, diligence and perseverance on students' part. These activities are precisely what television, radio, computers, calculators, graphing software and many other forms of technology often circumvent. In doing so, the most pervasive outcome is a change in students' attitudes—that learning should not be a struggle and that good teaching will make learning enjoyable and easy.

The practical outcome of this is that teachers at all levels are increasingly incorporating technologies into their classrooms to catch students' attention, but are doing so at the expense of serious study. Postman (1985), addressing the efforts to make classrooms more entertaining, writes:

> And in the end, what will the students have learned? They will, to be sure, have learned something about [the content in question], most of which they could have learned just as well by other means. Mainly, they will have learned that learning is a form of entertainment or, more precisely, that anything worth learning can take the form of an entertainment, and ought to. (p. 154)

Of course, effective teaching is mentally stimulating and often enjoyable. However, meaningful learning is just as often a discomforting struggle that is rewarding only after much cognitive and emotional effort. The point is that one unintended cognitive bias and social effect of technology has been its battering of habits congruent with the nature of serious learning.

Another Black Box

Actual Question to High School Students: Assume you drive your vehicle 10,000 miles/yr. Gasoline averages $1.29/gallon and your average fuel economy is 23 mpg. If you owned a vehicle with an average fuel economy of 40 mpg, how much money would you save each year in fuel costs?

> Answer from typically a quarter of students: $219,300
>
> Teacher: How much sense does that make?
>
> Student: That's the number my calculator gave me.
>
> Teacher: Do you believe everything your calculator tells you?
>
> Student: (Look of bewilderment)
>
> Teacher: If I could save $219,300 each year by buying a more fuel-efficient vehicle, I'd use public transportation and retire a millionaire in five years.

Paulos (1988) laments that the way mathematics is taught in schools contributes to what he terms "innumeracy"—the lack of a basic sense of numbers and what they mean. Technology also often impedes scientific literacy in the same way it can contribute to innumeracy. Despite recommendations to make technology "transparent," technology is often a "black box" that either misleads students into thinking they need not understand conceptually what the technology is doing for them, or worse, it promotes serious misconceptions about the concept under investigation.

For instance, researchers (Annenberg/CPB, 1997) found that even the brightest students in a high school physics classroom did not understand the basic concept of an electrical circuit despite two months of instruction on electricity. When asked how to make the bulb light, one student thought a bulb holder was a necessary part of a circuit. When trying to light the bulb, the student asks the interviewer, "Can I use the little piece we used in class?" When asked why she needed the bulb holder, she states, "It carries the charge or something.... I don't think it will light without it." The presence of this rudimentary piece of technology and the "black box" nature of it not only clouded the purpose of the bulb-holder, it created a misconception regarding the basic concept of a circuit, upon which many other concepts are built.

Equipment is often used before students have seriously grappled with the concepts under study. As a result, they can perceive the technology to be a necessary part of the concept, or worse, have little understanding about what they are doing. The first author recently interacted with a group of middle school students working on a water quality project. Equipped with expensive calculator-based laboratory (CBL) equipment, the students were excitedly putting probes into the water and recording results. The students were motivated to do the activity and the teacher was receiving district acclaim for being at the "cutting edge" of science education. When asked what they were doing, a group of students answered, "We're putting this probe in the water, reading the number, and writing it down in the box." "What does the number tell you?" The student looked at her paper and read her answer. "Dissolved oxygen concentration." "What's that?" "I don't know," the student responded, "I just write down the number." The second author observed a tenth grade class performing gel electrophoresis and had the following dialogue with a group of students:

Observer: "So what are you doing in this lab activity?"

Student: "Gel electrophoresis."

Observer: "What's that?"

Student: (pointing at apparatus) "Watching the blue dot move through the gel."

Observer: "What's the blue dot?"

Student: "I don't know."

Observer: "Why is it moving?"

Student: "Electricity. Current."

Observer: "How does the current cause the blue dot to move?"

Student: "We don't know." (nervous laughter)

Almy (1966) warned teachers that having students engage in manipulative or verbal operations which they cannot engage in mentally tends to erect knowledge super-structures which crumble under the slightest cognitive stress. Incorporating technology before students mentally grasp the underlying fundamental concepts does just that. Of course, technology can at times motivate students to learn the underlying concepts, but in the examples provided above, the technology was so far beyond students' conceptual understanding that it could not serve this motivating role.

Technology's inherent labor-saving bias encourages students and teachers to skip conceptual understanding. Many champions of technology in the classroom speak glowingly of projects that involve students in using multiple media to complete tasks formerly done by hand. Pearlman (1989) enthusiastically describes such a project:

> Last year kids at 200 schools throughout the country, working in teams, took water samples from local rivers, lakes, ponds, open fields, and water taps. Back in their classrooms the teams measured the pH levels of the water, recorded the results, and took averages for the samples. Then each team entered their results into a specially designed software program that allowed the class to average their results and then telecommunicate them, via modem, to a national computer.
>
> The next day the results of all sites were available for download from the national computer to the classroom computer, where they could be printed out and where special mapping software could generate color-coded maps of acid rain levels. Students then discussed the findings and communicated their analysis, again via modem to an expert at the National Oceanic and Atmospheric Administration, who wrote back and compared their findings to current scientific analyses. (pp. 13–14)

What is tragically missing from this account is any description of the teacher's role, how students were engaged in making meaning from this experience, whether they understood what they were reporting, or why they were even doing the activity. In our excitement for using technology, we have to remind ourselves that students' use of technology often hides their misunderstanding and interferes with learning and teaching.

TEACHERS ARE THE KEY!

These examples underscore the importance of the teacher. In the presence of technology, however, the teacher's role becomes more difficult as the "black box" nature of the technology works to help conceal students' thinking about the fundamental concepts. Overemphasizing technology, like overemphasizing curricula, neglects the teacher in favor of changing activities that students do. Despite the pervasive and critical role of curricula, the evidence is clear and overwhelming that teachers are the most influential factor in educational change (Duffee and Aikenhead, 1992; Fullan, 1991; Good and Brophy, 1994; Shymansky and Penick, 1981). The bottom line is that teachers, not technology, make exemplary science programs (Berliner, 1989; Penick, Yager, and Bonnstetter, 1986). Langer and Applebee (1987), after observing how teachers assimilated new writing activities into their old ways of thinking, wrote:

> For those who wish to reform education through the introduction of new curricula, the results suggest a different message. We are unlikely to make fundamental changes in instruction simply by changing curricula and activities without attention to the purposes the activities serve for the teacher as well as for the student. It may be much more important to give teachers new frameworks for understanding what to count as learning than it is to give them new activities or curricula. ...(T)o summarize bluntly, given traditional notions of instruction, it may be impossible to implement successfully the approaches we have championed. (pp. 87 & 139)

Teachers translate curriculum into a form ready for classroom application and decide what, how and why to learn. As Eisner (1985) writes, "In the final analysis, what teachers do in the classroom and what students experience define the educational process" (p. 59).

The principles of effective teaching are not changed by the presence or absence of technology. Understanding how individuals learn means teaching for conceptual understanding is an extremely interactive undertaking, requiring teachers to know their students' ideas and engaging students in meaning-making based on those ideas. This requires effective questioning, wait time, supportive non-verbal behaviors, active listening, responding to students in ways that further their thinking, and structuring activities to keep students mentally engaged.

ISSUES WHEN CONSIDERING TECHNOLOGY IN THE CLASSROOM

Be Wary of Introducing Technology Too Far Removed From Students' Conceptual Understanding

Teachers should carefully consider whether students' prior knowledge is sufficient to understand what the technology is doing for them and whether the novel concepts introduced by the technology can be linked to their prior understanding. If the distance between students' prior knowledge and the technology is too great for mental engagement to occur, then the technology becomes a "black box." Moreover, students may be perplexed regarding why they need to understand the concepts, since the technology does the work for them. In this case, the technology actually reduces motivation to learn complex content.

Technology Should Not Determine the Content or Activity

The availability of technology does not justify its inclusion in the classroom. Unfortunately, as technology becomes more widely available at lower cost, teachers are increasingly including activities in their curriculum that are neither developmentally appropriate nor coherent with other topics. Introductory high school biology and even middle school life science students are now performing gel electrophoresis and polymerase chain reactions (PCR) simply because the technology has recently become available at reasonable expense. Never mind that most students have little or no prior chemistry experiences and have significant difficulties conceptually understanding the molecular structure and function of DNA and proteins. Requiring students to utilize such technology may seem interesting and "cutting edge," but how well students can really understand what they are doing is questionable given their difficulties with more fundamental concepts that such technology is based upon. Simply because technology permits particular activities to be done doesn't mean they should be done.

Consider How the Technology Will Promote Desired Student Goals

Although comprehending the fundamental ideas in science is important, other student goals are just as critical to teachers (Goodlad, 1983) and their students' understanding of science (NRC, 1996). All goals need not be facilitated in every classroom activity, but teachers do need to consider how the introduction of a particular technology will help or hinder each goal. As we have argued, technologies may promote student interest, but retard conceptual understanding. Some technologies promote particular ways of communication while trivializing others. All technologies have biases, and these biases must be discerned by the teacher when deciding what to incorporate and how and when it should be used.

Consider Your Rationale for Using Technology. If Entertainment is the Primary Purpose, Consider Other Options

If the primary advantage of using the technology is that it will be fun for students or more "motivating," seriously consider why this is so. What we think you will find is technology often diminishes the need to seriously attend to prior knowledge, utilize metacognitive strategies, question prior ideas, generate examples, compare alternative solutions, grapple with new experiences, make sense of those new experiences, make new connections, and analyze whether prior connections continue to make sense. If the primary advantage of the technology is student interest, what are students being motivated to deliberately study and how effectively does the technology motivate them to do this? However, if the technology can do most all these things while being enjoyable, then incorporating it is appropriate.

Consider What is Gained and What is Lost By Using the Technology

Because all technology is a Faustian bargain, teachers are wise to consider both the gains and the losses when introducing technology in the classroom. Students who engage in e-mail activities with a distant class may gain in keyboarding skills and print-based communication skills, but they may lose in verbal skills such as speaking and listening as well as handwriting skills. Teachers who understand the gains as well as the losses can better make informed decisions about if, when and how technology should be used. In the above example, a teacher may value the skills gained through the e-mail project, but will need to engage students in additional activities that rely on handwriting as well as verbal speaking and listening skills.

How Does the Technology Promote or Inhibit Understanding Students' Thinking?

Because of the "black box" nature of most technology, student thinking is often hidden. When students use a calculator or computer to solve problems, their procedures are often not available for teacher diagnosis. Teachers can falsely assume students understand the concept because they have correctly solved a problem. Effective teachers work to understand students' thinking behind their answers and how technology helps or hinders this effort.

Any Technology Should Be A Means to A More Noble End, Not an End in Itself

Hawkridge, Jaworske and McMahon (1990) identify four main reasons cited for using technology in schools. First, a social rationale is that students need to know technology because technology is everywhere. A second reason is vocational; students need to learn technology because they may need it in their future careers. Third, a pedagogic rationale asserts students can learn better from a computer. Finally, some reformers assert that computers themselves can be a catalyst for systemic change in education. None of these reasons is adequately supported by research. The most careless rationale for including technology into the classroom is the pedagogic rationale that students can learn better from using computers, yet Ely (1995) found that research on student learning with technology is inadequate, contradictory and inconclusive.

These questionable rationales for including technology in the classroom, however, have been so widely and uncritically accepted that little serious discussion about them occurs. Technology is often included in schools as an end in itself. Technology is perceived as "good," and those who do not embrace it are given labels such as "resisters" and "Luddites." Cuban (1998) argues against those who use the above rationales and asserts that questioning the inclusion of computers in the classroom has received inadequate attention. As Postman argues, the inclusion of computers has received inadequate attention because:

> . . . educators confuse the teaching of how to use technology with technology education. . . . Technology education does not imply a negative attitude toward technology. It does imply a critical attitude. Technology education aims at students' learning about what technology helps us to do and what it hinders us from doing; it is about how technology uses us, for good or ill, and about how

> it has used people in the past, for good or ill. It is about how technology creates new worlds, for good or ill. (pp. 190–192)

Worse yet, the rhetoric of technology education as it often appears in the literature moves the dialogue away from such issues.

EXAMINING THE NATURE OF TECHNOLOGY

In science education, understanding the nature of science is a primary component of all reform documents (AAAS, 1989; AAAS, 1993; McComas & Olson, 1998; NRC, 1996). Science education is not simply about learning science content—its more important mission is to help students understand what science is, its limitations, how it impacts and is impacted by society, and how scientists and the scientific community work to generate knowledge (Clough, 2000). Likewise, meaningful technology education is far more than learning how to use technology to further other ends. It includes a rich understanding of what technology is, how and why technology is developed, how scientists and technologists operate as a social group and how society itself both directs, reacts to, and is unwittingly changed by new technologies. Education must move beyond its narrow focus of simply teaching students the mechanics of technology and blindly ignoring its more meaningful consequences. The concept of technology education must be broadened to include the nature of technology and confront questions like those raised by Postman (1995):

- For every advantage of technology, what is the corresponding disadvantage?
- How are the advantages and disadvantages of particular new technologies distributed unevenly?
- What is the underlying philosophy of particular technologies? For example, how do particular technologies change the way we think and act?
- How does new technology compete with older technology in regards to how we think of the world?
- What are the intellectual and emotional biases of particular technologies?
- What are the sensory, social, and content biases of particular technologies?
- What goals are promoted? What goals are ignored?
- How are other technologies impacted?
- How does this technology change the way we view schooling?
- How does the technology promote or inhibit thinking?

The cautionary perspective we take raises questions about the "cognitive biases and social effects" of technologies on education and how they often undermine what we know about effective teaching and learning. Our view from atop the fence has shown us that while technology has great potential to motivate and engage students, it can also change their fundamental ideas about the purposes of schools, potentially to their own detriment. Further, because of its black box nature, technology in classrooms often circumvents critical requirements of learning and can hide or even inhibit students' thinking. The teacher's role is critical, and attention must focus on how teachers engage students in making sense of their experiences. In addition, we need to address critical questions regarding the nature of technology, the inclusion of technology in classrooms, and carefully consider the Faustian bargain inherent in all technology.

REFERENCES

Almy, M. (1966). *Young Children's Thinking*. Teachers College Press, New York.

Annenberg/CPB (1997). *Minds of Our Own*. Harvard College President and Fellows, South Burlington, VT.

American Association for the Advancement of Science (1989). *Project 2061: Science For All Americans*. Author, Washington, D.C.

American Association for the Advancement of Science (1993). *Benchmarks For Science Literacy*. Oxford University Press, New York.

Berliner, D.C. (1989). Effective Schools: Teachers Make the Difference. *Instructor, 99*, 14–15.

Clough, Michael P. (2000). The Nature of Science: Understanding How the "Game" of Science is Played. *The Clearing House, 74*(1), 13–17.

Cuban, L. (1998). High-Tech Schools and Low-Tech Teaching. *Journal of Computing in Teacher Education, 14*, 6–7.

Duffee, L. & Aikenhead, G. (1992). Curriculum Change, Student Evaluation, and Teacher Practical Knowledge. *Science Education, 76*, 493–506.

Eisner, E. W. (1985). *The Educational Imagination: On the Design and Evaluation of School Programs, 2nd Edition*. Macmillan: New York.

Ely, D. P. (1995). Technology is the answer: But what was the question? Capstone College of Education Society James P. Curtis Distinguished Lecture Series. Syracuse University, April 14.

Fullan, M. G. (1991). *The New Meaning of Educational Change*, 2nd Edition. Teachers College Press, New York.

Good, T.L. & Brophy, J. E. (1994). *Looking in Classrooms*, 6th Edition. HarperCollins, New York.

Hawkridge, D. G, Jaworske, J., & McMahon, H. (1990). *Computers in third-world schools: Examples, Experiences and Issues*. Macmillan, London.

Langer, J.A., & Applebee, A.N. (1987). *How Writing Shapes Thinking: A Study of Teaching and Learning*. National Council of Teachers of English Research Report No. 22. NCTE: Urbana, pp. 87–37.

Matthews, P. (1997). Problems with Piagetian Constructivism. *Science & Education, 6*, 105–119.

McComas, W. F., & Olson, J. K. (1998). The Nature of Science in International Standards Documents. In McComas, W. F. (Ed.), *The Nature of Science in Science Education: Rationales and Strategies* (pp. 41-52). Kluwer Academic Press, The Netherlands.

National Research Council (1996). *The National Science Education Standards*. National Academy Press.

Nola, R. (1997). Constructivism in Science and in Science Education: A Philosophical Critique. *Science & Education, 6*, 55–83.

Ogborn, J. (1997). Constructivist Metaphors of Science Learning. *Science & Education, 6*, 121–133.

Osborne, J. (1996). Beyond Constructivism. *Science Education, 80*, 53–82.

Paulos, J.A. (1988). Innumeracy: Mathematical Illiteracy and its Consequences. Vintage, New York.

Pearlman, R. (1989). Technology's role in restructuring schools. *Electronic Learning*, June, 8–9, 12, 14–15, 56.

Penick, J.E., Yager, R.E., & Bonnstetter, R.J. (1986). Teachers make exemplary programs. *Educational Leadership, 44*, 14–20.

Phillips, D.C. (1997). Coming to Grips with Radical Social Constructivisms. *Science & Education, 6*, 85–104.

Phillips, D.C. (1995). The Good, the Bad and the Ugly: The Many Faces of Constructivism. *Educational Researcher, 24*, 5–12.

Postman, Neil (1995). *The End of Education: Redefining the Value of School.* Vintage, New York.
Postman, Neil (1985). *Amusing Ourselves to Death: Public Discourse in the Age of Show Business.* Penguin, New York.
Shymansky, J.A. & Penick, J.E. (1981). Teacher Behavior Does Make a Difference in Hands-on Science Classrooms. *School Science and Mathematics, 81*, 412–422.

This publication first appeared in Olson, J.K. & Clough, M.P. (2001). A Cautionary Note: Technology's Tendency to Undermine Serious Study and Teaching, The Clearing House, 75(1) 8–13. Copyright Heldref Publications. Reprinted with permission of Heldref Publications.

Joanne K. Olson
School of Education,
Iowa State University

Michael P. Clough
School of Education,
Iowa State University

CHAPTER 11

AMY NOELLE PARKS

SMART BOARDS, MONEY AND THE PEDAGOGY OF WATCHING

I became seriously interested in Smart Boards the day my daughter brought home an invitation asking me to attend a $500-per-family dinner at the home of a local minor celebrity. The purpose was to fund the installation of Smart Boards in every classroom at my daughter's public school. When I emailed the principal to express concern that this sort of invitation might be seen as exclusive, she said that she did not expect that all families could afford to attend, but that enough families would come to allow the school to buy Smart Boards for the last set of classrooms still using traditional white boards. She assured me that all children would benefit from this "fabulous" technology.

Interactive white boards (IWBs), which are sold most commonly under the names Smart Boards and Promethean Boards, have become ubiquitous in schools over the last decade. As of 2009, more than 75 percent of classrooms in Great Britain and more than 30 percent of classrooms in the United States have IWBs, and in the United States, that number has risen from just 5 percent of classrooms in 2004 (Corcoran, 2009). On their website, Smart Technologies (the company that sells roughly two-thirds of the IWBs in the United States) reports that more than 360,000 boards were sold in 2009, which shattered the previous sales record of 60,000 boards sold set in 2008. These sales represent no small investment of education funds in terms of dollars spent, with boards ranging in cost from roughly $2,000 to $7,000 per board, depending on the brand and features (T+D, 2008).

Despite the rapid proliferation of these boards and the increasing money spent on this technology, surprisingly little research has been done evaluating the pedagogical effectiveness of these boards, while a great deal has been written about them in the popular press. In a literature search using EBSCOhost with the search terms "interactive white board" or "Smart board" or "Promethean board" conducted in March 2011, I identified only 22 articles in academic journals. (In contrast, more than 2,000 news stories about Smart Boards were identified through Google News in just the first two months of 2011.) Of the articles identified on the scholarly search engine, four were editorials or columns; three were summaries of a variety

Michael P. Clough, Joanne K. Olson and Dale S. Niederhauser (Eds.), The Nature of Technology: Implications for Learning and Teaching, 201–216.

of technologies; another five were primarily examples of lessons or activities that teachers could use. One was a first person critique of IWBs by a British teacher educator. The remaining nine were empirical studies examining the use of IWBs in classrooms and reporting mostly positive effects. Five of these articles focused on research done in countries other than the United States, where IWBs have been in more classrooms for longer periods of time.

Much of the positive research on the use of IWBs (and nearly all of the research done in the United States) has come out of a group of researchers working in the field of special education. These studies have primarily examined the usefulness of IWBs in teaching reading skills to children with moderate intellectual disabilities. This body of work has demonstrated that children learn to read sight words from IWBs (including words intended for their peers to learn), that learning sight words from IWBs is more productive than learning them from flashcards, and that children productively learn letter sounds from IWBs (Campbell & Mechling, 2009; Gast & Thompson, 2008; Mechling, Gast & Thompson, 2008). Two of these articles (Mechling, Gast & Thompson, 2008; Campbell & Mechling, 2009) acknowledge the cost of boards as possible limitations, but none seriously raise the question of whether the better results demonstrated in the studies are worth the significantly greater costs of IWBs over traditional methods like flash cards. In terms of making generalizations to other fields, it also seems important to acknowledge that these studies have demonstrated increased efficiency in drill and practice pedagogies, but not in the kinds of teaching practices essential for engaging students in critical thinking and deep intellectual work around content.

A few studies *have* shown that IWBs can contribute to teachers' success in using reform-based pedagogies. For example, Schnittka & Bell (2009) reported that nine preservice teachers in biology were able to teach in reform-based ways using IWBs, such as by playing audio files of crickets chirping from which students could collect data. Similarly, Smith (2008) found that professional development on the use of IWBs helped secondary English teachers in Great Britain become more confident in teaching texts using multiple modes of instruction and multiple literacies. Preston and Mowbray (2008) found that IWBs enhanced both the teaching of basic skills, such as providing visualization for step-by-step instructions, and the teaching of analytic skills, such as the making of predictions based on simulations or models.

A few pieces of critical research on IWBs have been published, including a study of six primary school classrooms in Great Britain, which found that teachers tended to tightly control children's use of the boards and that teachers conceived of interactive lessons as ones where students interacted with the board, rather than lessons where students interacted with each other (Shenton & Pagett, 2007). The authors recommend new kinds of professional development to help teachers use these boards in different ways. Similarly, a study of IWB use in Turkey, a country that has invested heavily in educational technology including IWBs, found that lack of professional development, curricular support, and appropriate maintenance hindered the educational effectiveness of IWBs (Somyürek, Atasoy & Özdemir, 2009).

In light of the still slight research on the benefits of IWBs and the expense of the boards both in terms of dollars and opportunity costs, the purpose of this chapter is to undertake a conceptual analysis of texts about IWBs in the popular press and in advertisements. The goal is both to understand IWBs' rise in popularity as well as to examine some of the material and pedagogical consequences for U.S. schools. Given that the education research community has not produced a great deal of work advocating for the inclusion of IWBs in public school classrooms, the growing desire for the purchase of these tools by local school communities seems to be coming from other sources. Examining writing about IWBs in the popular press and in advertisements provides a way of understanding the circulating discourses of desire around this technology.

IDENTIFYING DISCOURSES

This analysis draws on poststructural theories of discourse, which see truth as produced through constant reiterations in spoken and written texts, the physical world, and social interactions (Fendler, 1999; Foucault, 1990). This is a broad definition of discourse that locates meaning-making not just in linguistic interactions, but also in physical objects (Latour, 2005). At the start of his book about teachers and technology, Cuban (1986) described a geography teacher showing a globe to a classroom of attentive students – all flying on an airplane above the landforms of the United States. Cuban pointed out the contradiction in this picture between the intended message – the promise of technology in education – and the unintentional one, which reinforced traditional images of show-and-tell instruction. This example demonstrates both the ways that discourses of truth are circulated and the many contradictory discourses available about teachers and technology. It also highlights the important role that Latour gives to physical objects in shaping social interactions. In the photo, the airplane and its windows make an entirely new kind of geography study possible, but the globe, the chalkboard, and the pointer call up other expectations about the roles of teachers and students.

One purpose of an analysis that draws on poststructural theories is to interrupt dominant discourses so that readers can see current truths as both unfamiliar and permeable. For example, instead of seeing the airplane as an advancement in teaching geography, readers can understand it as a new location for traditional pedagogies. In describing this kind of re-imagining, Foucault (1989/1996, p. 137) said: "It isn't possible not to think in terms of good and evil, true and false, but you have to say every time: and if it were opposite, what if the lines were elsewhere…." In the case of the current project, this means reconsidering the stories told about the necessity of IWBs.

In keeping with this chapter's theoretical perspective, I draw on research methods aligned with both rhetoric and social science, where documents, lived interactions, and physical spaces are read as texts available for interpretation (St. Pierre & Pillow, 2000). Drawing on strategies for textual analysis (Prior, 2003; Foucault, 1990), I read, sorted, classified, and examined particular phrases and words in a variety of texts related to IWBs, including:

- Promotional materials distributed by the two most prominent makers of IWBs: Smart Technologies and Promethean;
- Articles in the popular press identified through searches of the database, www.*googlenews.com*, using the search terms "smart boards," and "interactive white boards." For close analysis, I selected the first 50 articles listed as most relevant during the year 2011. I focused my analysis on SMART Boards, rather than their competitors, because of their large market share.
- Scholarly articles, identified using the database EBSCO with same search terms;
- Four video-taped lessons of elementary teachers using IWBs, which had originally been taped for analysis for other projects;
- Lesson plans for IWBs posted on the Smart Technologies and Promethean websites and in books written for teachers;
- An IWB training session at my university by Smart Technologies.

In analyzing these texts[1], I drew on strategies for content analysis, including identication of key themes, frequent words, and common narratives. In particular, I paid attention to writing about money, including amounts needed and spent, sources of funding, and opportunity costs described. In addition, I closely examined writing about particular pedagogies, including writing about teaching strategies or techniques, lessons, and curricula.

In doing this work, I turned all of the data into written documents. In other words, I transcribed and described the videos of the classroom and supplemented these with photographs. I also wrote fieldnotes and transcripts based on the audio recording of the IWB training session. These data, along with the articles and advertisements, were all entered into a qualitative software analysis program, which was used to identify frequent words as well as to code for particular themes. For example, I began by coding every mention of money in all of these texts and every description of a particular pedagogy. I then worked to generate more specific codes from these broad themes, an then read, wrote and thought about each set of texts identified with a different code. Although this work supported my deep thinking about the texts, it was not the heart of my analysis, which consisted of cycles of reading and writing about the ideas in the texts. While I did draw on the strengths of qualitative analysis software, in many ways my analysis was more in line with post-structural researchers (e.g., Lather & Smithies, 1997; Richardson & St. Pierre, 2005; Schmeichel, 2011) who reject coding as an unquestioned system of producing truth. Here the claim to credibility as researchers is *not* that any other researcher would produce the same claims in working with the same data, but instead that my own interpretation of the data is both valuable to the field and credible in relation to the evidence presented.

In doing this work of analysis my goal was not only to make an argument about the particular content of an individual text, but to look more broadly at the discourse surrounding IWBs. In Prior's words (2003, p. 67), "the common or garden office memo not only carries (mundane) information, but also gives expression to a set of (power) relations within an organizational setting. Indeed, memos give concrete expression to systems of hierarchy – of superordination and subordination – and certainly serve to define social networks." In other words, the justifications for the use of IWBs and the pedagogical practices described in written documents reveal some of the beliefs and practices of the broader communities in which the documents are situated. Given that there are as yet no comprehensive national studies about the impact of IWBs on pedagogical practices, examining national news stories about IWBs and other documents provides a window into how this technology is being viewed and used in a variety of particular contexts, including what problems these boards seem to solve, the pedagogies they promote, and the curricular and social issues they do and do not make apparent.

MONEY & SMART BOARDS: THE FIRST ONE IS ALWAYS FREE

The words most commonly associated with IWBs across virtually all of the texts I examined contributed to a dense discourse that associates new technology with progress, advancement, and getting ahead. A promotional magazine (Van Dusen, 2009) developed by Smart Technologies refers to the products as "advanced" (p. 25), "sophisticated" (p. 25), "cutting-edge" (p. 25), and "transformative" (p. 30). In the popular press, IWBs were described as "21st century" (Maynard, 2011), "up-to-date" (Aswell, 2011) and "top-of-the-line" ("New school gets ready to open," 2011). This kind of discourse works to sell IWBs in two ways: first, by enabling those who own the boards to feel as though their school setting is now at some "cutting edge", and second, by creating a fear of getting left behind for educators who do not have the boards. Often advocates for putting IWBs into classroom explicity drew on this discourse of keeping up. For example, one television station ("Willoughby: School fundraisers," 2011) reported a parent as saying "We want our students to be competitive" in her explantation of fund-raisers to purchase Smart Boards, which other local districts already had. Officials from another district bemoaned their inability to afford a $600,000 investment in Smart Boards to meet the level of buying done by neighboring districts (Davis, 2011). In my daughter's school, parents complained that the neighboring district, which had bought IWBs with Title I money, was "getting ahead." The Winona Daily News ("Winona schools lag," 2011) broadened the field of competition by reporting the comments of a representative of Smart Technologies who said: "Mexico put these interactive whiteboards in every elementary school classroom, and Europe is four years ahead of America." Like international comparisons of test scores, comments like this one suggest that keeping up with or surpassing other countries – whatever the measure – is imperative.

Through discourses like this one, parents come to see IWBs as an indicator of a high-quality school system, often while being entirely unfamiliar with the pedagogical opportunities and dangers these boards provide. This phenomenon could be seen in the number of PTOs and PTAs raising money for Smart Boards and also in the size of checks many of these organizations were writing, which in the news stories analyzed ranged from $3,000 to $200,000. Of the 50 news stories about Smart Boards analyzed, 21 of them dealt with issues related to money in some way, including eight about large donations, four about fund-raisers, and six about the difficulties of finding necessary monies.

As more and more districts put IWBs into their classrooms, other districts feel the need to follow suit so they are not perceived as traditional, dusty, or boring – all adjectives used by writers of these news stories to contrast older pedagogical tools with IWBs. Widespread desire for these products works to create new avenues for funneling money away from other priorities and toward IWBs. For example, as part of the economic recovery act that followed the 2008 recession in the United States, the federal government allocated $650 million to buy technology for schools. A number of private organizations also offer grants, and should these sources fail, parent teacher organizations turn toward local funding, such as in the anecdote that opened this proposal. In addition to funds from PTOs, schools in the stories analyzed received money from district foundations, individual donors, service clubs, and private businesses, such banks, restaurants, and energy companies.

The race to procure funds for buying enough IWBs to keep up with other districts contributes to an already dominant discourse that sees technology as the solution to all problems. In his discussion of neoliberal theories, Harvey (2007) linked this discourse to technology companies' needs to create more and more demand for existing products as well as to continually produce new products to sell to saturated markets. Harvey (2007, p. 68) wrote: "The neoliberal theory of technological change relies upon the coercive powers of competition to drive the search for new products, new production methods, and new organizational forms. This drive becomes so deeply embedded in entrepreneurial common sense, however, that it becomes a fetish belief: that there is a technological fix for each and every problem." In other words, IWBs are only the newest of educationally-marketed technologies, the most recent to become fetishized. But the viability of corporations depends on continually producing new demands, whether through updated versions of old products or through the introduction of new companion technologies, such as clickers, hand-held tablets, and Smart Tables.

SMART Technologies' Initial Public Offering (IPO) for investors to purchase stocks was successful, drawing in approximately $17.00 per share. Only months after the IPO, however, SMART Technologies was facing more than one class-action lawsuit accusing the company of misleading potential investors by not disclosing a decline in sales ("Smart technologies says lawsuit has no merit", 2011). With shares at approximately $9.00 in March, 2011, SMART Technologies must increase sales to meet investors' expectations. To do this, schools must be convinced to spend more and more money – both publicly and privately generated – on IWBs and related products.

In school districts where public dollars have been stretched to their limit, buying these new products will increasingly require a reliance on money from private sources, including fund-raising by parents and donations from local businesses. Both of these practices significantly impact the way schools engage with children and families. Because, as Vander Schee and Boyles (2010) point out in their analysis of sales to public schools of video exer-games (such as Dance, Dance, Revolution), private funding from individuals and companies necessarily makes public schools less public. Parents are reframed as consumers rather than as citizens, which means that families with a greater ability or willingness to pay are likely to feel more welcome both at school money-raising events and at other school functions where decisions about curriculum, school climate, and administrative matters are made. Similarly, companies that contribute money to schools will feel increasingly emboldened to market their own products through a variety of practices, such as by sending advertisements on stickers for children to wear home, placing banners in prominent places within schools, or providing basic school supplies such as paper or pencils with corporate logos on them. In this climate, it becomes more difficult for teachers to engage their students in critical reading practices of community institutions, such as critiquing a local restaurant's nutritional offerings or local banks' lending practices.

In addition, as companies seek to boost sales, teachers and teacher educators are themselves being increasingly enlisted in promoting the sales of IWBs and related products. For example, Promethean offers links on its website to various grants that teachers can apply for to buy IWBs, and Smart Technologies provides an article titled "What, not enough money?" (Knowlton, 2010) on its website with ten suggestions that educators could use to get needed funds; six of these suggestions involve building relationships with private companies.

SMART Technologies has also started reaching out to teacher education programs to market its products. My own college of education was recently given $40,000 of SMART products to use in preservice teacher education. As part of this "special product offer," a SMART Technologies memo asked that professors provide the company with lesson plans and activities for SMART Boards and other products, which the company can post on its website as part of its resources (personal communication, 2010). A company representative said that ten of the largest colleges of education were chosen for these partnerships. These alliances between colleges of education and the for-profit company serve a number of purposes. First, preservice teachers learn to use SMART boards as opposed to other kinds of IWBs so they are more likely to advocate for their purchase throughout their careers because of familiarity with that product. Second, the lessons created for and donated to SMART Technologies become part of the marketing package sold to other audiences. Sales materials trumpet "a comprehensive database of K-12 lesson activities" (ED Compass 2010, p. 2). Of course, to use these lessons, teachers must agree to use them only on SMART products.

Finally, the partnership produces a number of opportunities for the for-profit company to influence the research agenda of teacher educators. At my university, college administrators decided to give room-scheduling preference to professors who agree to require their students to produce plans for SMART Technologies, which ensures that proponents of IWBs are far more likely to introduce preservice teachers to the materials than critics. In addition, by supplying Smart Boards, the company makes it likely that research conducted around the use of IWBs will involve their own products. Strategies like this help to elevate the term "Smart Board" in the discourse to the equivalent of "interactive white board," much in the same way Kleenex often stands in for the term *tissue*.

Moreover, donating these products may indirectly influence the outcome of research by promoting positive images of the company among university researchers. Along with the products themselves, SMART Technologies occasionally supplies free meals on campus where products are demonstrated. A number of studies have demonstrated that even small gifts from sellers of pharmaceutical drugs influence physicians' prescribing practices (e.g., Brody, 2005; Wazan, 2000). Educators are unlikely to be immune to similar sorts of influence. For some time, for-profit corporations have sought to influence research at universities in a number of disciplines and many researchers in these disciplines have developed practices and codes of conduct designed to mitigate this influence. In education, these practices and codes of conduct have not routinely been a part of our work. However, there is some evidence of a need for education scholars to begin to have these conversations. Vander Schee and Boyles (2010) noted that some of the largest studies of exer-gaming are coming out of a university lab funded by the one of the companies that produces the product studied, while Saltman (2000) described the growing influence of curricula designed by corporations to promote the selling of their own products, such as counting activities that focus on particular kinds of candy or science experiments that require particular brands of soda.

Table 1: 10 Ways to spend $30,000 - $40,000

12 IWBs
550 children's violins
50 computers
30-40 Ipads
2,500 books
1 fully-equipped science laboratory
60 easels and $30,000 endowment fund for art supplies ($600/year at 2% interest)
80 complete sets of Froebel's Gifts
150 Digital cameras and 150 digital audio recorders
1 playground structure, 1 swing set, 25 soccer balls, 25 hula hoops, 25 jump ropes

In addition to the concerns about commercializing education schools, the cost of these products creates substantial opportunity costs for all buyers. Even at a relatively conservative estimate of $3,000 per classroom, many schools could spend $30,000 to $40,000 on IWBs without seriously asking what else that money might buy. (See Table 1 for some possibilities.) And, in addition to the original expense, money must be continually spent to keep these machines functioning. One district administrator noted that his district pays $150/hour to technicians to maintain the district's equipment ("Argus: Smartboards require," 2011).

In addition to maintenance costs for the original boards, schools are often pressured to buy new products to accompany IWBs because the market for the original boards has grown smaller as more and more schools buy boards for each of their classrooms. Over the last few years, a number of products related to IWBs have emerged, including clickers, which allow students to individually answer multiple-choice and short answer questions by pointing a remote control device at the screen; tablets, which allow students to write answers on the boards without leaving their seats; and touch tables, which allow children to play electronic games. School districts in the news stories I analyzed described spending $300,000 to $1,000,000 on technology, with funding for IWBs as a significant percentage of that expense. The spending of this money at a time when there is so much that we "cannot afford" in education has significant material and discursive consequences. Materially, many of the items listed in Table 1 (and others) go unpurchased. Discursively, this substantive spending allows the public to claim that we keep pouring money into education with little result, without noting that much of this money is pouring right through schools and ending up with for-profit companies. (Of course, IWB companies are not the only beneficiaries of this "educational" spending. Testing companies are also doing quite well.) As the price for a technologically-sophisticated classroom continues to climb, it seems important to stop and examine the kind of pedagogy we're getting for our money.

THE PEDAGOGY OF WATCHING: ALL EYES ON ME

One striking characteristic of the early academic literature on IWBs is the emphasis on rote learning (e.g., Mechling, Gast & Thompson, 2008; Campbell & Mechling, 2009). In contrast to the marketing claims about the boards' "interactive" and "transformative" possibilities, many of the lessons described across multiple texts do little to promote students' deep engagement with meaningful academic content. For example, a lesson on butterflies, which was highlighted as part of SMART Technologies' promotional materials (Van Dusen, 2009, p. 31), consists of a series of slides with elaborate graphics. During the lesson, students watch a movie, uncover the four stages of the life cycle, tap a climate where butterflies are not found, circle food that butterflies would eat, and draw lines from words to parts of a butterfly's body. This lesson, while quite bright and colorful, does not ask the students to do any intellectual work that they could not do on a worksheet, and unlike a worksheet, students must take turns and spend the majority of their time watching their classmates work. Perhaps most disturbingly, this lesson was produced as part of a four-week "intensive" training session on the use of the SMART Board, which raises questions about the quality of lessons produced under less leisurely and supportive circumstances.

Across a variety of texts, lessons produced for IWBs share the features of the lesson described above: emphasizing visual graphics, minimizing intellectual and emotional engagement, limiting access to materials, and promoting the watching of others as they interact with the technology. In the summer of 2010, there were 81 available lessons from the SMART Technologies website (www.exchange.smarttech.com) related to the Georgia fourth-grade social studies standards; 14 of these (or more than 15 percent) were about flags. Typically in these lessons, students name countries in a continent and then identify the appropriate flags. In a fourth-grade language arts lesson on literary characters, students watch a 13-slide PowerPoint presentation on defining character (round or flat; static or dynamic, etc.), take a short multiple-choice quiz, and then fill out an onscreen worksheet about a character they have read about. As with the butterfly lesson, students must take turns passing around the electronic pen. Again, as a teacher educator, there is little in this lesson to get excited about. Few elementary educators would recommend a 15-minute lecture as the optimum pedagogical tool to teach fourth-graders about character; and most could imagine a more productive and engaging use of that time. The slides designed for the lesson certainly have amusing pictures and brightly colored graphics, and it is equally the case that students could interact with the screen by writing on it or by using individual clickers (if available) to answer the multiple-choice questions. However, these are hardly the pedagogical practices called to mind with the words "interactive" or "transformative." In many of these lessons, the emphasis, for both the teachers and the children, is on the presentation of the content rather than on the big ideas. As a music teacher writing in praise of IWBs said: "The hardest part was deciding things like where to start, which background to use, and how many graphics to include, since the possibilities seemed endless" (Baker, 2007, p. 18). Here the intellectual energy of the teacher, like that of the children, is directed toward the mode of presentation rather than the content. In other words, the focus of both the children and the teacher was on the visual appearance of the display, not on the disciplinary content or on children's own experiences and interests.

The teaching practices described in the news stories I analyzed offered a similar vision of pedagogy. When talking in general terms, teachers, parents, and children described a vision of IWB teaching that might be worth the price tag. They called it "amazing" (Amo, 2011a), "interactive" (Smartboards help Willowville, 2011), and "fun" (Gillhoolley, 2011). However, when describing particular activities students have actually done on the IWBs, a different picture emerged. According to these news stories, IWBs allowed students to:

- Sort "words into categories by clicking and dragging words with their fingers" ("Smartboards help Willowville, 2011).
- "Write and draw using digital markers and other tools" (Yeatman, 2011).

- Do a Weekly Reader (Lindsey, 2011).
- Answer "multiple-choice questions" (Las Vegas Sun, 2011)
- "Watch videos" (Lindsey, 2011)
- "Write-in the answers" (Maynard, 2011)
- "Try out the new technology with some interactive activities like hangman" (Amo, 2011b).

Meanwhile, with IWBs teachers:

- "Put problems on the board" (Gillhoolley, 2011).
- "Write on the Smart Board and highlight things" (Gillhoolley, 2011).
- "Download worksheets or games from the Internet" (Maynard, 2011).
- "Load their lessons straight from the computer onto the Smart Board ... to keep the same lesson there for next year to use again." (Aswell, 2011).

These practices, which teachers and children describe actually using in their classrooms, hardly seem to be practices made possible only by a $3,000-per-classroom investment. Of course, some IWB activities described were more engaging. For example, one student described watching the teacher electronically dissect a frog (Maynard, 2011) and a teacher described using Google Maps to show the polar ice caps melting in real time ("Winona schools lag," 2011). However, these sorts of activities were quite rare in the total data set and tended to be described in the hypothetical when talking about what *could* be done, as opposed to what actually occurred. And even these practices place most students in the position of watching someone else (either a teacher or peer) engage with the technology.

These brief descriptions of practice in the news media resembled the teaching I observed in the four IWB lessons captured on video. Each of these videos was made as part of larger projects examining teaching in mathematics. Teachers volunteered to participate in the projects and were video-recorded with the goal of documenting typical teaching in their mathematics classrooms, rather than of assessing any particular program, pedagogy, or tool, including IWBs. Across these four lessons, the IWB activities positioned children as passive viewers rather than as active learners. In each classroom, IWBs took up prime real estate at the front of the classroom, relegating open space for "traditional" drawing and writing to the margins of the classroom. Student desks tended to be positioned to ensure that all students had a clear view of the board and frequently in primary classrooms space was cleared in front of the board for students to sit on the floor.

One of the classroom videos showed a second grade teacher beginning a mathematics lesson on the IWB. The children were seated on the carpet in front of the board where the teacher was standing. The following was a typical interaction:

> The screen on the Promethean Board showed a brightly-colored screwdriver, wrench, and hammer. Below was a giant picture of a ruler, which could be dragged around on the screen using the board's tools. Ms. Smith began the lesson by asking Nina to estimate the length of the screwdriver. Nina estimated "six."

Ms. Smith: Six what?
Nina: Five.
Ms. Smith: Five what?
Nina: Inches. (Ms. Smith wrote the estimate with her electronic pen.)

Ms. Smith placed the ruler under the screwdriver, noting: "See, this ruler does not start all the way at the end." She pointed to the space between the edge of the ruler and the first line.

Ms. Smith pointed to the end of the screwdriver on the ruler.

Ms. Smith: It comes out to about what?
Nina: Four.

Ms. Smith: So you had a pretty good estimate. What do you think about that wrench, Kyle?

Ms. Smith then engaged in a similar interaction with Kyle. Sometimes during these interactions, Ms. Smith invited students to come to the front of the room to write on or otherwise interact with the board.

This excerpt reveals a relatively typical interaction in an elementary classroom around measurement. The Promethean Board offers some advantages because of its clear projection of the ruler, which allowed Ms. Smith to model how to read and interpret the lines on the ruler. However, the board also worked to promote passivity among students which many educators seek to minimize. As in a lesson with a traditional white board or an overhead projector, the majority of students in this lesson sit silently as individual class members interact with the teacher and the technology. This means that all students spend the majority of their time sitting and watching others, and if students are invited up to the screen to write and move objects around, the time spent watching is increased.

In many ways, IWBs exacerbate this problem because rather than seek to minimize this part of the lesson, teachers using this "sophisticated" technology may be tempted to prolong it. This can happen for a variety of reasons. First, with the amount of money spent on IWBs and the amount of time spent on professional development for teachers, use of these boards is encouraged both explicitly and implicitly. At Ms. Smith's school, she had access to IWB activities assigned to each day of the school year that she was expected to use. Because of this library of activities, she did very little additional planning of other kinds of activities in mathematics. As she told me in an interview: "This year I have it so easy. I'm the luckiest first-year teacher in the world." Second, because of the graphics, music, and novelty of IWBs, students often will sit calmly in front of these screens longer than in front of traditional white boards, giving teachers less motivation to engage students in more active work. Finally, with their connection to the internet, IWBs provide teachers with easy access to movies, cartoons, and songs, which prolong the time children spend watching the screen without adding significantly to their intellectual engagement.

Ms. Smith did not stand out among the teachers videotaped in her use of the IWB. Three of the four teachers spent the majority of their mathematics time in front of the board, even when it was used in kindergarten. This instructional decision has profound consequences for children because in a school day with limited hours, children who spend more time in front of the screen will necessarily spend less time measuring objects, reading books, conducting science investigations, and problem solving with classmates. This is a feature of IWB use that is unlikely to be eliminated through professional development. Unlike other popular new technologies, like netbooks and tablets, IWBs by the nature of what they are will always steer teachers toward lessons where most students are watching.

BAIT AND SWITCH

One of the goals of this paper was to explore reasons why IWBs have grown so popular and in the conclusion section, I argue that the passive pegagogies promoted by IWBs are actually a key component of their popularity. In his essay on why teaching practice in the United States has remained remarkably constant over the years, Cohen (1988, p. 10) argued that lay narratives of teaching – that is, ideas of teaching in the popular culture – are deeply embeded in our collective psyches as a result of the long tradition of teaching in Western cultures where "learners are relatively passive; students are accumulators of material who listen, read, and perform perscribed exercises. And knowledge is objective and stable." Reform practices – where we recognize that students actively construct their knowledge in sometimes idiosyncratic ways, requiring the teacher to do far more than tell information and expect recall – is a "radical departure" from this lay tradition. One of the reasons that teaching practice stays the same, Cohen argued, is that it is almost impossible for reformers to subvert this narrative. At the same time, there is a strong narrative within the United States about progress – particularly in relation to technology (e.g., Cuban, 1986; Harvey, 2007). From our schools and ourselves, we desire "continual improvement" and "annual yearly progress." IWBs are popular, in part, because they feed both these narratives at the same time. Following Latour, the physical characteristics of the boards themselves contribute to both discourses of technological progress and of teacher demonstration. People can visit classrooms and be amazed at children dragging numbers across the board with their fingers, answering test questions with remote controls, and doing worksheets complete with animation, while at the same time having all of their traditional ideas about teaching confirmed. IWBs are popular because they masquerade as educational change without actually causing any. And, of course, it doesn't hurt that there's plenty of money to be made in the selling of them.

Teacher educators committed to reform pedagogies of all kinds must begin to interrupt the dominant discourse that insists that newer technologies are inherently better for students. This will be difficult work. As a field, we've made substantial financial investments in IWBs and their counterparts. Psychologists and economists call this money "sunk costs" (Arkes & Blumer, 1983) and note that even in the face of a lack of evidence or negative evidence; people are more likely to persevere with a choice when they have invested time and money. For this reason, amid the current discourses and having already invested considerable dollars for IWB, simply raising cost-benefit concerns is cognitively demanding. Advocates of IWBs will assert that the problem is with the individual teacher, not the board, but the board itself and all of its associated resources favor passive pedagogical practices, and reduce the number of students who have to participate in the lesson. Thus, going beyond calls for more or higher-quality professional development is an even greater challenge than admitting that money on IWBs is money not well-spent. However, if we're going to be prepared for the next technological imperative, then we need to question the foundations of the pressure to place these boards (as well as other expensive technologies) in our schools. Who is benefiting most from this pressure? Schools, teachers, students, or private companies? What kinds of pedagogies are they designed to promote? In the history of U.S. public schooling, a number of promising and relatively inexpensive pedagogies have been pushed to the margins of our schools while many questionable practices developed by for-profit companies have been taken up and promoted enthusiastically. In addition to exploring what does and does not work for students, our research needs to increasingly investigate the relationship between the expansion of certain pedagogical practices and who profits from the adoption of those practices—and ask important questions about the appropriateness of that relationship.

That IWBs are being marketed as a desirable new technology while embodying a pedagogy that has been present in U.S. schools since they began is a sad irony. In *The School and Society* (1956, p. 31) Dewey told a story about shopping for student desks. He wrote:

> We had a great deal of difficulty in finding what we needed, and finally one dealer, more intelligent than the rest, made this remark: 'I am afraid we have not what you want. You want something at which the children may work; these are all for listening.' That tells the story of the traditional education. Just as the biologist can take a bone or two and reconstruct the whole animal, so, if we put before the mind's eye the ordinary schoolroom … we can reconstruct the only educational activity that can possibly go on in such a place. It is all made 'for listening.'

Our schools still lack the workshops and laboratories that Dewey called for a century ago, but we've found hundreds of new – and more expensive – ways for students to listen.

ACKNOWLEDGEMENTS

I want to thank my colleague Stephanie Jones for her substantive feedback throughout the drafting of this manuscript.

This material is based in part upon work supported by the National Science Foundation under Grant No. 0546164 and Grant No. 844445. Any opinions, findings, and conclusions or recommendations expressed in this material are those of the author(s) and do not necessarily reflect the views of the National Science Foundation.

NOTE

[1] In this theoretical perspective, the word "text," is conceived of broadly, including both written documents and other representations of social meaning, such as videos and transcripts of lived interactions.

REFERENCES

Arkes, H.R. & Blumer, C. (1986). The psychology of sunk cost. *Organizational Behavior & Human Decision Processes, 35*(1), 124–140.

Amo, D. (2011a, Feb. 24). *Smart boards installed at JWP. Waseca County News*. Retrieved February 24, 2011 from Google News.

Amo, D. (2011b, Jan. 28). *JWP elementary receives $14000 donation from PTA. Waseca County News*. Retrieved February 24, 2011 from Google News.

Argus: Smartboards require maintenance. (2011, Feb. 18). *The Daily Argus*. Available at www.my.hsj.org. Retrieved February 24, 2011 from Google News.

Aswell, T. (2011, Feb. 13). *School enjoys gift from governor's wife. Natchez Democrat*. Retrieved February 24, 2011 from Google News.

Baker, J. (2007). Smart board in the music classroom. *Music Educators Journal, 93*(5), 18–19.

Brody, H. (2005). The company we keep. Why physicians should refuse to see pharmaceutical representatives. *Annals of Family Medicine, 3*, 82–85.

Campbell, M.L. & Mechling, L.C. (2009). Small group computer-assisted instruction with smart board technology: *An investigation of observational and incidental learning of nontarget information. Remedial & Special Education, 30*(1), 47–57.

Corcoran, E. (2009). Getting to the top of the class. *Forbes*, 184(6), 40–41.

Cuban, L. (1986). Teachers and machines: *The classroom use of technology since 1920*. New York: Teachers College Press.

Dewey, J. (1902). The school and the life of the child. In J. *Dewey The Child and the Curriculum/The School and Society*, 30–62. Chicago: University of Chicago Press.

EDCompass newsletter. (2010). EDCompass newsletter: *News and resources for educators using SMART products*, February 2010. Available online at: http://education.smarttech.com/ste/en-US/CommunityB/Newsletter/Previous+issues.htm (accessed July 2, 2010).

Fairclough, N. (2003). Analyzing discourse: *Textual analysis for social research*. New York: Routledge.

Fendler, L. (1999). Making trouble: prediction, agency, and critical intellectual. In T.S. Popkewitz and L. Fendler (Eds.), Critical Theories in Education: *Changing Terrains of Knowledge and Politics*. New York: Routledge, 169–188.

Foucault, M. (1990). The History of Sexuality: *An Introduction*. New York: Vintage Books.

Foucault, M. (1989/1996). Talk Show. In S. Lotringer (Ed.), *Foucault Live*: Interviews 1961–1984, 133–145. New York: Semiotext(e).

Gillhoolley, A. (2011, Feb. 19). The Skype's the limit. *Lebanon Daily News*. Retrieved February 24, 2011 from Google News.

Knowlton, N. (2010). What? Not enough money? *Smart Technologies website. Available online at*: http://www2.smarttech.com/NR/rdonlyres/3CFDF4D8–45A3–4595-A463–21FEB9182FE0/0/WhatNotEnoughMoney.pdf. Accessed July 2, 2010.

Lather, P. & Smithies, C. (1997). Troubling the angels: Women living with HIV/AIDS. *Boulder, CO*: Westview Press.

Latour, B. (2005). Re-assembling the social. *An introduction to actor-network theory*. Oxford: Oxford University Press.

Lindsey, J. (2011, Feb. 6). Smart technology transforms classrooms. *Daily Ardmoreite*. Retrieved

February 24, 2011 from Google News.
Maynard, J. (2011, Feb. 22). New: Education: Beal City: *Small school with big technology*. The Morning Sun. Retrieved February 24, 2011 from Google News.
Mechling, L.C., Gast, D.L. & Thompson, K.L. (2008). Comparison of the effects of smart board technology and flash card instruction on sight word recognition and observational learning. *Journal of Special Education Technology, 23*(1), 34–46.
New school gets ready to open. (2011, Feb. 16.) *WLBZ-TV*. Retrieved February 24, 2011 from Google News.
Preston, C. & Mobray, L. (2008). *Use of Smart Boards for teaching, learning and assessment in kindergarten science*. Teaching Science – the Journal of the Australian Science Teachers Association, *54*(2), 50–53.
Prior, L. (2003). *Using documents in social research. London*: Sage.
Richardson, L. & St. Pierre, E.A. (2005). Writing: A method of inquiry. In N.K. Denzin & Y.S. Lincoln (Eds.), *Handbook of Qualitative Research (3rd ed., pp. 959–978)*. Thousand Oaks, CA: Sage.
Saltman, K.J. (2000). Coca-Cola and the commercialization of public schools. In (K.J. Saltman) Collateral damage: Corporatizing public schools – a threat to democracy, 57–75. Lanham: Rowman & Littlefield Publishers.
Schnittka, C.G. & Bell, R.L. (2009). Preservice biology teachers' use of interactive display systems to support reforms-based science instruction. Contemporary *Issues in Technology and Teacher Education, 9*(2), 131–159.
Schmeichel, M. (2011). Feminism, neoliberalism, and social studies. *Theory and Research in Social Education, 39*(1), 6–30.
Shenton, A. & Pagett, L. (2007). From 'bored' to screen: *The use of the interactive whiteboard for literacy in six primary classrooms in England. Literacy, 41*(3), 129–136.
Smart Technologies website. (2010). *Teachers Hub*. Available online at: http://www.education.smarttech.com/ste/enUS/Ed+Resource/Teachers+Hub/getting+started/. Accessed July 2, 2010.
Somyürek, S., Atasoy, B., & Özdemir, S. (2009). Board's IQ: What makes a board smart? *Computers & Education, 53*(2), 368–374.
Sowell, E.J. (1989). Effects of manipulative materials in mathematics instruction. *Journal for Research in Mathematics Education, 20*, 498–505.
T+D (2008). *What does it cost to purchase whiteboards? T+D*, 62(2), 96.
Vander Schee, C.J. & Boyles, D. (2010). 'Exergaming,' corporate interests and the crisis discourse of childhood obesity. Sport, *Education & Society, 15*(2), 169–185.
Van Dusen (2009). eSN Special Report. June, 25–32. SMART Technologies.
Wazan (2000) Physicians and the pharmaceutical industry. Is a gift ever just a gift? *Journal of the American Medical Association, 283*, 373–380.
Willoughby: School fundraisers needed for classroom basics. (2011, Feb. 15). *WKYC-TV*. Retrieved February 24, 2011 from Google News.
Winona schools lag in the use of the technology. (2011, Feb. 23). *Winona Daily News*. Retrieved February 24, 2011 from Google News.
Yeatman, J. (2011, Feb. 18). *Aiming higher, teaching smarter. So Md News*. Retrieved February 24, 2011 from Google News.

Amy Noelle Parks
Department of Elementary and Social Studies Education,
University of Georgia

CHAPTER 12

JOANNE K. OLSON

THE PURPOSES OF SCHOOLING AND THE NATURE OF TECHNOLOGY

The End of Education?

INTRODUCTION

Early in my career, I taught middle school science and mathematics in South Central Los Angeles. One day, a sixth-grader, Marcus, was doing what he had done every day since school began—nothing. Each day, he arrived on time and sat in his chair, with no paper, no pencil, no homework, and no books. As the year progressed, I became increasingly frustrated with Marcus' refusal to do anything despite my multiple attempts to engage him in the class. Before the reader assumes that I simply needed to make the content of the course better, I will state that I am an award-winning science teacher, and I was a doctoral student at the time with several years of teaching experience. I was reading the professional literature, using what I was learning about how people learn and how to create engaging classroom environments. My class was a flurry of hands-on activities, deep thinking about science concepts, and field trips to accomplish what we could not do within the four walls of our classroom. On this particular day, out of exasperation with the continued lack of engagement by Marcus, I said to him, "Marcus, why are you here?" This was not my best pedagogical move, but I was unprepared for what he said. "I don't know. You tell me." I looked at him and realized that I had no compelling answer. Why was he in school? Why should he be here?

My standard rationale for schooling was college preparation. I wanted to open as many doors as possible for all of my students to pursue college and eventually careers so that they could be economically stable and successful. However, I also understood that not all students go to college, and not all students are concerned with what they will be doing 6–7 years in the future. My science class needed to provide students with a purpose now, not an uncompelling vision of a distant future of college-readiness. In fact, teachers in the United States graduate more criminals than scientists (Leyden, 1984). While I would never use those sobering statistics to label my individual students or project their individual futures, I was bothered by the

Michael P. Clough, Joanne K. Olson and Dale S. Niederhauser (Eds.), The Nature of Technology: Implications for Learning and Teaching, 217–248.

fact that if my students did things other than go to college (for whatever reason—economic, personal choice, etc.), what they were doing in my class needed a purpose and I did not have one that included anything other than "go to college." Schooling focused primarily on college preparation, when students end up doing other things (many of which are very valuable to society!), appeared to me to be a milder version of a prison system—we take children against their will and send them to school, telling them to exert effort for 13 years so they can go to college when that may not be what they wish to do or can do.

Why do we have schools? What compelling purpose can we provide to students so that they will exert the effort we request from them for 13 years? Surprisingly, very little explicit attention is paid to this matter in the modern education community (Kohn, 2011a; Postman, 1995). We discuss curriculum and pedagogy matters *ad nauseum*—arguing over standards, reading programs, simulations, algebra, phonics, inquiry, etc. Postman (1995) notes that we exert far too much energy discussing the technical matters of schooling without being clear on the purpose of schooling. "The truth is that school cannot exist without *some* reason for its being…" (Postman, 1995, p. 27). Conversations about methods and effectiveness make little sense if we do not know where we are going—methods to accomplish what? Effectiveness at what? Any curriculum and pedagogy decision is utterly dependent upon the desired ends (Kohn, 2011a). As Lewis Carroll (1865) so wisely noted in *Alice and Wonderland*, when Alice is lost and asks the Cheshire cat which way she should go, the cat says, "That depends a good deal on where you want to get to." When Alice answers that she does not care where she goes, the cat says, "Then it doesn't matter which way you go" (p. 513). Without a purpose for schooling, any route will do.

WHY DO WE HAVE SCHOOLS? THE PUBLIC PERSPECTIVE

I teach curriculum theory, and each semester the graduate students in the course interview fifteen individuals, with the restriction that only five can be from the professional education community. My students ask these participants a series of questions designed to elicit their underlying purpose for schools. The responses are nearly unanimous that schools exist so that students will get a good job when they graduate, either by being well prepared for college, or well prepared to immediately enter the workforce. Postman has noted this same tendency of the public to define individuals by what they do. We tell students a narrative that goes like this: "If you will pay attention in school, and do your homework, and score well on tests, and behave yourself, you will be rewarded with a well-paying job when you are done" (Postman, 1995, p. 27). Postman refers to this narrative for schools as Economic Utility. People are seen as workers, with varying degrees of status assigned to particular occupations, and thus to people's worth to society. Related to this narrative of Economic Utility is Consumership—that we are what we own. When children are asked why they need a good job, they often say that jobs are needed so they can buy things.

This view so permeates the United States that references to this narrative go virtually unchallenged. For example, the president of the United States gave a speech directed toward the nation's youth in 2009, and the speech was shown in most public schools. Many members of the opposition party opposed the concept of the schools showing a speech from the president, accusing him of having a "direct channel" to children, but interestingly, the content of the speech was largely met with silence (McKinley & Dillon, 2009). Specifically, the president stated, "You want to be a doctor, or a teacher, or a police officer? You want to be a nurse or an architect, a lawyer or a member of our military? You're going to need a good education for every single one of those careers" (Obama, in Hardin, 2009). In summary, "It takes hard work in school to prepare for a rewarding career" (Hardin, 2009).

Postman notes that the Economic Utility narrative is fundamentally flawed, for a number of reasons. First, the assumption that the economy is linked to the quality of schooling is faulty. "There is no correlation between achievement, as measured with test scores, and international market competitiveness as defined by the World Economic Forum" (Bracey, 2003, p. 160). Second, "Economic Utility is rarely believed by students and certainly has almost no power to inspire them. Generally, young people have too much curiosity about the world and far too much vitality to be attracted to an idea that reduces them to a single variable" (Postman, 1995, p. 30). Most children in schools are many years away from employment, and the concept of adult work is so distant that it is meaningless. They have no idea what they want to do in the workplace at a level of specificity to which a school could respond. When one considers what most of us do for our employment, the expectation for children to be compelled by visions of a distant future sitting in an office cubicle, typing on a computer, or other tasks of most adult work seems rather ridiculous. Such future work is hardly a compelling or developmentally appropriate reason for students to exert effort for 13 years or more. Third, given the current and foreseeable economic situation in many countries, the promise of "work hard and get a good job" is not a promise that can be delivered upon. The national unemployment rate for those aged 18–29 in the U.S. is, at the time I write this, 12% (twice the national average); in Spain and Greece, 50%; Italy, 35%, and France and the United Kingdom, 22% (Kotkin, 2012). These data include a large number of college graduates who are unable to find work. Half of U.S. law school graduates in 2011 could not find jobs in the legal field nine months after graduation (Kotkin, 2012). Thus, not surprising is the fact that over 43% of recent U.S. college graduates are employed in positions that do not require a college education. Sixteen percent of parking attendants have a bachelor's degree or more (Kotkin, 2012). The promise of this narrative is faltering, indeed. "Any education that is mainly about economic utility is far too limited to be useful, and, in any case, so diminishes the world that it mocks one's humanity" (Postman, 1995, p. 31). Thus, students need a reason to go to school that has meaning for them now, as well as in the future. Childhood is far too precious to waste, and human beings are far too complex to be reduced to cogs in an economic machine.

NARRATIVES, METAPHORS, AND SCHOOLING

Why is Economic Utility so widely held as a purpose for schools when it is so faulty? Eisner (2002) asserts that metaphors are frequently used for complex endeavors such as education. Metaphors often frame our conception of education because

> education is one of those normative, qualitative abstractions like "justice" or "beauty…". While we seem to share a sense of the concept, a core idea sufficient for its meaningful use in everyday concepts, we disagree ceaselessly about the full-fledged and proper account of the concept, about its "essential" qualities. (DeNicola, 2012, p. 2)

When dealing with complex concepts, metaphors tend to be created to capture the essence of the concept using a more familiar example. These metaphors are adopted through acculturation and socialization, and create a language and set of images that serve to define complex endeavors such as education and schooling. These metaphors, according to Eisner (2002), are usually absorbed without conscious attention to the values and perspectives inherent in them.

> When we talk about learners rather than children, competencies rather than understanding, behavior rather than experience, entry skills rather than development, instruction rather than teaching, responses rather than action, we make salient certain images: our language promotes a view, a way of looking at things, as well as a content to be observed. (p. 360)

We all need some conception of education, and metaphors shape those conceptions, making particular ideas and solutions appear plausible; importantly, other conceptions fail to appear at all because they do not fit with the metaphor. The narrative of Economic Utility discussed earlier is so widely-held and plausible because it fits well with the dominant metaphor used for schooling in the United States. Eisner describes this metaphor as follows: "The dominant image of schooling in America has been the factory and the dominant image of teaching and learning the assembly line" (p. 361).

If schools are an assembly line for job preparation, then employers and businesses are the "end user" of the output of the factory—little more than skilled workers. Evidence of this metaphor permeates discussions of education at all levels. Corporations have lamented for some time that new hires lack specific job skills and have demanded that education institutions provide such training. For instance, law schools have been accused of not providing enough "practical training" (Vukowich, 1971) and business schools similarly accused of not providing sufficient practical skills (Rubin & Dierdorff, 2009). Critics have also targeted K-12 public schooling. A headline in 1994 read: "Clinton tells Educators Youths are not getting Practical Skills for Jobs" (Postman, 1995, p. 31). Recent criticisms claim education is an inefficient use of tax dollars. *Forbes* magazine called education the most inefficient enterprise in the country, citing as evidence "more and more money spent on declining results"—"results" being defined as international ranking on test scores (Noer, 2012). The assumptions here are worth noting: the goal of schooling appears to be increased test scores (in language arts and mathematics), and an increased investment should yield increased output. These are business assumptions based on a factory model of inputs and outputs. What goes unquestioned is whether such business concepts can and should be applied to schools. This is a limitation of using metaphors.

If the goal of schooling is specialized training for jobs, then the current structure of schools is in many ways mismatched to this desired outcome. Certain corporate interests would probably like to see schools designed around narrow training programs that prepare students with specific skills to enter specific jobs. Student interests could be assessed and students could be tracked into curricula that prepare them for particular jobs, eliminating any time spent "wasted" on unrelated coursework. Several online degree-granting schools advertise their programs in just this way. "Why waste time with general education courses I'm never going to use?" says a young man in one commercial for such a program. By enrolling, "I can get my degree in two years and go straight to work without taking unnecessary classes." According to this argument, why take math when we can take accounting instead and more efficiently prepare accountants? Why take humanities when we could use the time to teach technical job skills?

One simple answer for why we do not narrowly specialize schooling and turn it into job training is that we cannot possibly anticipate what professions students aspire to, and schools cannot provide the specialized competence that would be needed. In addition, career changes are commonplace. The precise number of career changes the average adult will make is unknown, but is estimated to be around five. Whatever the exact number, "most people can expect to make many changes during their working lives and that flexibility begins with a quality core education…" (Lain Kennedy, 2008). Inherently, schools will be woefully behind in narrowly preparing students for unknown career changes and jobs that do not yet exist. Another reason we do not highly specialize schooling into job training is noted by Postman (1995):

> If we knew, for example, that all our students wished to be corporate executives, would we train them to be good readers of memos, quarterly reports, and stock quotations, and not bother their heads with poetry, science, history? I think not. Everyone who thinks, thinks not. Specialized competence can only come through a more generalized competence, which is to say that economic utility is a by-product of a good education. (p. 31)

Eisner (2002) has sharply criticized efforts to describe schools, and what we want from schools, in factory-related training-oriented terms. "These images misconceive and underestimate the complexities of teaching and neglect the differences between education and training" (p. 361). The distinction between training and education is important. A person who is *trained* uses knowledge and skills in predictable circumstances. The goal of training is for the individual to use those skills in the same way efficiently and effectively every time. In contrast, a person who is *educated* uses a broad knowledge base flexibly in unpredictable circumstances. The goal of education is for the individual to exercise sound decision-making to use in shifting contexts. Unfortunately, the metaphor of the factory has made training efforts at the expense of education not only plausible, but desirable. Eisner argues that several terms (and associated concepts) have entered our vernacular unannounced, and because they fit the dominant metaphor,

> such terms become ubiquitous, their conceptual implications are taken for granted, they become a part of our way of educational life without the benefit of critical analysis. The consequences of such concepts and the images they imply are, in my view, devastating. They breed the illusion of a level of precision in practice that is likely to be achieved only by reducing education to training. The assembly-line mentality that was so persuasively described by Callahan in his study of the scientific management movement in education between 1913 and 1930 is still with us. Such an image of education requires that schools be organized to prescribe, control, and predict the consequences of their actions, that those consequences be immediate and empirically manifest, and that they be measureable. In such a school, the exploitation of the adventitious, the cultivation of surprise, and the use of ingenuity are regarded as "noise." (p. 361)

The factory metaphor is seen in Cuban's (1993) description of a pervasive desire "to make teaching and learning productive, that is, better, more of it, and faster. The core values sought are efficiency in the use of limited school resources and enhanced individual productivity so that future workers will make U.S. businesses internationally competitive" (p. 18). As evidence of this metaphor and the solutions it allows, Eisner (2001) notes how we describe schools with terms that show a fixation on efficiency, control, productivity, standardized outcomes, measureable products, supervision and evaluation of teachers, quality assurance and quality control, performance contracts, payment by results, etc. The metaphor has made these concepts acceptable, despite research showing that such factory concepts do not translate well to school settings. For example, educational psychologists have repeatedly found that "an overemphasis on assessment can actually undermine the pursuit of excellence" (Maehr & Midgley, 1996), yet education reform efforts in the United States ignore these and other research findings and emphasize increased assessment, using factory-oriented terms such as "accountability" to the investors (taxpayers) to justify the practice.

Kohn (2011a) asserts that educators rarely make the purpose of schools explicit, because if we did, we would be forced to confront whether our purpose (and associated methods) is really our paramount goal. Is Economic Utility what we want from our schools? How content are we with children being viewed as little more than future workers? When we are so embedded in a factory metaphor with Economic Utility seen as the goal for individuals who complete schooling, we can easily overlook the existence of other narratives and metaphors. We may also fail to see the folly of many advocated practices that are suited for factories but not for schools. Furthermore, as evidenced by those who seek more job training in schools, we may be confused about why remnants of other purposes of schooling continue to exist in schools and appear to be in direct opposition to efficient job preparation.

THE DISAPPEARING GOAL OF A LIBERAL EDUCATION

A tension exists between training students so that they can be successful in specific jobs, or educating students more broadly, with the desire that they can apply that knowledge in many aspects of their lives. This tension is a long-standing issue, occurring in very early discussions of compulsory schooling. A liberal education has historically been a purpose for schooling, and reflects a different metaphor and narrative than a factory and Economic Utility. I view the metaphor of a liberal education as a cognitive apprenticeship, and the narrative as creating a critically-thinking, engaged public out of diverse individuals. The desire to create a critically-thinking public is why broad subjects such as literature, mathematics, science, art, and history have been central parts of the curriculum, rather than accounting, engineering, and court reporting.

The word *liberal* shares the same Latin root (*liber*) as *liberty* and *liberate*. The notion of a liberal education is an education that frees the individual from ignorance, prejudice, and intolerance. Adam Smith argued for such an education and noted that "a man without the proper use of the intellectual faculties of a man, is, if possible, more contemptible than even a coward, and seems to be mutilated and deformed in a still more essential part of the character of human nature" (Smith, 1776, p. 343). Furthermore, he asserts that this is the unfortunate condition of "the great body of the people," who are confined by their employment status

> to a few very simple operations, of which the effects are perhaps always the same, or very nearly the same, has no occasion to exert his understanding, or to exercise his invention in finding out expedients for removing difficulties which never occur. He naturally loses, therefore, the habit of such exertion.... [As a result,] the torpor of his mind renders him, not only incapable of relishing or bearing a part in any rational conversation, but of conceiving any generous, noble, or tender sentiment, and consequently of forming any just judgment concerning many even of the ordinary duties of private life. (Smith, 1776, p. 340)

For Smith, the very nature of most human work was stupefying to the intellect needed to engage in civic and private life. In contrast, a liberally educated public has the advantage of providing stability for a free society.

> The more they are instructed, the less liable they are to the delusions of enthusiasm and superstition, which, among ignorant nations, frequently occasion the most dreadful disorders....They are more disposed to examine,

> and more capable of seeing through, the interested complaints of faction and sedition, and they are upon that account, less apt to be misled....In free countries, where the safety of government depends very much upon the favourable judgment which the people may form of its conduct, it must surely be of the highest importance that they should not be disposed to judge rashly or capriciously concerning it. (Smith, 1776, p. 343)

In a democracy, an educated public was considered necessary for the defense of freedoms, thoughtful civic decision-making, and the avoidance of extremism and divisive factions. Adler and Gorman (1952) note,

> When the vocation of man is thus understood, a general or liberal education is vocational in that it prepares each man for the common conditions and callings of human life. In this sense specialized training, which by implication at least seems to be the object of Smith's criticism, is not vocational. It fits a man only for some specialized function, according to which he or his social class is differentiated from some other man or class. (p. 378)

Aristotle divided education similarly into education that is liberal, and that which is illiberal. "Certain subjects are illiberal by nature, namely, 'any occupation, art, or science, which makes the body or soul of the freeman less fit for the practice or exercise of virtue.' In this category Aristotle includes 'those arts which tend to deform the body, *and likewise all paid employments, for they absorb and degrade the mind*'" (Aristotle, in Adler & Gorman, 1952, p. 378, italics added). Aristotle's view of a liberal education is that it must treat humans and their growth as an end, not as a means to be used by others or by the state. In other words, we are more than what we do for a living.

Liberal education has three general purposes: 1) a broad transmission of culture, 2) self-actualization, and 3) promotion of an understanding of the world and the forces that shape the individual (DeNicola, 2012). These purposes are viewed as liberating the individual to rise above the daily tasks of employment to engage in a fuller life with meaning, a sense of public responsibility and internal character, and the capacity to understand the larger issues of humanity. How students are to achieve these goals has varied widely over the centuries (DeNicola, 2012), but one or more of these three aims for a liberal education can be traced to the earliest writings on education. For example, Plato viewed education as drawing "the soul toward being" (self-actualization).

Unfortunately, the Economic Utility narrative has become so entrenched in our culture that we seem largely unaware of the power this narrative exerts in shaping what we view as appropriate in schools. Views of education as fostering virtue, civic responsibility, bodily health, and ethical behavior are often seen as quaint relics of history rather than more currently valued goals such as high scores on reading comprehension tests. Our desire to manage, control, and measure has caused noble goals of education that are difficult to measure to fade from our stated goals for schools. "The result is that teachers may become more adept at measuring how well students have mastered a collection of facts and skills whose value is questionable—and never questioned" (Kohn, 2011b, p. 30).

The remainder of this chapter is devoted to an exploration of how the factory metaphor and its progeny, Economic Utility, have resulted in the largely unquestioned acceptance of dramatic changes in science education in the United States. This is certainly not the only instance of this metaphor's influence, but this case is illustrative of the need for education rather than training, and the nature of the disciplines (such as the nature of science, and nature of technology) to have a prominent place in that education.

SCIENCE EDUCATION AND THE DISPLACEMENT OF SCIENCE

The United States has witnessed several decades of increased national standardization of schooling, reflecting the factory metaphor. The United States has historically considered K-12 schooling to be outside of federal control, leaving the decision of subject matter to states, where such issues could be decided at the state level or even locally. Specified "outcomes" or "standards" are relatively recent in the United States, becoming commonplace among states in the 1990s. In 2009, governors in forty-five of the fifty states agreed to adopt and mandate a single set of national standards, beginning with English/language arts and mathematics, followed by science. The narrative underlying these "Common Core" standards is clear: "Preparing America's students for college and career" is the slogan featured prominently on materials prepared by this unprecedented initiative to standardize U.S. schooling (CCSSI, 2012).

The development of the first nationally-mandated science education standards in the U.S. is an interesting example of the depth and power of the factory metaphor and Economic Utility narrative. Federal control and standardized outcomes, now accepted as inevitable and necessary, would have been seen several decades ago as a tremendous waste of time and resources while stripping states and local school districts of the authority to make decisions about the education of their children.

Unlike previous "recommended" standards that were developed by the National Academy of Sciences, the new *Framework for Science Education* and the *Next Generation Science Standards* were jointly developed by the National Academy of Sciences and the National Academy of Engineering (NRC, 2012). A group calling themselves "Achieve, Inc." was also involved in the writing of the standards, and this group is associated with the Business Roundtable (a group representing business interests with an explicit focus on influencing federal political policy) (Bracey, 2003). The result of this business/engineering/science-influenced writing effort is that the subject of science for all children in public schools in 45 states now includes engineering content. The argument commonly used for its inclusion and displacement of one-fourth to one-third of science content is that engineering uses science. Arguments can be made that other disciplines "use science" (such as agriculture) and yet are excluded, but the larger issue is what this says about the influence of the Economic Utility narrative, and the concomitant power of the factory metaphor.

Many reasons may underlie the fundamental change in science education to include engineering. Perhaps this is simply a political maneuver to increase the job pipeline for engineering careers. If so, the country's preoccupation with Economic Utility likely accounts for the lack of open questioning of the wisdom of this decision. Another reason for the inclusion of engineering may be due to the ways that the nature of technology and the nature of science are understood.

THE NATURE OF TECHNOLOGY IN U.S. REFORM EFFORTS

Technology is portrayed by developers of the U.S. science education standards in glowing Economic Utility terms. One developer of the *Framework for Science Education* (NRC, 2012) described the purpose of the standards as follows:

> Giving each child an opportunity to hear the story that is science, appreciate what makes science unique, and experience how engineering helps us solve both our own and society's problems will result in all students having greater options and opportunities when they leave high school.... All students...will be able to access science and be far better prepared for their chosen field, regardless of what it is. (Pruitt, 2012, p. 3)

This is consistent with the foreword of the *Framework for Science Education* (2012, p. x) that refers to a need for a greater "percentage of students who are motivated... to pursue careers in these fields" and "an ever larger number of jobs require skills in these areas..." as rationales for the document. The stated goal of these standards is not to create a public who can peacefully co-exist and defend its freedoms. It is not to create well-educated, thoughtful people who care for others and the environment. It is not about creating moral or ethical people. It is about creating workers for "their chosen field." If these other goals are valued at all, they are implicit and ancillary to the explicit goal of job preparation.

The *Framework for Science Education* (NRC, 2012) makes clear efforts to distinguish science from engineering, and to articulate what engineers do and how the field works. A dominant focus is on the engineering design process. However, the nature of technology and engineering (NOT/E) in the *Framework* completely ignores or downplays significant issues in NOT/E that citizens need to understand in order to be informed and to make wise technology decisions.

As exemplified by Pruitt's comment above, the *Framework for Science Education* (NRC, 2012, hereafter called the *Framework*) conveys engineers as solving personal and societal problems. The engineering design process is described in considerable detail, and students are expected to learn, for example, that engineers ask questions to define a problem, they determine criteria for a solution, they identify constraints, and they communicate their design solutions (NRC, 2012, p. 54). A typical standard for engineering states:

> By the end of grade 8: Engineering advances have led to important discoveries in virtually every field of science, and scientific discoveries have led to the development of entire industries and engineered systems. In order to design better technologies, new science may need to be explored (e.g., materials research prompted by desire for better batteries or solar cells, biological questions raised by medical problems). Technologies in turn extend the measurement, exploration, modeling, and computational capacity of scientific investigations. (NRC, 2012, p. 211)

Students who meet the expectations described in the *Framework* will have a very positive perspective of engineers as solving our problems, of technology as progressive, of technology providing solutions to problems, of an engineering design process that is rather straightforward and assures that constraints and costs have been appropriately considered, and of life as getting better because of our technology. This positive view is consistent with broader societal misconceptions about technology. "In the Anglo-American empiricist, French enlightenment, and European positivist traditions, technology is widely depicted as an unproblematically beneficial force for human progress. For these traditions, technology needs only the proper association with modern science to fulfill its promise…" (Scharff & Dusek, 2003, p. 3).

THE IGNORED AND DOWNPLAYED NATURE OF TECHNOLOGY

Such a positive view of engineering likely appears accurate to those unfamiliar with the nature of technology, including inherent limitations of the engineering design process, and consequences resulting from technology. In this section, I highlight issues in the nature of technology that are overlooked or downplayed in the *Framework*. To those unfamiliar with the nature of technology, the list below may appear to be deliberately negative. This is not my intent. In addition, my intent is not to advocate an unrealistic abandonment of modern technology. I am not in this chapter proposing solutions to the limitations of technology. The *Framework* highlights positive aspects of engineering and technology, yet ignores many concepts that are fundamental to technology and the engineering design process. As noted by Bunge (2003), "the conceptual side of technology is neglected or even ignored by those who equate technology with its practice or even with its material outputs" (p. 174). To more fully understand technology, we must examine its philosophical underpinnings rather than simply define it as a process and the products that result. Thus, my purpose here is to identify issues overlooked or downplayed in the *Framework* that engineers and philosophers of technology have raised so that educators can consider what issues may be appropriate for students to learn at varying levels of education. Ignoring these issues, as currently occurs in the *Framework*, will likely result in ignorant consumers of technology rather than liberally educated individuals who can understand the world around them and the consequences of modern technologies, and make informed, conscious decisions about technology.

The following list highlights aspects of the nature of technology downplayed or absent in the *Framework*.

1. *The distinction between "the natural world" and "the human-built world" is a myth.* Unfortunately, the *Framework* perpetuates the concept of "the human-built world" as separate from "the natural world" in an attempt to distinguish science from engineering (p. 55). While this distinction is useful in terms of understanding human-made and natural objects, a more dangerous misconception is simultaneously created. In terms of the consequences of engineering, we cannot separate the objects that we create from the natural world. The production of any physical technology requires natural resources, many of them nonrenewable, and produces waste products. Nonrenewable natural resources are finite. Many of the waste products of our technologies contain synthetic compounds that do not biodegrade, creating long-term impacts on the environment. Furthermore, many technologies have disrupted the natural cycling of many elements, such as carbon and nitrogen. Describing technology as "the human-built world," as if it is a literal world apart from nature, enables us to more easily ignore the impacts of technologies on the environment and the sustainability of wide scale production, distribution, consumption, and disposal.

   > All this leads us to believe that we have made our own environment and no longer depend on the one provided by nature. In the eager search for the benefits of modern science and technology we have become enticed into a nearly fatal illusion: that through our machines we have at last escaped from dependence on the natural environment. (Commoner, 1971, p. 15)

2. *All technologies have unavoidable and unintended consequences.* "Given the extreme complexity of industrialized societies, nearly everyone, including highly educated scientists and engineers, who are essentially laypeople in any field other than that in which they specialize, is ignorant of the intricacies and possible side effects of most advanced technologies outside their narrow fields of expertise. In the absence of knowledge, it is easy to succumb to technological optimism" (Huesemann & Huesemann, 2011, p. 155). We simply do not have the ability to predict many of the consequences that occur from technologies. The inventor of the machine gun had the positive outlook that its presence and capability would be a deterrent to warfare (Johnson, 1984). Instead, carnage resulted when the machine gun was used. The developer of the automobile could not have foreseen the significant impacts on the environment, distribution of the population, layout and structure of cities, political and economic impacts of fossil fuel dependence, social isolation, and other consequences of this technology (Huesemann & Huesemann, 2011, p. 31).
3. *We cannot design away the negative consequences of technology.* "Both positive and negative effects are necessarily produced by any technology. It

is impossible for technologies to provide only benefits without any costs in a world in which everything is connected…" (Huesemann & Huesemann, 2011, p. 73). The *Framework* mentions negative consequences, but uses qualifiers that make negative consequences appear rare, rather than inevitable. For example, the *Framework* states: "Technologies that are beneficial for a certain purpose may later be seen to have impacts (e.g., health-related, environmental) that were not foreseen" (p. 213). The solution to such problems is described by additional engineering efforts to "increase their benefits...decrease known risks, and to meet societal demands…" (p. 213). Unfortunately, this is unrealistic and misportrays the nature of technology.

> The technical phenomenon cannot be broken down in such a way as to retain the good and reject the bad…. It is an illusion, a perfectly understandable one, to hope to be able to suppress the "bad" side of technique and preserve the "good." This belief means that the essence of the technical phenomenon has not been grasped. (Ellul, 1976, p. 111)

We often see very real and immediate gains from technologies, such as gains in efficiency, convenience, or pleasure. However, technology inherently also has negative consequences. This is because we live in an ecological system. The planet is an interconnected system, and "anything extracted from it by human effort must be replaced. Payment of this price cannot be avoided. It can only be delayed" (Commoner, 1971, p. 46). When an assembly line is automated, the product may be more consistent and produced faster. However, people who worked on that assembly line are now unemployed. Physical products require natural resources and produce pollutants and waste during production and disposal. Negative consequences of technology frequently occur in unanticipated ways, in distant locations, or at some time in the distant future, and thus we can easily dissociate the consequences from their cause, thereby downplaying or ignoring their existence.

4. *Absolute technological progress is a myth.* We have certainly benefited in many ways from technologies, and this leads to the perception that absolute technological progress—that the world is getting better all the time due to our technology—is an undisputable axiom. Much of this myth depends upon personal perspective. I have family members whose very existence is dependent upon technological devices and procedures, so I do not make this claim flippantly. However, when one views the overall system and takes a macro-approach rather than the micro "what has technology done for me lately" approach, we see a very different picture emerge. We must keep in mind that, as readers of a book such as this, we are likely of an economic status that directly experiences benefits from technology and do not have direct contact with many of its consequences, and may not consider many other consequences that have impacted us. This does not mean that those consequences do not exist.

"There is no absolute technical progress. With each advance we can see a certain setback" (Ellul, 1990, p. 40). Not only do technologies require the investment of capital and resources, but we also have to consider other losses when such technologies are produced. Ellul (1990) cites an observation made by Kolm: "The United States would seem to have double the national revenue of France, but would much of this gap remain if one took into account social values and deducted a measure of the ugliness of American cities?" (p. 40). Some readers may find this statement humorous; however, physical and mental health are profoundly impacted by our visual environment (Orians & Heerwagen, 1992). Many technologies promise increased leisure time and happiness. Fortunately, we can study whether these promises have been met. For instance, the automobile promised to decrease time spent traveling. This was a false promise; time spent traveling has remained constant, because we now travel greater distances, a phenomenon known as Hupke's Constant (Huesemann & Huesemann, 2011). Rather than gaining leisure time, we have changed the locations and ways we live, and gained more material goods. As Brende (2004) notes, technology designed to produce leisure time

> does not so much *save* labor as *separate* it out in time and place and thereby make it *harder* to obtain—physical exercise in the gym, moneymaking in the office, education in the school, and "quality time" with the family in the national park. Rather than an integrated whole, life becomes a temporal and geographical obstacle course. (p. 3)

The average woman today has one hour less leisure time per day than women did in the early 1900s, and this does not include time spent commuting (Huesemann & Huesemann, 2011). Modern life is characterized by increased demands on the individual (Kegan, 1994), higher stress, higher rates of depression, less leisure time (Louv, 2005), greater rates of obesity and related diseases, less job security, and a host of other difficulties. We have greater access to information, but less ability to make decisions due to information overload (Bauerlein, 2008). Dramatic negative impacts have occurred to the environment and to the ways of life of entire cultures.

In our focus on the short-term gains from our technologies, we often overlook the fact that "a complex mechanical entity readily overwhelms or subverts the very purpose for which it was deployed" (Brende, 2004, p. 230). "Mechanized farming may produce more food, but we have to consider that a calorie of food now demands the consumption of a calorie of fuel. A team of oxen or horses would use less" (Ellul, 1990, p. 41). The average speed of cars in rush hour traffic in major cities, often less than 10 km per hour, is lower than that of the average speed of a horse drawn carriage (Graedel & Allenby, 1998; Myers & Kent; 2001, in Huesemann & Huesemann, 2011). My purpose here is not to advocate a return to the horse. But this is the reality of our technology whether or not we like to hear about it.

5. *Technologies can change the way we think and act, often without our conscious awareness.* How many of us would have willingly agreed to be accessible to our employers seven days a week, and during evenings and even nighttime hours? Yet the very nature of e-mail has created such working conditions for many of us to which we would have never consciously assented. E-mail is an instantaneous inbox, filled with tasks that arrive at all hours, and accompanying expectations for the employee to respond. The cell phone has changed family dynamics as well. How many of us would have consciously made the decision to avoid interacting with our children or our partners? Yet looking at families dining at restaurants indicates that this is the case—children and adults can be seen at tables, sitting together but each one silently absorbed in an electronic device. Cars are being sold with DVD players in the back seats to "entertain" the children. The result is certainly entertained and rather quiet children. The result is also reduced social interaction between parents and children, and reduced exposure to the world beyond the back seat of the car. Technologies are created to be used in certain ways, and when they are used in such ways, consequences result that impact our behavior. Keyboards, cell phones, and many other devices are designed for a single user. Thus, they foster solitary behavior, including a lack of social interaction with others who are physically present. Conversation occurring while using solitary devices is more often than not a distraction. These issues are addressed in far more detail by Bugeja in Chapter 3 and Parks in Chapter 11.
6. *The benefits and negative consequences of technologies are distributed unequally.* History is filled with examples of technological decisions that had the benefits reaped by wealthy individuals, while the negative consequences were passed to those who are "poor and relatively powerless members of society, whose burdens in polluted and little-visited neighborhoods can be easily buried in the 'reasonable and objective' 'bottom lines' of cost-benefit calculations" (Ferre, 1995, p. 84). For example, the United States has outsourced the manufacture of a number of technologies, where working conditions and environmental impacts are not subjected to the same regulations, thus resulting in exploitation of people and the environment elsewhere while U.S. corporations and consumers benefit. Within the United States, computers in schools were touted as providing more equitable access to information. However, Cuban (1993) reported an Office of Technology Assessment study that found that access to computers in schools is greater for children of wealthy families. In addition, students whose native language is not English have less access than other students to computers in schools, and poor children are more likely to use computers in schools for drill and practice exercises rather than reasoning and problem solving activities. Even within a wealthy country, the distribution of technologies is unequal and may further exacerbate inequalities.
7. *Modern technologies create physical, emotional, and temporal barriers that make it less likely for us to see negative consequences of those technologies.* Electricity

generation, the extraction of natural resources, working conditions in factories, waste produced by factories, and many other aspects of technology development usually occur out of sight of those who use the technologies. Thus, the exploitation of workers, nature, and other consequences are often ignored by the public and quietly promoted by those who profit because they are geographically passed somewhere else or to future generations.

8. *Technology decisions have consequences that are extremely long-term.* A classic example of long-term consequences can be seen in our use of DDT. What was seen as a highly safe and effective method of pest control has remained in the environment, becoming stored in the tissues of organisms throughout the planet. The disposal of synthetic compounds and nuclear waste created now must be addressed for timespans that extend far beyond our lifetimes. Once dispersed, we have no solution for removing such substances from the environment.
9. *Efficiency improvements alone cannot solve our problems.* An assumption is often made that we can increase benefits, and decrease the use of limited resources, by making technology more efficient. Limits exist in efficiency efforts, as described by the second law of thermodynamics—energy loss always exists in any energy conversion effort, and 100% efficiency can never be reached. While efficiency efforts certainly can result in a single technology requiring fewer natural resources, the result of such efficiency improvements when they are adopted in the society is counterintuitive. Consumption and demand increases. This is because population is increasing, material affluence has risen, new products are available, and the cost of goods and services decreases. This leads to what is known as the Jevons paradox, or the "rebound effect"—efficiency efforts designed to decrease the use of limited resources actually increases their consumption (Huesemann & Huesemann, 2011).
10. *Technology and engineering have a value orientation, and reflect a value system.* A common myth exists that technologies are neutral, or have only a positive potential if used well. The phrase "technology is just a tool; the problem resides in how we use it" has taken on almost mantra-like status.

> This idea would be merely preposterous if it were not so widely accepted, and so dangerous. In believing this, however, we allow technology to develop without analyzing its actual bias. And then we are surprised when certain technologies turn out to be useful or beneficial only for certain segments of society. (Mander, 1991, p. 35)

When developing technologies, decisions are made during the process that reflect specific values. This occurs with any creative process, including engineering design, yet these values "are either unrecognized, or simply taken for granted" (Cotgrove, 1982, p. 68).

Engineering as it has occurred in industrial societies operates on several assumptions: 1) humans are "separate from and more valuable than nature" (Bunge, 2003, p. 180); 2) nature is something to be mechanized; 3) humans have a right (or even duty) to alter or subdue nature to the benefit of humans (Ferre, 1995); 4) "the ultimate task of technology is the fullest exploitation of natural and human resources (the unlimited increase in GNP) at the lowest cost without regard for anything else" (Bunge, 2003, p. 180); and 5) moral and ethical decisions about technology are largely the responsibility of policy makers (Bunge, 2003).

Philosophers and engineers have raised concerns about these assumptions that underlie our current technological practices. Ferre (1995) notes that when we view nature as something to be mechanized, we view technology as serving as an "anti-entropic ordering function" (Ferre, 1995, p. 66). However, as noted by basic principles of ecology, negative entropy in one part of the system creates entropy elsewhere. Bunge (2003) notes a growing distrust of the assumptions listed above because they condone harmful technologies and practices, but he also notes, "As yet, we have not offered an alternative ethical code" (p. 180). Thus, while we may not like these assumptions, they continue to undergird engineering practice.

In addition to the assumptions that underlie engineering, Huesemann and Huesemann (2012, p. 236–237) assert that values are also embedded in the produced technologies. Some of these values include: cost-effectiveness, efficiency, marketability, power, control, exploitation, profit-maximization, speed, uniformity, mass production, repetitiveness, quantification, precision, standardization, dependency, materialism, consumerism, and individualism. These engineers note, "In general, the more specifically a technology has been designed for a particular use, the more completely it will embody the values of the designers, and the less it will be of use for purposes reflecting different values. Thus, the more specific the design, the less human choice and control there will be regarding the final use of the respective technology" (p. 237).

Important to note is that the technology is not at all neutral. Any technology was developed with particular assumptions about humans, nature, and our right to alter our environment. The resulting technologies reflect these assumptions and also have biases toward particular sensory systems or uses, and not others.

> A modern automatic machine is no mere inert tool. It is a complex fuel-consuming being with needs of its own. It gobbles up energy; it demands care and maintenance; it even has bouts of temperament. In many cases no diaper will contain its mess. And all this on top of the initial chunk of cash it bites—its purchase price—which often amounts to a king's ransom. For these reasons, it not only serves but must be served. But it is more than another mouth to feed; as it becomes more involved and involving, it can easily invade the living space we formerly reserved for ourselves, taking on functions once our own. (Brende, 2004, p. 230)

As Huesemann and Huesemann note, "A distinct value-orientation can be demonstrated for any technology" (2011, p. 238). For example, the television is often perceived as a positive entertainment device or a neutral tool that is dependent on how we use it, but numerous authors have written about its impacts that are a natural result of its bias toward sedentary, solitary, silent viewing of mass-distributed messages produced elsewhere by people unknown to the viewer and with motives that span beyond entertainment. For instance, Mander (1978) has noted that television favors images that are larger because subtle emotions, detail, and complexity cannot be well conveyed via that medium. Thus, due to the design of the technology, the content of all television programming is affected and is biased toward sports, conflict, drama, violence, and "the grosser end of the human emotional spectrum" (Mander, 1978, p. 269). The amount of time young children watch television is directly correlated with concentration difficulties (Healey, 2004). Because conversation with others during television viewing is not conducive to comprehension of what is being viewed, the television reduces communication and promotes isolation. Given profit motives of those who own the broadcasting stations, advertisements have pushed an agenda on the public to be docile consumers, who watch television an average of 4–5 hours or more per day in industrialized nations (Kivel, 2004). Modern entertainment and communication technologies "grant us unprecedented powers to associate with whom we want, when we want, to the degree we want, under the terms we want, finessing and filtering out those we don't want—and thin out the possibilities of social growth accordingly" (Brende, 2004, p. 80).

The "technology is just a neutral tool" myth is dangerous for several reasons. First, by making technology appear neutral and objective, the fact that technology can be used to exploit and control people is overlooked (Proctor, 1991). Second, when the user can be conveniently blamed, technology and the developers of it avoid criticism. Third, the public is removed from decision-making about technology because there are no decisions to be made if something is simply neutral and objective. Fourth, the profit motive of technology can be conveniently overlooked if technology is portrayed as neutral and value-free (Huesemann & Huesemann, 2011).

11. *Profit is often a motive of technology.* Technology's focus on utility and profit is in stark contrast to basic science (Bunge, 2003). The *Framework* conveys technology as seeking to "solve human problems" while ignoring the reality that many, if not most, technologies are entangled with issues of profits, markets, marketing, patents, and "consumers."

> We are told constantly that innovation occurs only in response to human needs, that the desires for improved technology are the desires of the great buying public. In my view, these statements are humbug. The desires are those of the engineers and scientists, ambitious to achieve ever more elegant solutions to self-imposed problems. The desires are also those of the entrepreneurs, eager to carve out a niche for themselves

> and make a good profit. The desires are those of the manufacturers, eager to stimulate new waves of purchase for new products when markets are saturated. (Braun, 1995, p. 190)

When the primary motive is one of profit, many questions need to be raised about whether such technologies that result are addressing legitimate human needs or simply supporting corporate interests. For example, how many consumer products have created and encouraged human over-consumption? Are human needs really being addressed, and to what extent are needs and wants fabricated or promoted by the technology or the corporations who will profit from it? How do earnings responsibilities to shareholders bias decisions toward profits, at the expense of other factors such as health, well-being, and the environment?

12. *The engineering design process is biased toward positive perspectives of technology while downplaying or ignoring other important considerations.* An important part of engineering design is to assess costs, benefits, and risks, and make decisions about whether and how to pursue the development of a particular technology. The basic underlying philosophical position used is the Principle of Utilitarianism.

> Utilitarianism offers a relatively straightforward method for deciding the morally right course of action for any particular situation we may find ourselves in. To discover what we ought to do in any situation, we first identify the various courses of action that we could perform. Second, we determine all of the foreseeable benefits and harms that would result from each course of action for everyone affected by the action. And third, we choose the course of action that provides the greatest benefits after the costs have been taken into account. (Andre & Velasquez, 1989)

This perspective is congruent with the descriptions of engineering design in the *Framework*. Utilitarianism is usually summarized as "the greatest good for the greatest number." We are so used to utilitarianism that we often are unaware of other positions, such as egalitarianism, the "common good," or a position based on rights. Utilitarianism has been criticized for its limitations and the resultant negative effects on minority groups, the environment, and future generations (e.g. Hadjilambrinos, 2000). The utilitarian position results in biases in several aspects of the engineering design process related to cost-benefit analysis and risk-benefit analysis. These biases are described in more detail below.

12a.*Cost-benefit analyses are biased toward short-term gains and easily measurable impacts due to the practice of discounting.* Fundamental to utilitarianism is the calculation of costs and benefits, because decision-making is based on the greatest good for the greatest number. Thus, impacts must be quantified or monetized. Because consequences develop and continue over time and cannot be accurately calculated, a process called discounting occurs in which future benefits and costs are reduced relative to those benefits and costs in the present. For example, the negative consequences of a technology such as a nuclear waste storage facility to individuals more than a few decades in the future are significantly discounted to negligible levels or even zero (Huesemann & Huesemann, 2011). This results in a positive bias that often greatly underestimates intangibles and future impacts.

12b. *Engineering design is inherently biased toward anthropocentrism and does not adequately address considerations of ecological justice.* Human wants and needs are a centerpiece of engineering design. Animals and nature are considered "resources" or commodities to be used by humans rather than entities that exist for some other reason than to serve humans. Because impacts on the environment are often long-term and difficult to determine (e.g., what is the "value" of an undiscovered species that may go extinct?), calculations of risks and costs cannot address the full cost to the ecosystem of species extinction, alterations of carbon and nitrogen cycles, etc., and thus, they are often downplayed or ignored. This results in a bias toward "present exploitation of opportunities and against the values of conservation for the future" (Ferre, 1995, p. 82).

12c. *Cost- and risk-benefit analyses use aggregating methods and by doing so, they "tend to be defective in concern for distributive justice"* (Ferre, 1995, p. 84). Calculations of costs, risks, and benefits are not designed to address ethical concerns that deal with issues of fairness, equity, and justice. "Cost-benefit analysis is not designed to pay attention to the ethically crucial question: 'Who pays the costs, who gets the benefit?'" (Ferre, 1995, p. 83). For example, if a power plant is to be built next to a residence, cost-benefit analysis determines the net good; this means that the thousands of more distant customers who benefit from the power plant outweigh the impacts to the resident next to the plant in the calculation. If the net benefit results in a positive number after the resident's projected increased medical costs and property devaluation are subtracted from the "greater good," the Principle of Utilitarianism has been met (the greatest good for the greatest number), but the Principle of Justice is ignored. Building the power plant while expecting the resident to suffer from the negative consequences simply due to an accident of geography is not fair, since in a just society all individuals should be considered equally deserving. Ferre points out that in a just world, at a minimum, the individual bearing the negative consequences should be compensated, "but the world approved by cost-benefit analysis is not a just one" (p. 84). Importantly, how can aesthetics, health, and life itself be accurately quantified and subsidized?

12d. *Cost- and risk-benefit analyses downplay or ignore intergenerational justice.* Intergenerational justice refers to the passing of consequences of technology decisions to future generations. Because future generations are not current people, they cannot vote or advocate for themselves. Yet many important technology decisions have been made for current benefits while the costs and risks are passed to future generations (Huesemann & Huesemann, 2011). For example, the proliferation and storage of nuclear waste, the extraction of nonrenewable natural resources, and the development of an automobile-based transportation system create very real consequences for future generations while present generations reap the benefits. Because many of the actual costs involved in living with the consequences cannot be determined, they are reduced through the process of discounting when performing cost-benefit analyses.

13. *Technology optimism is inversely proportional to our knowledge about the technology.* As new technologies are spread and adopted throughout society, both positive and negative impacts begin to emerge. Therefore, we tend to see most technological optimism when such technologies are new and relatively poorly understood (Huesemann & Huesemann, 2011). For instance, the internet has been widely adopted for some time and with much fanfare, yet we are only beginning to understand the powerful impacts on people, including adult depression and loneliness (Kraut *et al.*, 1998, in Louv, 2005), internet addiction (Louv, 2005), decreased reading, decreased written skills, and increased social problems in children (Bauerlein, 2008).
14. *Many technologies can be considered counter-technologies and social fixes. Counter-technologies and social fixes rarely, if ever, deliver on their promises but instead, they create additional problems.* Counter-technologies are technologies intended to offset or neutralize the negative effects caused by other technologies. Environmental technologies such as those designed to remove pollution (e.g., wastewater treatment plants, carbon dioxide sequestration proposals) and many military technologies (e.g., nuclear weapons, anti-aircraft missiles) are examples of counter-technologies. These technologies are intended to address the negative consequences of other technologies, yet while doing so, they also create unintended consequences. For instance, attempts to remove pollutants often use strategies of concentration (thus reducing environmental dispersal, such as a landfill or nuclear waste containment), dilution (spreading the contaminant broadly via smokestacks or other means), transference (e.g., placing the contaminant in the air rather than in groundwater), incineration, and bioremediation (e.g., using bacteria to metabolize the targeted pollutant) (Huesemann & Huesemann, 2011). Each of these strategies has severe limitations. These pollutants go somewhere. When pollutants are concentrated, we must create additional technologies to monitor, contain, and prevent their accidental spread. When pollutants are dispersed, we simply pass the problem elsewhere in the ecosystem where we face more difficulties in removing such pollutants from the environment, and we know less about the long term impacts. (For example, moving pollutants into the air has consequences we often do not consider, and we have not removed DDT from the environment.) Even the process of bioremediation can create toxic byproducts (Huesemann & Huesemann, 2011). The result of our decision-making about counter-technologies is what Tenner (1996) terms the "Rearranging Effect"—we consider counter-technologies effective because we view the short term consequences while we are unable to foresee the risk being transferred in space and time. This is either because we do

not fully know what the risk is, or we have little concern about future or distant consequences.

Social fixes are technologies designed to solve social, cultural, economic, and political problems. A familiar example of a social fix is the attempt to create weight loss medicines that people will gladly take rather than address the underlying problem of inappropriate eating habits and inactivity.

A fundamental problem of counter-technologies and social fixes is that they both fail to address what created the problem in the first place. Thus, technologies that have their own set of unintended consequences are added to the consequences of other technologies and to societal problems. We may benefit in the short-term by masking the symptoms, but the underlying problems are overlooked (allowing them to continue), and risk and consequences are often passed elsewhere.

> The weakness of the technological fix, however, is not its short-term efficacy but its seductiveness, the temptation to see technology as an autonomous reality that can be manipulated outside the system that produced it, making it unnecessary for people and societies to change their behavior and reorder priorities. Technological fixes can and do encourage undisciplined consumption and waste, putting off a socially and economically disruptive day of reckoning. (Stunkel & Sarsar, 1994, p. 82)

Ellul (1990) refers to this as "the law that problems grow with the growth of techniques" (p. 50).

15. *Technological systems tend toward increased centralization, standardization, and a loss of privacy, shared control, personal dignity, and freedom.* Technology optimists often tout the increased freedom that will be gained with the adoption of new technologies. Individuals will be able to choose where to gain their information, when and where to travel, or what music they wish to listen to. However, systems become nationally and globally connected and interdependent (Braun, 1995). The freedom promised by the distribution of the automobile turned into the opposite: we now have a vast network existing only to support the car: oil refineries, pipelines and tankers, millions of miles of roads, gasoline stations, repair shops, and the auto insurance industry. Those living in the suburbs are dependent upon the car for carrying out almost all daily activities outside the home. Families often need both adults to have employment outside the home in order to afford the cost of the car, insurance, gasoline, and repairs, a situation that often leads to additional costs such as childcare and the loss of parental time spent raising children (Brende, 2004). This is in addition to social isolation, reduced quality of life in cities, air pollution, and other impacts. Our individual freedom of mobility is a benefit; the costs are many, and even restrict other freedoms.

The proliferation of traffic cameras, computer information gathering, the permanency of anything typed on a computer (even after "deleting" it), and the use of cell phone records and even recordings in the court system are just a few of the losses in privacy due to modern technologies (Sassower, 1990). The founder of Sun Microsystems, Scott McNeely is quoted as saying, "You have zero privacy. Get over it" (Manes, 2000, p. 312). Is the loss of privacy simply something we should "get over" because laws regarding personal privacy have not kept pace with the capabilities of modern technology?

Solar and wind generation of electricity are optimally suited to be used on a local basis (Schumacher, 1997). However, Huesemann and Huesemann (2011) note that because the engineering of these technologies is inextricably connected to profit motives and the need for research and development investment capital to be offset by profits from a successful product, not surprising is the use of vast amounts of land for centralized wind farms and solar panel arrays. The result of these grand scale operations is to maintain centralized control over energy production and distribution, and continue consumer dependence upon corporations.

Thus, despite the promises of freedoms, technology's consequences tend to create restrictions on other freedoms, in often unforeseen ways.

REFLECTIONS ON THE *FRAMEWORK*

Broadly speaking, the *Framework for Science Education* (NRC, 2012) ignores or downplays central issues in the philosophy of science and engineering, and replaces both with an emphasis on *practices*. Students are expected to learn what scientists and engineers do. These sociological practices are reductionistically divided into specific skills, such as *develop models, make predictive inferences*, and *test hypotheses*. According to the *Framework*, "The idea of science *as a set of practices* has emerged from the work of historians, philosophers, psychologists, and sociologists over the past 60 years" (NRC, 2012, p. 43, italics added). This statement is problematic on many levels. Any rudimentary examination of the philosophy of science indicates that issues of ontology and the fundamental assumptions of science are an inherent part of the discipline, not simply practices. The psychology of science deals with issues that go beyond practices, such as inherent bias in observations. Science as a set of practices is not a complete portrayal of what is represented in the research on the nature of science in science education over the last several decades, either. As McComas and I noted in 1998, international standards documents concur that the nature of science for K-12 students includes issues related to the philosophy and psychology of science—issues that examine the ontological and epistemological basis of science. The field is in agreement that students deserve to know, for example, that observations are theory-laden, that scientific knowledge has a tentative yet durable character, and that theories are not tentative guesses nor do they "grow up" into laws. As reflected in the *Framework*, both science and engineering are portrayed as if these issues are unimportant—that students only need to know and engage in the "practices" that scientists and engineers do to reach conclusions. That these practices, and even the conclusions reached, might have biases, limitations, or rest on particular assumptions is apparently considered irrelevant for U.S. students by the writers of the *Framework*. This overarching focus on practices is dangerous. "Some of the major disasters of mankind have been produced by the narrowness of men with a good methodology" (Whitehead, 1929, p. 8).

Consider the following statements from the *Framework* and how they misportray the nature of technology.

> The engineering design process begins with the identification of a problem to solve and the specification of clear goals, or criteria, that the final product or system must meet. Criteria, which typically reflect the needs of the expected end-user of a technology or process, address such things as how the product or system will function (what job it will perform and how), its durability, and its cost. *Criteria should be quantifiable whenever possible* and stated so that one can tell if a given design meets them. (p. 204, italics added)

> By the end of grade 12: ...Criteria and constraints also include satisfying any *requirements set by society*, such as *taking issues of risk mitigation into account*, and they should be *quantified to the extent possible* and *stated in such a way that one can tell if a given design meets them*. Humanity faces major global challenges today, such as the need for supplies of clean water and food or for energy sources that minimize pollution, *which can be addressed through engineering*. These global challenges also may have manifestations in local communities. But whatever the scale, the first thing that engineers do is define the problem and specify the criteria and constraints for potential solutions. (p. 205, italics added)

The bias toward quantifiable criteria and risk is an example of how consequences of an intangible, long-term, environmental, and/or ethical nature are downplayed or ignored. How quantifiable are species extinction, quality of life, illness, death, natural resource depletion, or even changes to family interactions or working conditions? These issues are not raised with students. The assumption is made that engineers simply quantify the risks and work toward a solution. The standard also makes the claim that society has set requirements for constraints, yet Marchant (2011) and others have argued that society, and the legal system in particular, is unable to keep up with the ethical and legal issues raised by the exponential growth of technology, creating a vast and growing gap between emerging technologies and the law. It is impossible to follow mandates that don't exist, and thus, engineers themselves need to deeply account for ethics and issues of justice—yet these are not adequately addressed because of the limitations of the engineering design process (Ferre, 1995), nor are ethical, legal, and environmental issues sufficiently described for students in the *Framework*.

In this engineering description and standard, "major global challenges" are portrayed as being solvable by engineering. Our inadequate supply of clean water and increasing energy needs are consequences of other technologies that have produced pollution and enabled unprecedented population growth. Thus, the engineering "solutions" that will be employed to solve these problems are counter-technologies and social fixes, which have historically created more problems than they have solved. Keep in mind that counter-technologies and social fixes do not address the initial cause of the problems, but they try to mitigate the negative consequences that result. In doing so, the initial cause of the problems continues, while the counter-technology and/or social fix creates its own set of unintended consequences. As noted by Dubos (1959),

> Developing counter-technologies to correct the new kinds of damage constantly being created by technological innovations is a policy of despair. If we follow this course we shall increasingly behave like hunted creatures, fleeing from one protective device to another, each more costly, more complex, and more undependable than the one before. (in Ehrenfeld, 1981, p. 108)

Perhaps the strongest case made in the *Framework* (2012) that technology has consequences is found on page 214:

> By the end of grade 12: Modern civilization depends on major technological systems, including those related to agriculture, health, water, energy, transportation, manufacturing, construction, and communications. Engineers continuously modify these technological systems by applying scientific knowledge and engineering design practices to increase benefits while decreasing costs and risks. Widespread adoption of technological innovations often depends on market forces or other societal demands, but it may also be subject to evaluation by scientists and engineers and to eventual government regulation. New technologies can have deep impacts on society and the environment, including some that were not anticipated or that may build up over time to a level that requires attention or mitigation. Analysis of costs, environmental impacts, and risks, as well as of expected benefits, is a critical aspect of decisions about technology use.

Many points are worth noting regarding this standard. First, the myth of efficiency ("increase benefits while decreasing costs and risks") is conveyed to students, as if we can simply design technologies to avoid negative consequences. Second, unintended consequences are addressed as a minor afterthought ("including *some* that were not anticipated") and are then described as being addressed by possibly requiring "attention" or "mitigation." The notion that we can engineer our way out of the consequences of technologies is present here, and misleading. Many of the greatest challenges that modern humans must face are consequences of our technologies: global climate change, pollution, species extinction, and depletion of natural resources. Proposed engineering "fixes" to these consequences further disturb natural cycles and create high costs and risks, and counter-technologies have a strong history of lackluster results. Finally, the myth of the technological imperative is present in this standard. This myth asserts that individuals are helpless—that the genie is out of the bottle, and therefore, whatever can be done technologically, should be done. The only individuals involved in making decisions about technology in the standard listed above are engineers, scientists, and the government. Do we really intend to convey to children that they have no voice in technological matters and that the engineers, scientists, and government will make the best decisions for us?

The answer to this question depends on the metaphor and narrative for schooling. If schools are factories to produce workers, then moving the responsibility for decision-making to engineers, scientists, and the government is logical. However, simply because a position is logical does not make it wise. Given our track record of exploiting the environment and other people, depleting nonrenewable natural resources, and unleashing technologies such as DDT and leaded gasoline that continue to pollute the environment and the organisms in it, we must question how well ethical and complex environmental issues have been considered in our current technological decision-making process. Marcuse (1964) noted that when the public is ignorant of the consequences of technology and remains passively engaged in stupefying work and entertainment, the "technological veil conceals the reproduction of inequality" (p. 32), leaving those who control the technologies free to continue unquestioned exploitative or unjust practices. Marcuse has asserted that it is only through educating the public that we stand a chance to make technological decisions that are more equitable and sustainable.

THE NATURE OF TECHNOLOGY FOR A LIBERAL EDUCATION

Given the very real consequences of modern technologies that impact all of us and future generations, denying students access to the nature of technology, including its assumptions, limitations, and consequences, is nothing short of educational malpractice. However, given that the Common Core standards initiative makes overt claims that job preparation is its paramount goal, rather than education, perhaps we have in fact reached the End of Education as posited by Postman (1995).

If we have reached the End of Education, a rather bleak future likely awaits. We will have become a version of what Huxley (1932) predicted in *Brave New World*—a society where individuals are little more than workers in an economic machine, tracked early in life and destined for a narrowly-defined future. Our method to reach this state, however, is far more like Bradbury's (1953) description in *Fahrenheit 451*—rather than some external government or corporation forcing its wishes upon us, we have passively done this to ourselves. We happily consume material goods, blissfully ignoring the consequences that others bear for our overly-consumptive lifestyle. As long as we do not see the impacts on the environment, other people (including future generations), or even the impacts on ourselves, our desires for entertainment can conveniently continue while we pursue and identify ourselves with our jobs. Consider this description of modern society made by Marcuse in 1964:

> The distinguishing feature of advanced industrial society is its effective suffocation of those needs which demand liberation—liberation also from

> that which is tolerable and rewarding and comfortable—while it sustains and absolves the destructive power and repressive function of the affluent society. Here, the social controls exact the overwhelming need for the production and consumption of waste; the need for stupefying work where it is no longer a real necessity; the need for modes of relaxation which soothe and prolong this stupefication; the need for maintaining such deceptive liberties as free competition at administered prices, a free press which censors itself, free choice between brands and gadgets. (p. 7)

Ferre (1995) summarizes Marcuse's position as follows: "When everything present is affirmed, when everyone is 'happy,' then imagination itself is crippled in its power to take account of the absent, to long for what *is not*" (p. 71). As written, the *Framework* conveys a happy world of engineers solving our problems with technology solutions—distant experts who have our needs and wants in mind with no need for us to question any of it or provide input. The student is thus expected to become a complacent consumer of the products of engineering, while consequences are mentioned only in passing and conveyed as fixable with simple "mitigation." The engineer is conveyed as seeking efficient solutions, with risks being quantified and compared to benefits in a rather straightforward fashion.

We must ask why students are denied a more complete education regarding technology and engineering. Why do we fail to address the assumptions of technology, the unintended consequences of all technologies, profit motives, the biases and values inherent in technologies, the passing of consequences to future generations, the limitations of counter-technologies, the impacts of natural resource extraction methods (and consequences of them), and the limitations of our methods of waste disposal?

> There is a tendency to divide up the danger.... This is a technocratic mistake. We must look at the ecological question in its entirety, with all the interactions and implications, without reductionism. We then see that the problem raised is a thousand times more vast and complex.... The danger is so great that people prefer to ignore it. After a period of awareness between 1955 and 1970, the public has lost interest and governments do their best to deny the danger. Those who have really studied the situation regard the danger as such that immediate measures are necessary on a global scale if we are to restore ecological balance to our environment, since we are dealing with an ecology that is now socio-agro-industrial. (Ellul, 1990, p. 51)

Why do we fail to address the impacts of technology on the intangible aspects of human life and dignity? The technological bias toward efficiency has spread via the factory metaphor from the assembly line to the school. Schools are supposed to be an efficient use of taxpayer "investment" with narrowly defined "outcomes" that fail to consider unquantifiable aspects of our existence, such as art, ethics, faith, creativity, liberty, equity, justice, and aesthetics.

> In a world dominated by efficiency, each development would serve only narrow and practical purposes. Beauty, creativity, fantasy, enjoyment, inspiration, and poetry would fall by the wayside, creating an unappealing world indeed. (McDonough & Braungart, 2002, p.65; in Huesemann & Huesemann, 2011, p. 114)

If we expect to live sustainably on a planet with limited resources, we need educated citizens who know about these issues and can work toward solutions—seen not solely through limited technological fixes. These issues should be an essential part of a science and technology education. "We cannot afford to ignore the nature of technology, let alone despise it, if we want to gain full control over technology in order to check its dark side" (Bunge, 2003, p. 181), and wisely use technology. This education must situate itself between the unproductive extremes of technophilia (promoting unquestioned consumership of technologies and unwarranted belief in technological progress) and technophobia (promoting fear and disengagement) and focus on making technology an object of inquiry, including our gains and losses (Sassower, 1990).

When the public is ignorant of what is lost because of the technological decisions being made, the ability to exercise reason and make changes is suppressed (Marcuse, 1964). Yet the free exercise of reason is perhaps the only way out of our present dilemmas. In summarizing Marcuse, Ferre (1995) notes:

> If the totalitarianism of modern technological society is to be fought, it will first of all require the re-stimulation of the lost dimension of imagination and negative critique. The gadfly Reason of Plato will need to be released on all domains of life. Then the machine could be recaptured for the fulfillment of the human rather than *vice versa.* (p. 72)

This sentiment continues in discussions of educational philosophy in the present day:

> Such a world is not the place for simplistic solutions; now is not the time for simple-mindedness. Yet many people, in coping with such issues, grasping for common sense, searching for anchoring, find them attractive. Some act as though ethical commitment requires closed-mindedness: ignorance, willful and proud, is celebrated.... Technological advances and sophisticated consumer information tracking have made it possible for people to stay comfortably cocooned in their prejudices, encountering only the preferred (which is to say the familiar and self-reinforcing) music and entertainment, products, and even news. What is needed is an education for complexity, for globalization, for autonomy, for self-reflection and continuing self-development, for dialogue with other viewpoints, for critical reason and informed judgment, for wholeness and integrity. What is needed is an education that opens us to the possible as well as the actual and the necessary. What is needed is a liberal education. (DeNicola, 2012, p. 245)

If we wish to change our current course, a better curriculum in science education that addresses the nature of technology is necessary, but insufficient. We must also make the narrative of Economic Utility and the metaphor of the factory into objects of inquiry and critique. Only when this metaphor and narrative are questioned and changed will other solutions appear plausible. We must make more clear what education, as opposed to training, involves, and why the development of reason must assume a place of primacy. When the narrative changes, as Postman noted, employment can be seen as a by-product of a good education, rather than the sole goal.

Now, more than ever, we need a liberally educated public that understands the Faustian bargain of technology. Technology makes positive promises; but it delivers both benefits and negative consequences. This is not a negative position; it is simply an informed one that acknowledges the multiple facets of technology. The nature of technology as conveyed in the *Framework* ignores or significantly downplays fundamental issues in the philosophy of technology that are necessary for an educated public. The result of the current *Framework* will be students who are trained to admire technology while being utterly ignorant of the serious consequences that result from it. Yet these students will have to live with the consequences that resulted from decisions made by our generation and those before us, and trying to solve technological problems with more technologies is hardly a complete and sufficient answer. To solve these problems, they will need an education in ethics, the philosophy of technology, ecology, and complex geological and biological systems, to name a few (Ellul, 1990). Rather than simply becoming motivated to pursue engineering careers (a stated goal of the *Framework*), we need an educated public that may have to take greater control over engineering decision-making, put a halt to some of our current practices, change its utilitarian biases, and make difficult decisions about our current course of action.

Instead, sadly we are viewing students as individual workers who will fit into a job at some point in the future, adding to the economic machine. Such a narrow training perspective dehumanizes the individual, replacing education with training and stupefying the very people that we need to help us take a different look at our unsustainable technological path.

> But do we have to stop teaching the rudiments to children? If we reduce these rudiments to useful, technical knowledge (which does not develop the intelligence but is essential for entry into this society), we have demented programs that simply serve to crush children's personalities and sensibilities. (Ellul, 1990, p. 57)

Our children deserve better than to be treated as trained employees and ignorant consumers of technologies foisted upon them by those who desire a profit. Understanding the nature of technology and engineering, including its very real, but less pleasant side, should be a crucial part of a modern education.

My attempt here is to begin the conversation about these issues. Much work needs to be done to determine what nature of technology concepts are appropriate at what age and how we can teach students such issues without displacing or downplaying other subject matters of importance in a liberal education, or creating a sense of utter hopelessness. Ignoring these issues is not an option.

> As civilized human beings, we are the inheritors, neither of an inquiry about ourselves and the world, nor of an accumulating body of information, but of a conversation, begun in the primeval forests and extended and made more articulate in the course of centuries. It is a conversation that goes on both in public and within each of ourselves.... Conversation is not an enterprise designed to yield an intrinsic profit, a contest where a winner gets a prize, nor is it an activity of exegesis; it is an unrehearsed intellectual adventure....
>
> ...Education, properly speaking, is an initiation into the skill and partnership of this conversation in which we learn to recognize the voices, to distinguish the proper occasions of utterance, and in which we acquire the intellectual and moral habits appropriate to conversation. And it is this conversation which, in the end, gives place and character to every human activity and utterance. (Oakeshott, 1962, p. 199)

When our vision for children and their future is broadened to account for their humanity rather than simply their future employment potential, perhaps we can provide children like my former student Marcus a compelling reason to be in school and a meaningful role in the conversation that must occur to solve our most pressing problems. We need more from this next generation than high test scores, because as Einstein noted, "No problem can be solved at the same level of consciousness that created it." We need a broader, systems perspective to complex challenges that can only result when we develop the ability to reason, think critically, and exercise creativity, and when we prepare our children to more fully understand and participate in the world around them. Such a perspective must include the nature of technology, as our very future may be dependent upon that understanding.

REFERENCES

Adler, M. J., & Gorman, W. (1952). The Great ideas: A Synopticon of great books of the western world. In Hutchins, R. M. (Ed.) (1952). *Great books of the western world.* Volume I. Chicago: Encyclopedia Britannica.

Andre, C., & Velasquez, M. (1989). Calculating consequences: The utilitarian approach to ethics. *Issues in Ethics, 2*(1), Accessed December 21, 2012 from http://www.scu.edu/ethics/publications/iie/v2n1/calculating.html

Arthur, W. B. (2009). *The nature of technology: What it is and how it evolves.* New York: Free Press.

Bauerlein, M. (2008). *The dumbest generation: How the digital age stupefies young Americans and jeopardizes our future.* New York: Penguin Group.

Bracey, G. W. (2003). *On the death of childhood and the destruction of public schools.* Portsmouth, NH: Heinemann.

Bradbury, R. (1953). *Fahrenheit 451.* New York: Ballantine Books.

Braun, E. (1995). *Futile progress—Technology's empty promise.* London, UK: Earthscan Publications Ltd.

Brende, E. (2004). *Better off: Flipping the switch on technology.* New York: HarperCollins.

Bunge, M. (2003). Philosophical inputs and outputs of technology. In R. C. Scharff & V. Dusek (Eds.). *Philosophy of technology: The technological condition* (pp. 172–181). Malden, MA: Blackwell Publishing Ltd. Originally appeared in G. Bugliarello & D. B. Doner (Eds.) (1979). *The history and philosophy of technology* (pp. 262–281). Urbana: University of Illinois Press.

Carroll, L. (1865). *Alice's adventures in Wonderland.* In Griffith, J. W., & Frey, C. H. (Eds.) (1987). *Classics of children's literature,* pp. 487–542. New York: Macmillan, Inc.

CCSSO (2012). *Common Core State Standards Initiative*. Council of Chief State School Officers. Accessed November 12, 2012 from www.corestandards.org.
Commoner, B. (1971). *The closing circle—nature, man, and technology*. New York: Alfred A. Knopf.
Cotgrove, S. (1982). *Catastrophe or cornucopia*. New York: John Wiley and Sons.
Cuban, L. (1993). Computers meet classroom: Classroom wins. *Teachers College Record, 95*(2), 185–210.
DeNicola, D. R. (2012). *Learning to flourish: A philosophical exploration of liberal education*. New York: Continuum International.
Dubos, R. (1959). *Mirage of health—Utopias, progress, and biological change*. New York: Harper and Brothers.
Ehrenfeld, D. (1981). *The arrogance of humanism*. Oxford, UK: Oxford University Press.
Eisner, E. W. (2002). *The educational imagination: On the design and evaluation of school programs*, 3rd Ed. Upper Saddle River, NJ: Pearson Education, Inc.
Eisner, E. W. (2001). What does it mean to say a school is doing well? *Phi Delta Kappan, 82*(5), 367–72.
Ellul, J. (1990). *The technological bluff*. Translated by Geoffrey W. Bromiley. Grand Rapids, MI: William B. Eerdmans Publishing Co.
Ellul, J. (1976). *The technological society*. New York: Alfred A. Knopf.
Ferre, F. (1995). *Philosophy of technology*. Athens, GA: University of Georgia Press.
Graedel, T. E., & Allenby, B. R. (1998). *Industrial ecology and the automobile*. Upper Saddle River, NJ: Prentice Hall.
Hadjilambrinos, C. (2000). An egalitarian response to utilitarian analysis of long-lived pollution: The case of high-level radioactive waste. *Environmental Ethics, 22*(1), 43–62.
Hardin, M. (2009). Obama's speech to students: Work hard in school for a good career. *Denver Business Journal*. Accessed September 8, 2009 from www.bizjournals.com/denverstories/2009/09/07/daily4.html
Healey, J. M. (2004). Early television exposure and subsequent attention problems in children. *Pediatrics, 113*(4), 917–918.
Huesemann, M., & Huesemann, J. (2011). *TechNo-Fix: Why technology won't save us or the environment*. Gabriola Island, BC, Canada: New Society Publishers.
Huxley, A. (1932). *Brave new world*. New York: Harper & Row.
Johnson, J. T. (1984). *Can modern war be just?* Westford, MA: Murray Printing Company.
Kegan, R. (1994). *In over our heads: The mental demands of modern life*. Boston, MA: Harvard University Press.
Kivel, P. (2004). *You call this a democracy? Who benefits, who pays, and who really decides*. New York: The Apex Press.
Kohn, A. (2011a). Teaching strategies that work! But for what? *The Washington Post Answer Sheet*. Accessed August 10, 2011 from http://www.washingtonpost.com/blogs/answer-sheet/post/teaching-strategies-that-work-but-for-what/2011/08/10/gIQAw3Yt6I_blog.html?wprss=answer-sheet
Kohn, A. (2011b). The case against grades. *Educational Leadership, 69*(3), 28–33.
Kotkin, J. (2012). Generation screwed. *Newsweek*. Accessed November 1, 2012 from http://magazine-directory.com/Newsweek.htm
Kraut, R., Lundmark, V., Patterson, M., Kiesler, S., Mukopadhyay, T., & Scherlis, W. (1998). Internet paradox: A social technology that reduces social involvement and psychological well-being? *American Psychologist, 53*(9), 1017–1031.
Lain Kennedy, J. (2008). Careers now: How many career changes in a lifetime? *Tribune Media Services*. Accessed November 12, 2012 from http://www.starnewsonline.com/article/20080608/CAREERS03/709742305
Leyden, M. B. (1984). You graduate more criminals than scientists. *The Science Teacher, 51*(3), 26–30.
Louv, R. (2005). *Last child in the woods: Saving our children from nature deficit disorder*. Chapel Hill, NC: Algonquin Books.
Maehr, M.L., & Midgley, C. (1996). *Transforming school cultures*. Boulder, CO: Westview.
Mander, J. (1991). *In the absence of the sacred*. San Francisco, CA: Sierra Club Books.
Mander, J. (1978). *Four arguments for the elimination of television*. New York: HarperCollins.
Manes, S. (2000). Private lives? Not ours. *PC World, 18*(6), 312.
Marchant, G. E. (2011). The growing gap between emerging technologies and the law. In Marchant, G. E., Allenby, B. R., & Heckert, J. R. (Eds.). *The growing gap between emerging technologies and*

legal-ethical oversight: The pacing problem (pp. 19–33). The International Library of Ethics, Law, and Technology. Volume 7.

Marcuse, H. (1964). *One-dimensional man: Studies in the ideology of advanced industrial society*. Boston, MA: Beacon Press.

McComas, W. F. & Olson, J. K. (1998). The nature of science in international standards documents. In W. F. McComas (Ed.). *The nature of science in science education: Rationales and strategies* (pp. 41–52). The Netherlands: Kluwer Academic Publishers.

McDonough, W., & Braungart, M. (2002). *Cradle to cradle—Remaking the way we make things*. New York: North Point Press.

McKinley Jr., J. C., & Dillon, S. (2009). Some parents oppose Obama school speech. *The New York Times*. Retrieved September 4, 2009 from www.nytimes.com/2009/09/04/us/04school.html

Myers, N. & Kent, J. (2001). *Perverse subsidies—How tax dollars can undercut the environment and the economy*. Washington, DC: Island Press.

National Research Council (2012). *A framework for K-12 science education. Washington,* DC: National Academies Press.

Noer, M. (2012, November 19). One man, one computer, 10 million students: How Khan Academy is reinventing education. *Forbes*. Accessed December 21, 2012 from http://www.forbes.com/sites/michaelnoer/2012/11/02/one-man-one-computer-10-million-students-how-khan-academy-is-reinventing-education/

Oakeshott, M. (1962). The voice of poetry in the conversation of mankind. In *Rationalism in politics and other essays*. London: Methuen.

Orians, G., & Heerwagen, J. (1992). Evolved responses to landscapes. In J. Barkow, L. Cosmides, & J. Tooby (Eds.). *The adapted mind: Evolutionary psychology and the generation of culture, 7*(1), 555–579. Oxford, UK: Oxford University Press.

Postman, N. (1995). *The End of Education*. New York: Vintage Books.

Proctor, R. N. (1991). *Value-free science? Purity and power in modern knowledge*. Cambridge, MA: Harvard University Press.

Pruitt, S. (2012). Next generation science standards—Our standards, our opportunity. *NSTA Reports, 24*(4), 3. Arlington, VA: National Science Teachers Association.

Rubin, R. S., & Dierdorff, E. C. (2009). How relevant is the MBA? Assessing the alignment of required curricula and required management competencies. *Academy of Management Learning & Education, 8*(2), 208–224.

Sassower, R. (1990). The alarmist view of technology. In P. T. Durbin (Ed.). *Broad and narrow interpretations of philosophy of technology* (pp. 167–180). Dordrecht, The Netherlands: Kluwer Academic Publishers.

Scharff, R. C., & Dusek, V. (2003). *The philosophy of technology: The technological condition*. Oxford, UK: Blackwell.

Schumacher, E. F. (1997). *This I Believe and Other Essays*. Dartington, UK: Green Books Ltd.

Smith, A. (1776). An inquiry into the nature and causes of the wealth of nations. Republished in Hutchins, R. M. (Ed.) (1952). *Great books of the western world. Adam Smith.* Volume 39. Chicago: Encyclopedia Britannica.

Stunkel, K. R., & Sarsar, S. (1994). *Ideological values and technology in political life*. Lanham, MD: University Press of America.

Tenner, E. (1996). *Why things bite back—Technology and the revenge of unintended consequences*. New York: Alfred Knopf.

Vukowich, W. T. (1971). The lack of practical training in law schools: Criticisms, causes, and programs for change. *Case Western Reserve Law Review, 23*, 140–152.

Whitehead, A. N. (1929). *The function of reason*. Princeton, NJ: Princeton University Press.

Joanne K. Olson
School of Education,
Iowa State University

CHAPTER 13

DALE S. NIEDERHAUSER

LEARNING *FROM* TECHNOLOGY OR LEARNING *WITH* TECHNOLOGY

Theoretical Perspectives on the Nature of Using Technology in the Classroom

> ". . . with teaching machines and programmed instruction one could teach what is now taught in American schools in half the time with half the effort."
> –B. F. Skinner

> "We focus on technology and ask whether its use is improving student achievement, but it is educational practices that determine how well students learn, and technology is not a process but a tool through which educational practices are mediated."
> –Steve Rappaport

INTRODUCTION

When modern *instructional technologies*[1] became widely available and affordable in the 1980s and 90s, visionaries and pundits claimed that these new tools would prompt a dramatic change in schooling practices and the nature of education in the United States (c.f., David, 1991; Papert, 1980; Sheingold, 1991; Skinner, 1984). However, the form that this change might take were open to a vigorous, and often contentious, debate. Some saw the change as a matter of degree—technology would help teachers more efficiently and effectively deliver the kinds of didactic and drill-and-practice-based forms of instruction that had become the norm in the first half of the 20th century (c.f., Skinner, 1984). Others saw the change as a matter of kind—technology would push teachers to fundamentally adapt their instructional practices in ways that were consistent with a more child-centered project-based instructional view (Sheingold, 1991). The purpose of this chapter is to examine some of the theoretical perspectives about learning that underlie technology use in schools.

In *The Computer in the School: Tutor, Tool, Tutee*, Taylor (1980) proposed three orientations that exemplified differing theoretical perspectives toward the

Michael P. Clough, Joanne K. Olson and Dale S. Niederhauser (Eds.), The Nature of Technology: Implications for Learning and Teaching, 249–268.

ways that instructional technologies might best be used with students. In the *tutor* mode, rooted in medieval Judaic and Christian traditions that teaching is telling (Cohen, 1990) and behaviorist conceptualizations of programmed instruction and operant conditioning (see Skinner, 1981), the role of instructional technology was conceptualized as a *teaching machine*. From this perspective, computers allowed the use of dynamic graphics and integrated instructional supports that prompted learners to more actively engage and interact as they learned, and be guided through the carefully structured instructional interaction presented through the machine. Performance data on learning (collected while students completed activities) was used to assess learning and provide data that could be used to determine an appropriately challenging sequence for presentation of content and activities. In this view knowledge is transferred from the teaching machine to the student and students learn skills through feedback-guided practice.

In *tool* mode, Taylor saw instructional technologies as devices that could be used to support teachers and students by transferring tasks of a tedious or mechanical kind to the computer (e.g., using a calculator to solve arithmetic problems, using a database to organize information, using word processers in writing workshops, etc.). In this view the technology supports the learner by accomplishing some of the more mundane tasks when he or she engages in educational activity. Jonassen and his colleagues (Jonassen, 1996; Jonassen & Carr, 2000; Jonassen & Reeves, 1996) expanded on Taylor's tool mode through the introduction of the *Mindtools* concept. Mindtools ". . . refer to technologies, tangible or intangible, that enhance the cognitive powers of human beings during thinking, problem solving, and learning" (Jonassen & Reeves, 1996, p. 693). Through Mindtools Jonassen extends Taylor's tool mode in terms of potential learning benefits associated with freeing up cognitive capacity through the use of technological tools (which I link to cognitive load theory in this chapter; see Sweller, 1988; Sweller, Van Merrienboer, & Paas, 1998), and situating the nature of those learning benefits in a Piagetian constructivist theoretical framework.

In the *tutee* mode, students teach (program) the computer. In so doing, proponents claim the child learns more deeply, and learns more about the process of learning, than he or she would through being tutored by software developed by others (diSessa, 1985; Papert, 1980, 1993; Resnick, 2007). Benefits include: (1) learners developing deep and connected understanding because the student has to understand the content such that he or she can create a representation of that content (e.g., teach it to the computer), and (2) teaching (programming) the computer to do something using the narrow capabilities of computer logic will help the learner develop *computational thinking*, while promoting a richer understanding of his or her own conceptual models and thought processes. While Taylor and his colleagues proposed computer programming as the vehicle for the tutee mode, Jonassen and his colleagues broadened this conception to include a variety of technology-based modeling applications.

From the tutee perspective the computer provides a virtual environment in which a learner's mental model can be translated into a working computer-based model;

and, using that working model, the learner (and others—teacher, peers, etc.) can examine, test and probe this external representation. This can benefit the learner in two basic ways that draw on what Piaget (1971) identified as fundamental learning processes: *organization* and *adaptation*. First, learners must organize their thoughts and ideas about a complex process, procedure, or principle into a coherent meaningful and internally consistent mental model. Striving to represent that mental model using the computer forces the learner to refine and reframe the mental model as he or she works within the constraints imposed by limitations of the hardware and software systems. Further, when the computer-based working model is complete, examining and probing the representation can reveal anomalies. When anomalies are encountered (i.e., the working model does not accurately reflect the mental model) the learner experiences what Piaget referred to as *disequilibration* which can prompt reorganization of the mental model through the process of *accommodation*—Piaget's key adaptive process associated with changing schemata—triggering what Jonassen and his colleagues, refer to as *meaningful learning* (Howland, Jonassen, & Marra, 2011).

These scholars have conceptualized models that illuminate roles for technology in enhancing student learning. In the following sections I will explain these ideas and examine how these perspectives provide theoretical grounding for ways that teachers use technology with their students.

TECHNOLOGY AS TUTOR: TEACHING MACHINES

From the tutor perspective, instructional technology-based materials were thought to be necessary so that the instructional process "could not be corrupted by 'the average teacher,' who was [viewed as] less competent . . . than a 'teaching machine.'" (Lagemann, 2000). Computer-based teaching machines were viewed as "a cost-effective albeit capital-intensive way of individualizing education," as a way to "simplify the extensive recordkeeping required for individualized instruction," and as a "salable skill" that provides a "strong vocational advantage" for those entering the workforce (Nesbitt, 1982, p.33). Examples that Nesbitt provided of the ways computers were being used in the classroom included:

1. Students in the Alaskan Bush study Alaskan history and English using microcomputers hooked up to television sets.
2. Harrisburg, Pennsylvania students use computers for instruction and to learn programming,
3. Neighborhood centers in Wilmington, Delaware offer after-school tutoring on computers hooked up to a data bank at the University of Delaware. (p.34)

These examples suggest a view of the role of instructional technology that seems rather naïve from 30 years away. Much of the writing at that time was grounded in an objectivist orientation toward teaching and learning. From an objectivist perspective one key role for instructional technology is to *present* or *transmit* information (e.g., about the Alaskan Bush, Alaskan history and English). The audio-visual aids provided by "microcomputers hooked up to television sets" can be used to display material for students, fulfilling one important function of the teacher and, when successful, make content so clear and interesting that students learn (Nesbitt, 1982). Those who hold this objectivist view assume that learning involves passive transmission of information to the learner (Clements & Battista, 1990).

From an objectivist perspective "The role of education is to help students learn about the real world. Students are not encouraged to make their own interpretations of what they perceive; it is the role of the teacher or the instruction to interpret events for them. Learners are told about the world and are expected to replicate its content and structure in their thinking." (Jonassen, 1991, p. 10). Educational media associated with these types of uses were typically developed by teams of educational technologists (instructional designers, media producers, media managers, etc.) working with subject-matter experts and teachers and using an instructional systems design model (cf. Dick & Carey, 1996; Gagne, Briggs and Wager, 1987) that is grounded in effective and efficient presentation of content in a hierarchically structured format. The other key objectivist position on the role of technology in the classroom goes beyond this passive information transmission view to promote a more interactive role for technology aimed at more active instructional interactions. This perspective is informed by behaviorist ideals associated with operant conditioning principles (see Skinner, 1958).

Leading figure in operant conditioning, B. F. Skinner advocated a more active role for the learner and cautioned against the widespread use of "equipment designed simply to present material," fearing that the student would become "more and more a mere passive receiver of information." (1968; p. 30) From Skinner's perspective, the essential role of the *teaching machine* (see Fry, Bryan & Rigney, 1960 for more on teaching machines) was to promote an efficient and effective *instructional interchange* between student and teacher (or teaching machine). Based on early devices developed for the automatic testing of intelligence and information retention initially developed by Pressey (1926), Skinner (1968, 1984) envisioned and designed mechanical teaching machines that drew on his conception of learning as the proper arranging of *reinforcement contingencies* around *programmed materials*. That is, the target information or skills to be learned must be analyzed and broken into component parts which can then be presented sequentially such that students' build up to the target skill. Each step in the process can be systematically *shaped* through a series of stimulus-response-reinforcement loops until the target behavior (performing the skill or retrieving the information from memory) becomes automatic. Skinner repeatedly demonstrated the efficacy of his methods with a variety of animals (primarily rats and pigeons) and worked to demonstrate that his success training these types of animals had direct implications for the teaching of schoolchildren.

As computers became more readily available in schools mechanical teaching machines gave way to newer more powerful computer-based instructional technologies that were well-suited to Skinnerian views on teaching and learning.

This modern version of the teaching machine allowed easy and efficient structuring of a productive instructional interchange. "The computer displays a problem for the student (stimulus) who, in turn, responds with an answer (response). The computer then provides feedback to the student regarding whether he or she has provided the 'right' answer (reinforcement)." (Niederhauser & Stoddart, 2001, p. 18). Through systematic presentation of programmed materials, and use of the stimulus-response-feedback loops for monitoring performance, computer-based teaching machines enabled an instructional interchange that provided:

1. constant interchange between program and the machine,
2. assurance that a given point is thoroughly understood before the student is allowed to move on,
3. exposure to just that material for which the student is ready—and asks the learner to take only that step which he is at the moment best equipped and most likely to take,
4. support for the student to arrive at the right answer through orderly construction of the programmed materials and using techniques like hinting, prompting, and suggesting that are derived from an analysis of verbal behavior, and,
5. immediate reinforcement for every correct response. (Skinner, 1968)

Ultimately, Skinner saw the potential for an instructional technology, which was grounded in the use of teaching machines based on operant conditioning principles (programmed instruction, stimulus-response-feedback loops, and schedules of reinforcement) that would be more efficient and effective than human teachers: "The simple fact is that, as a reinforcing mechanism, the teacher is out of date. This would be true even if a single teacher devoted all her time to a single child, but her inadequacy is multiplied many-fold when she must serve as a reinforcing device to many children at once." (Skinner, 1968, p. 22). He later claimed "With teaching machines and programmed instruction one could teach what is now taught in American schools in half the time with half the effort" (Skinner, 1984, p. 948). The tutor perspective was the driving metaphor behind much of the initial tutorial and drill-and-practice software developed in the 1980s, and the Integrated Learning Systems developed during the 1990s.

Current evidence of the tutor perspective can be found in a wide variety of drill-and-practice educational apps developed for the iPad, as well as instructional software developed for interactive whiteboards that mirror the "come to the blackboard and solve the problem" activities that have been present in US classrooms for the better part of the last century. Current evidence of presenting information can be seen in the vast amount of media (that can be broadly construed as educational) available through the World Wide Web—at a level probably unimaginable when Nesbitt provided his examples of educational computer use in 1982.

Thus, proponents of teaching machines have championed the potential for instructional technology to change schooling by making the existing instructional practices and methods more effective and efficient. Change based on a tutor-based view of instructional technology was more evolutionary in nature because use of technology did not challenge the traditional didactic instructional paradigm, it simply provided devices that made more traditional instructional practices more effective and efficient. However, instructional technology was also viewed as having the potential to promote a deeper, more revolutionary change in instruction and schooling by supporting a shift in teachers' perspectives and practices to reflect a more progressive student-centered view of curriculum and pedagogy.

TECHNOLOGY AS TOOL: MINDTOOLS FOR LEARNING

Taylor's conception of the tool mode was grounded in the idea that technological tools could be used to help learners accomplish the more mundane mechanical tasks that students encounter in many school-based activities. He saw the tool mode as the primary mode of computer use by those outside of education; however, the computing-in-education visionaries who contributed to his book did not tend to view tool use as central to their work: "All assume heavy use must be made in education of tool mode computing; none advocates it as most important or focuses his own major interest upon it. All five [authors] have advocated the use of the computer as a calculator and a word processor, and all have advocated various other tool uses as well." (Taylor, 1980, p. 9). Availability of technological tools was seen as necessary for learners engaged in both tutor and tool use, although "Use of the computer in tool mode may teach the user something during use, but any such teaching is most likely accidental and not the result of any design to teach." (p. 8)

Jonassen and his colleagues (c.f., Jonassen, 1996; Jonassen & Carr, 2000; Jonassen & Reeves, 1996) moved beyond the notion of using technology to offload tedious and mechanical tasks to propose that using technology as a Mindtool provides ways for using technology to prompt and support meaningful learning. The Mindtool perspective is grounded in the idea that learners actively construct knowledge—and in this essay I argue that these knowledge construction processes require the application of various forms of a finite amount of cognitive resources (see Miller, 1956; Sweller, 1988; Sweller, Van Merrienboer, & Paas, 1998). Using technological tools can help reduce the need to employ cognitive resources for completing routine tasks, thereby allowing learners to allocate freed up cognitive resources to develop deeper, more sophisticated understanding of the content. I will begin this section with an introduction to the Mindtools concept, then propose a theoretical rationale for the value of offloading tedious or mechanical tasks to technology that is grounded in cognitive load theory.

Mindtools

In developing the Mindtools construct Jonassen and his colleagues looked at using technologies in a way that focused more directly on student cognition, specifically on promoting higher order thinking. In this view instructional technologies are described as *cognitive tools*, or *Mindtools* ". . . that enhance the cognitive powers of human beings during thinking, problem-solving and learning" (Jonassen & Reeves, 1996, p. 693). Central to this idea is the notion that using these tools can promote active learner engagement in higher-order thinking as they classify information, analyze and break-down content, organize information in personally meaningful ways, and integrate and elaborate their conceptual understanding of the information at hand.

From this perspective, the definition of learning inherent in the tutor model—or, as Jonassen characterized it, learning *from* technology—is problematic because learners are not acknowledged as "active constructors of knowledge" (Duffy & Jonassen, 1992). From a *constructivist* perspective, one learns *with* technology. The learner uses technology as a tool for analyzing the world, accessing information, interpreting and organizing their personal knowledge, and, rather than simply *consuming* media (as in the tutor mode), *creating* media to express and represent what they know—thereby developing ". . . critical-thinking skills as authors, designers, and constructors of knowledge and learn more in the process than they do as the recipients of knowledge prepackaged in educational [media]" (Jonassen & Reeves, 1996, p. 713).

From a cognitive constructivist viewpoint (see Duffy & Jonassen, 1992; von Glasersfeld, 1995), technology can be used to support knowledge construction by helping learners interpret, integrate and organize the new objects and events that they encounter through their experiences into their existing knowledge structures. Using ". . . cognitive tools, and the goals, tasks, culture, resources, and human collaboration integral to their use, enable learners to engage in active, mindful, and purposeful interpretation and reflection" (Jonassen & Reeves, 1996, p. 695). Further, there is a powerful relationship between designing and learning. The belief being that students will benefit more from designing and producing media to reveal their own knowledge representations than they would as consumers of media designed by others. With appropriate pedagogy, curriculum, and learning environment; having students represent their knowledge with technology can promote reflective thinking (the kind of careful deliberate thinking that helps us make sense of what we have experienced) and support construction of new meaning as students mindfully engage in an "'intellectual partnership' between the learner's mind and various cognitive tools" (p. 696).

Contrary to the "effective and efficient" teaching machine model, through which, proponents claim, content can be easily learned by students; from the Mindtools perspective meaningful learning is not easy. Cognitive tools are not designed to make learning easier by providing a streamlined, optimized instructional interaction that makes learning easier, rather, cognitive tools promote and enable reflection that amplifies, extends and perhaps even prompts humans to reorganize their mental powers in ways that help learners construct their own knowledge base and complete challenging tasks—especially when they are pursuing investigations that are relevant to their own lives (Jonassen and Reeves, 1996). Rather than making learning easier, meaningful learning using technology-based tools is viewed as a difficult and complicated (though inherently worthwhile) process.

So, while teaching machines can be thought of as "intelligent" devices (because they are based on expert and student models that drive instructional decisions about what students need to know, how much and what kind of instruction learners need, the instructional sequence that will best benefit a given learner, and whether or not the student has learned the material); Mindtools rely on the learner to provide the intelligence (because Mindtools merely support the learner in representing his or her knowledge and understandings without making judgments about what the student should learn, how it should be learned, or whether the student has learned—responsibility which, falls on learners and teachers rather than programmed materials as was the case in the tutor mode). From a tool perspective the learner, rather than the technology, takes responsibility for key instructional elements like planning, decision-making, and self-regulation. The instructional technology serves as an intellectual partner, with learners assuming responsibility for tasks they do best, while allocating to the technology tasks for which it is better suited.

Offloading Extraneous Cognitive Load

Tasks that computers do well include mundane tedious computational and algorithmic processes. The basic idea is that technology can take responsibility for completing some of those tasks, allowing the learner to offload a portion of the effort required to complete that aspect of the activity so that he or she can focus on building higher-level conceptual understandings. Cognitive load theory provides insights into this process.

Cognitive load is a construct representing the burden that engaging in a learning task imposes on the cognitive system—particularly on *working memory* (Paas & VanMerrienboer, 1994; Sweller, 1988; Sweller, Chandler, Tierney, & Cooper, 1990; Sweller, Van Merrienboer, & Paas, 1998) The nature of working memory suggests that only a limited number of elements can be held at any given time; however, elements can be "chunked," such that information can be remembered regardless of whether the chunks are letters or words, single digits or larger numbers (Miller, 1956). Further, if the information has to be processed (organized, contrasted, compared, etc.), only two or three items of information can be dealt with simultaneously. Learners use working memory as a kind of processing space to mediate the integration of new information into existing knowledge. That is, they draw on *schemata* in long-term memory to make sense of incoming information, and incorporate that new information into existing schemata. Cognitive load falls into three categories: Intrinsic, extrinsic, and germane (Sweller, Van Merrienboer, & Paas, 1998).

Intrinsic load on working memory is associated with the nature of the material that is being learned (Sweller, 1994). Some material has low "element interactivity." For example, learning that Fe is the symbol for iron can be accomplished without learning that Pb is the symbol for lead. The elements that are learned do not interact and can be learned in isolation. Learning this type of material is associated with a low cognitive load—it does not require extensive use of working memory resources. Material that has high element interactivity, however, entails substantial allocation of working memory resources. High element interactivity is involved when the learner tries to understand complex processes, systems, or concepts (Sweller, 1994). For example, comparing and contrasting the relative benefits of different economic systems (like capitalism and Marxism) would involve holding several interacting elements in working memory at once.

Extraneous cognitive load is the demand placed on working memory due to things like the manner in which material is presented to learners, and/or the tasks and activities required of the learner (Sweller, Van Merrienboer, & Paas, 1998). For example, if a student were engaged in a learning activity and the software was new to him, learning to use the software would require use of some of the available cognitive resources. Further, cognitive processes become automated over time and put less load on the system. The proficient reader experiences less cognitive load from the act of reading than does a beginning reader who must purposefully decode each word (Reynolds, 2000). Extraneous cognitive load is, therefore, apparent in the effort that the learner expends through participating in the learning activity.

Finally, germane load is associated with cognitive effort related to the sense-making process. As was described in the previous section, when using technology as a Mindtool learners must take responsibility for their own learning—planning, decision-making, and self-regulation—while making meaning through assimilation, accommodation and equilibration. Germane load is the aspect of cognitive load that is associated with the construction and processing of schemata. Since cognitive load is additive in nature (the sum of intrinsic, extraneous, and germane load), increases in intrinsic load (arising from trying to learn more complex content) and extrinsic load (brought on by tasks that are unfamiliar, difficult, complex or confusing) tend to make learning more difficult (Sweller, Van Merrienboer, & Paas, 1998). Thus, Jonassen's conceptualization of Mindtools is consistent with cognitive load theory in that providing learners with tools for addressing tedious and mechanical tasks would reduce extrinsic load, and allow more cognitive resources to be allocated to the sense-making process.

Mindtools in the Classroom

Jonassen and Reeves (1996) provided examples of how the Mindtools perspective can be enacted in classrooms through the use of technologies like semantic networks, hypermedia and multimedia authoring tools, databases and spreadsheets, and expert systems. The first two examples, semantic networks and hypermedia/multimedia authoring tools, allow learners to represent their knowledge while providing a window that allows learner, peers, teachers, parents, etc. to view that knowledge representation, question and challenge it when appropriate, and modify it as learning progresses. Spreadsheets and databases provide tools for conceptualizing, organizing and analyzing subject-matter-content. Expert systems allow learners to model a process and to test the model to determine whether it performs as expected, or, perhaps reveals flaws in the model (and, therefore, in the learners conceptual model that was represented using the expert system), indicating a need to go back and re-conceptualize.

Semantic networks. Semantic networks, also known as concept maps, help learners organize and represent structural and conceptual knowledge. Software tools that were designed to help students construct concept maps, like Inspiration (Inspiration, 1988–2010) and SemNet (Fisher, 1990, 1992), provide a flexible environment in which learners develop a graphic representation of their understanding of complex ideas using a nodes and link structure that allows them to represent key concepts and the relationships among them. Knowledge and intelligence resides in the learner, and the technology serves as a tool that learners use to organize and represent their knowledge and understanding. Creating concept maps engages learners in analyzing the structural relationships among the content being studied, while the created map provides a concrete snapshot of the learner's knowledge at a given time. This concrete representation allows users to reflect on their knowledge, teachers to assess their students' knowledge, and provides a way to examine knowledge acquisition, or learning, when students construct multiple iterations of concept maps on a given topic that can be compared over time. Using these tools allows the learner to quickly and easily develop a graphical representation of their conceptual understanding, and modify and update that representation by adding, removing, and restructuring the representation as an ongoing process—allowing the user to focus attention on increasing sophistication of his or her understanding of the concept rather than getting bogged down in the drawing and redrawing of the representation.

Hypermedia and multimedia authoring tools. While concept maps can provide an outline of a learner's knowledge and understandings, hypermedia and multimedia authoring tools support users in designing and developing richer, more complex representations. Although hypermedia created by others can provide excellent resources, students can learn more by constructing hypermedia instructional materials than by studying hypermedia created by others. The value of hypermedia and multimedia construction is grounded in the idea of *knowledge as design* (Perkins, 1986), which shifts the instructional process away from the view of knowledge as information and the teacher as transmitter to one in which teachers and students are collaborators in the knowledge construction process (Jonassen & Reeves, 1996). Students define and refine the nature of the identified problem, reconstruct their knowledge-base to address the problem, and represent their solution using hypermedia (Lehrer, 1993). This entails the use and development of major cognitive skills including: Project management skills, research skills, organization and representation skills, presentation skills and reflection skills. These technologies provide learners with the tools to easily design and develop sophisticated multimedia representations, freeing them to focus on the content they are presenting rather than the mechanics of producing it.

Databases and spreadsheets. Databases and spreadsheets can also serve as cognitive tools. A database is a computerized record-keeping system in which information is structured such that the same type of information is organized and stored across a series of records. The key functionality of a database is that it allows users to search and sort the information in the database. For example, a database of census data for a given city might include information on number of residents, average household income, ethnic makeup, types of industry, etc. By organizing the information in the database, and providing ways to search and sort it, the user is able to easily and efficiently review and reorganize the data to identify patterns and relationships, thereby offloading some of the cognitive demands associated with accomplishing these tasks without the technology. Performing these analytical and research tasks help learners develop a deeper understanding of structural relationships within the content. The tool allows data to be searched and sorted to answer specific questions about the content, or to reorganize the data in ways that help the learner manipulate the dataset and make inferences about patterns and relationships among the data. The technological tool allows the learner to offload organizing, sorting and searching tasks and focus on the higher-order learning associated with making sense of patterns and relationships in the data.

Computer spreadsheets also provide opportunities for students to develop higher order thinking skills. Spreadsheets have three primary functions: storing, calculating, and presenting (typically numerical) information. Set up as a grid of columns and rows, the intersection of a particular row and column constitutes a cell. Values, formulae and functions can be entered into a cell and formulae and functions can reference data stored in other cells to perform mathematical and logical calculations. The information in the spreadsheet (both entered and calculated values) can then be presented through charts and graphs. Thus, like databases, the value of using spreadsheets as a Mindtool comes through both the design and use of the spreadsheet.

Spreadsheets are rule-using tools that require users to become rule-makers (Vockell & Van Deusen, 1989). When designing a spreadsheet, users identify patterns and relationships among the data and use rules (algorithms, functions, etc.) to mathematically model those patterns and relationships. In accomplishing this learners develop an understanding of the ways that mathematical models are used to represent content domains. Students begin to develop deeper understanding of calculations (antecedents and consequents) because they are focused on identifying and specifying the interrelationships among calculation components. Spreadsheet construction necessitates that leaners demonstrate all steps of problem solutions, showing the progression of calculations as they are performed. Creation of the spreadsheet models forces students to internalize the mathematical logic implied by the calculations.

When using a completed spreadsheet, learners can easily manipulate values in the spreadsheet to answer "what if?" questions. For example, if students are using a spreadsheet that computes compound interest, learners can quickly examine the affect of various interest rates (e.g., What if the rate were 3%?, what if it were 5%?, what about 7%?) on the total amount paid over the course of the loan without tediously computing the compound interest for each rate. This allows students to engage higher-order thinking skills to reflect on the affect of interest rates on amount paid, rather than spending time on the procedural tasks associated with computing a series of compound interest rates.

Use of computer tools continues to be one of the most common uses of technology in schools. In a recent Project Tomorrow report (2008), a large percentage of students reported that their primary use of technology consisted of using word processors to complete writing assignments (74%) and using web browsers to conduct online research (72%). Further, in a large scale interview study with technology-using teachers, researchers concluded that, although creating multimedia projects was a prominent activity, (multimedia presentations, word processed documents and graphics (charts and graphs for the presentations and documents) constituted 74% of products developed by students) many of the activities did not push students to work at higher cognitive levels and few teachers took full advantage of the potential for computers to promote problem-solving capabilities in their students. (Niederhauser & Lindstrom, 2006). Thus, while tool-use still appears to be the most pervasive use of technology in the classroom, use seems to be more in line with Taylor's original conception of productivity tools, rather than Jonassen's more elaborate Mindtools vision.

TECHNOLOGY AS TUTEE: CREATING DYNAMIC REPRESENTATIONS

In the tool-based examples discussed above, technology served as a way to structure, represent and reflect on one's conceptual understandings (through tools like semantic networks and authoring tools), and to construct organizational systems for conceptualizing and exploring relationships among data (through tools like spreadsheets and databases). Using technology as a tutee, through programming and use of expert systems, allows learners to create a dynamic representation of phenomena, processes and/or procedures—and to test their conceptual understanding of the systems and rules underlying the representation created using the technology.

A key distinction for tutee mode involves the self-correcting nature of this use. In tutor mode the teaching machine provides explicit feedback as to whether the learner has selected the right answer (necessary if one is to reinforce correct responses). In tool mode, evaluation of the products generated with the tools (e.g., the word processed paper, multimedia report, concept map, etc.) is typically completed by teacher and/or peers. That is, the technology is used to create the product, but tells the learner nothing about how his or her representation is consistent with, or differs from, the concept, process or procedure he or she is trying to represent. In tutee mode the learning experience is more like playing with a puzzle. The learner puts the puzzle together (creates the representation), then checks to see if the pieces all fit together as expected. The learner does not need an external authority (the program in the case of the tutor model or teacher in the case of the tool model) to tell him when the pieces fit or do not fit, flaws in the learner's thinking become apparent when the representation does not function as expected.

Some basic principles that underlie the tutee model include:

1. Computers can be used to create a dynamic, working, testable model that can serve as a concrete external representation of student thinking,
2. The learner can benefit from teaching (or tutoring or programming) the computer because it forces him or her to be clear, systematic, and explicit in his or her thinking,
3. The computers make a good 'tutee' because of its dumbness, its patience, its rigidity, and its capacity for being initialized and started over from scratch (Taylor, 1980, p. 4),
4. Problems that arise with the conceptual model itself, or the way the model is implemented using the programming language, present opportunities for further problem-solving and learning.

I now turn to a discussion of the nature of the theoretical orientations that underlie use of the Tutee mode.

Programming

The tutee mode has been most widely championed by proponents of a family of computer programming languages and applications that were designed specifically for educational purposes. Initially developed at the MIT Artificial Intelligence laboratories in the 1970s, "LOGO is the name of a philosophy of education and a growing number of computer languages that goes with it" (Papert, 1980, p. 217). More recently, applications like Boxer (diSessa, 1985; diSessa, Abelson, & Ploger, 1991) and Scratch (Resnick, 2007) were developed as programmable learning environments that enable beginners to experience learning in tutee mode without having to learn syntactically-based programming languages first. Scratch, developed by the Lifelong Kindergarten group at the MIT Media Lab, motivates learning through playfully experimenting and creating projects like interactive animations and games.

While recently developed intuitive systems like Scratch have seen more widespread use of late, LOGO continues to serve as the underlying philosophy for using instructional technology in the tutee mode. Designed as a powerful computer language, Logo provides entry into computer programming for beginners who have little or no prior mathematical knowledge or experience with programming. To that end a specific subset of primitive commands, called Turtle Talk, was developed that allowed learners to control the actions of the Turtle, which Papert (1980) describes as a "computational 'object to think with'" (p. 11). For the sake of this discussion, we can think of the Turtle as an object on the computer screen that responds to Turtle Talk instructions. For example, one primitive is the FORWARD command. If the learner types in FORWARD 50 the Turtle moves forward 50 steps. Another primitive, RIGHT, tells the Turtle to rotate in a clockwise direction—so if the learner types RIGHT 90 the Turtle will rotate clockwise 90 degrees. The power of LOGO lies in the ability to use these primitive commands to create more complex commands. The TO primitive is used to create 'procedures' which act like primitives. For example:

```
TO SQUARE
FORWARD 50
RIGHT 90
FORWARD 50
RIGHT 90
FORWARD 50
RIGHT 90
FORWARD 50
RIGHT 90
END
```

With square defined in this manner, typing SQUARE will cause the Turtle to follow the commands to draw a square with sides of 50 units on the computer screen. Further, SQUARE is now a defined procedure (essentially the same as a primitive) that can be used in defining other procedures. While turtle graphics is the most widely known application of LOGO, students can also use LOGO to program non-graphical solutions to logic problems like the *Tower of Hanoi* dilemma (for an explanation of the Tower of Hanoi problem see Kotovsky, Hayes, & Simon, 1985). Using LOGO, children can learn about geometry, logic, computational thinking, and their own mental models and thought processes, as they 'teach' the Turtle routines that create complex figures and perform a variety of logical functions.

Since the Turtle Talk lexicon includes only very basic and simple 'primitive' commands, learners must learn to organize their thoughts and teach the computer in ways that the computer will 'understand.' Learners can build sophisticated vocabularies and procedures using the TO command, but the Turtle cannot do much until those new vocabularies and procedures are taught. Teaching the computer promotes logical and systemic thinking, and necessitates the development of troubleshooting skills when the program does not behave as anticipated.

Drawing on aspects of Piaget's developmental model (c.f., Piaget & Inhelder, 1975), Papert (1980) claimed teaching the computer (creating a computer program that constitutes a representation of the learner's internal conceptual model) made it possible for "computers [to] concretize (and personalize) the formal," (p. 21) promoting the development of two essential aspects of the formal stage of intellectual development, combinatorial thinking and self-referential thinking. Further, in learning to program, students develop computational thinking, which involves:

1. Formulating problems in a way that enables us to use a computer and other tools to help solve them,
2. Logically organizing and analyzing data,
3. Representing data through abstractions such as models and simulations,
4. Automating solutions through algorithmic thinking (a series of ordered steps),
5. Identifying, analyzing, and implementing possible solutions with the goal of achieving the most efficient and effective combination of steps and resources,
6. Generalizing and transferring this problem solving process to a wide variety of problems. (ISTE & CSTA, 2011)

Finally, rather than the right/wrong dichotomy that drives much of student work in schools, when students serve as a tutee for the computer the focus shifts to isolating and correcting "bugs." The question is not whether they got it right or wrong, rather the question becomes whether the problem is fixable—helping to make the student more self-motivated, self-directed, and self-evaluative.

Expert Systems

While much of the tutee mode perspective has evolved from the use of computer programming languages, Jonassen & Reeves (1996) proposed expert systems as an alternative that is more relevant to solving the kinds of problems typically encountered in daily life. Expert systems typically include a *knowledge base* and an *inference engine*. The knowledge base includes factual information and rules that govern the ways the information is inter-related. For example:

> An expert system designed to diagnose cars that will not start might include facts and rules such as:
>
> Fact: Battery supplies voltage to ignition.
>
> Fact: Ignition routes voltage to solenoid.
>
> Rule: If ignition is on,
>
> AND solenoid is not engaged,
>
> THEN battery is dead,
>
> OR ignition switch is faulty. (p. 708)

The inference engine is a logic-driven component that acts on the knowledge base using data and parameters from the current problem to generate solutions. While often used as a job aid to help technicians diagnose and solve problems, expert systems best serve as a cognitive tool when a learner is allowed to develop an expert system that represents his or her own knowledge base in a given domain. Constructing the knowledge base requires that the learner explicitly articulate the expert-knowledge that the system requires in the form of facts, rules and causal relationships. Identifying and codifying the factual information, procedural knowledge and causal relationships underlying a knowledge domain necessarily engages expert systems designers in higher-order thinking.

Building an expert system requires that the developer models the knowledge of the expert in an explicit and concrete way (Starfield, Smith & Bleloch, 1990), which dramatically changes the roles of teachers and students. Students become "knowledge engineers" and assume a more active role in acquiring prerequisite knowledge and focusing and directing their interactions with the teacher. Students assume responsibility for analyzing the knowledge domain, and synthesizing rules and rule sequences—becoming expert. Teachers assume a supporting role, guiding and responding to student probing concerning the more demanding and interesting aspects of the content domain, and providing technical support with the technology (Morrelli, 1990).

Thus, with an expert system learners first develop a conceptual model of a phenomenon, process or procedure (supporting development of deep, connected, causal understandings of the domain), then uses instructional technology to create a dynamic working model of the learner's conceptual model. The working model can then be probed and tested to assess its' validity and reliability, and, when shortcomings are identified, the learner develops a new iteration of the working model that accounts for the shortcomings (which has direct impact on restructuring the development of the learners underlying conceptual model). Although use in classrooms has been limited, applications like STELLA (ISEE, 1985–2012) offer a practical way to dynamically visualize and communicate how complex systems and ideas work. Stella can been used to model answers to a variety of complex questions including: How does climate change influence an ecosystem over time? Would Hamlet's fate have changed if he'd killed Claudius earlier? And how do oil prices respond to shocks in supply and/or demand? Having students use STELLA to develop representations that model their responses to these kinds of questions can provide powerful learning opportunities and insights into students' deep understanding of the content.

CONCLUSION

This discussion provides a framework for conceptualizing the nature of technology use in instructional settings—particularly with respect to the roles that technology can play in mediating instructional interactions. Technology can be used to support various orientations toward learning that range from tutoring systems that are grounded in a behavioral paradigm, to tool and tutee models that are framed by cognitive- and social-constructivist perspectives. There does not appear to be anything in the nature of the technology that necessitates that it be used in ways that reflect a particular ontological, epistemic or pedagogical orientation; rather, teachers tend to use technology with their students in ways that are consistent with their (the teacher's) beliefs about learning and pedagogy (Niederhauser & Stoddart, 2001); and adapt instructional innovations to fit with their practices.

US schools have seen a massive infusion of funding and resources with estimates that overall spending on instructional technology will top $56 billion in 2012 (Nagel, 2008), yet widespread well-integrated use of instructional technologies remains unrealized—underutilized by teachers and students alike (Ertmer & Ottenbreit-Leftwich, 2010). When embarking on technology integration efforts, like the current one-to-one computing movement that is sweeping through many US states, pedagogical and curricular issues must be central to the discussion. The common practice of simply handing out devices to students and teachers, with some minimal training on how to operate it, is not enough. Embedding technology integration in a broader context of pedagogical and curricular reform, helping teachers make explicit their beliefs about teaching and learning, and encouraging and supporting broader discussions about the variety of ways that technology can support different pedagogical orientations and learning outcomes discussed here, would increase the likelihood that teachers might begin to integrate student use of technology in more substantive and meaningful ways.

NOTE

[1] In this chapter instructional technology refers to computer-based devices, software, and network infrastructure used for teaching and learning.

REFERENCES

Clements, D. H. & Battista, M. T. (1990). Constructivist learning and teaching. *Arithmetic Teacher, 38*(1), 34–35.

Cohen, D. K. (1990). Teaching practice: Plus ça change. In P. W. Jackson (Ed.), Contributing to educational change: Perspectives on research and practice (pp. 27–84). Berkley, CA: McCutcheon.

David, J. L. (1991). Restructuring and technology: Partners in change. *Phi Delta Kappan, 73*(1), 37–40,78–82.

Dick, W., & Carey, L. M. (1996). *The systematic design of instruction* (4th Ed.). New York: HarperCollins.

diSessa, A. A. (1985). A principled design for an integrated computational environment. *Human-Computer Interaction, 1*, 1–47.

diSessa, A. A., Abelson, H., & Ploger, D. (1991). An overview of boxer. *Journal of Mathematical Behavior, 10*, 3–15.

Duffy, T. M. & Jonassen, D. H. (1992). Constructivism: New implications for instructional technology. In T. Duffy & D. Jonassen (Eds.), *Constructivism and the technology of instruction: A conversation* (pp. 1–16). Hillsdale, NJ: Lawrence Erlbaum.

Ertmer, P. A., & Ottenbreit-Leftwich, A. T. (2010). Teacher Technology Change: How Knowledge, Confidence, Beliefs, and Culture Intersect. *Journal of Research on Technology in Education, 42*(3), 255–284.

Fisher, K. M. (1990). Semantic networking: New kid on the block. *Journal of Research on Science Teaching, 27*(10), 1001–1018.

Fisher, K. M. (1992). SemNet: A tool for personal knowledge construction. In P. Kommers, D. Jonassen

& T. Mayes (Eds.) *Cognitive tools for learning*. Berlin: Springer.
Fry, E. B., Bryan, G. L., & Rigney, J. W. (1960). Teaching machines: An annotated bibliography. *Audio Visual Communication Review, 8*(2), 1–2, 5–79.
Gagne, R. M., Briggs, L. J., & Wager, W. (1987). *Principles of instructional design* (3rd ed.). New York: Holt, Rinehart, & Winston.
Howland, J. L., Jonassen, D. H., & Marra, R. M. (2011). *Meaningful learning with technology (4th Ed)*. Columbus, OH: Pearson.
Inspiration (1988–2010). *Inspiration* [computer software). Beaverton, OR: Inspiration Software Inc.
ISEE (1985–2012). *STELLA* [computer software). Lebanon, NH: ISEE Systems.
International Society for Technology in Education (ISTE) and the Computer Science Teachers Association (CSTA), (2011). Operational definition of computational thinking for K-12 education. Available: http://csta.acm.org/Curriculum/sub/CurrFiles/CompThinkingFlyer.pdf
Jonassen, D. H. (1991). Objectivism versus constructivism: Do we need a new philosophical paradigm? *Educational Technology Research and Development, 39*(3), 5–14,
Jonassen, D. H. (1996). *Computers in the classroom: Mindtools for critical thinking*. Columbus, OH: Prentice Hall.
Jonassen, D. H., & Carr, C. S. (2000). Mindtools: Affording multiple knowledge representations for learning. In S.P. Lajoie (Ed.), *Computers as cognitive tools: No more walls* (Vol. 2, pp. 165–196). Mahwah, NJ: Lawrence Erlbaum.
Jonassen, D. H., & Reeves, T. C. (1996). Learning *with* technology: Using computers as cognitive tools. In D. Jonassen (Ed.) *Handbook of research on educational communications and technology* (pp. 693–719). New York: Macmillan.
Kotovsky, K., Hayes, J. R., & Simon, H. A. (1985). Why are some problems hard? Evidence from Tower of Hanoi. *Cognitive Psychology, 17*(2), 248–294.
Lageman, E. C. (2000). *An Elusive Science: The troubling history of educational research*. Chicago: University of Chicago Press.
Lehrer, R. (1993). Authors of knowledge: Patterns of hypermedia design. In S. Lajoie & S. Derry (Eds.) *Computers as cognitive tools,* 197–227. Hillsdale, NJ: Erlbaum.
Miller, G. A. (1956). The magical number seven, plus or minus two: Some limits on our capacity for processing information. *Psychological Review, 63*, 81–97.
Morrelli, R. (1990). The student as knowledge engineer: A constructivist model for science education. *Journal of Computing in Higher Education, 2*(1), 78–102.
Nagel, D. (2008). Education IT spending, fueled by telecom, to top $56 billion by 2012. Campus Technology. Available: http://campustechnology.com/articles/2008/09/education-it-spending-fueled-by-telecom-to-top-56-billion-by-2012.aspx
Nesbitt, J. (1982). *Megatrends: Ten new directions transforming our lives*. New York: Warner Books.
Niederhauser, D. S., & Lindstrom, D. L. (2006). Addressing the NETS for students through constructivist technology use in K-12 classrooms. *Journal of Educational Computing Research, 34*(1), 91–128.
Niederhauser, D. S., & Stoddart, T. (2001). Teachers' instructional perspectives and use of educational software. *Teaching and Teacher Education, 17*(1), 15–31.
Paas, F. G., & VanMerrienboer, J. G. (1994). Instructional control of cognitive load in the training of complex cognitive tasks. *Educational Psychology Review, 6*, 351–371.
Papert, S. (1980). *Children, computers, and powerful ideas*. New York: Basic Books.
Papert, S. (1993). *The children's machine: Rethinking school in the age of the computer*. New York: Basic Books.
Perkins, D. N. (1986). *Knowledge as design*. Hillsdale, NJ: Erlbaum.
Piaget, J. (1971). *The origins of intelligence in children*. (M. Cook, Trans.). New York, NY: W. W. Norton & Co. (Original work published 1952).
Piaget, J. & Inhelder, B. (1975). *The origin of the idea of chance in children*. (L. Leake, P. Burrell & H. Fishbein Trans.). New York: WW Norton & Co.
Pressey, S. L. (1926). A simple apparatus which gives tests and scores-and teaches. *School and Society, 23*, 373–76.

Project Tomorrow (2008). *21st century learners deserve a 21st century education.* Selected National Findings of the Speak Up 2007 Survey. Retrieved October 23, 2012, from http://www.tomorrow.org/speakup/speakup_congress_2007.html

Resnick, M. (2007). Sowing the seeds for a more creative society. *Learning and Leading with Technology, 35*(4), 18.

Reynolds, R. E. (2000). Attentional resource emancipation: Toward understanding the interaction of word identification and comprehension processes in reading. *Scientific Studies of Reading, 4,* 169–196.

Sheingold, K. (1991). Restructuring for learning with technology: The potential for synergy. *Phi Delta Kappan, 73*(1), 17–26.

Skinner, B. F. (1958). Reinforcement today. *American Psychologist, 13*(3), 94.

Skinner, B. F. (1968). *The technology of teaching.* Des Moines, IA: Meredith.

Skinner, B. F. (1981). Selection by consequences. *Science, 213*(4507), 501–504.

Skinner, B. F. (1984). The shame of American education. *American Psychologist, 39*(9), 947–954.

Starfield, A. M., Smith, K. A., & Bleloch, A. L. (1990). *How to model it: Problem solving for the computer age.* New York: McGraw-Hill.

Sweller, J. (1988). Cognitive load during problem solving: Effects on learning. *Cognitive Science, 12,* 257–285.

Sweller, J. (1994). Cognitive load theory, learning difficulty, and instructional design. *Learning and Instruction, 4,* 295–312.

Sweller, J., Chandler, P., Tierney, P., & Cooper, M. (1990). Cognitive Load as a Factor in the Structuring of Technical Material. *Journal of Experimental Psychology: General, 119,* 176–192.

Sweller, J., Van Merrienboer, J. G., & Paas, F. G. (1998). Cognitive architecture and instructional design. *Educational Psychology Review, 10,* 251–296.

Taylor, R. P. (1980). *The computer in the school: Tutor, tool, tutee.* New York: Teachers College Press.

Vockell, E., & van Deusen, R. M. (1989). *The computer and higher-order thinking skills.* Watsonville, CA: Mitchell.

Von Glasersfeld, E. (1995). *Radical constructivism: A way of knowing and learning.* New York: RoutledgeFalmer.

Dale S. Neiderhauser
School of Education,
Iowa State University

CHAPTER 14

WARREN J. BLUMENFELD

THE NATURE OF SOCIAL COMMUNICATION TECHNOLOGIES AND CYBERBULLYING

Filtered Cues and Disinhibited Actions

Pheobe Prince moved with her family from Ireland, their country of origin, to begin a new life in the United States, settling in the beautiful western Massachusetts town of South Hadley. When she began attending South Hadley High School, she made a few friends who commented how happy and well-adjusted she appeared. Under the surface, though, she was concealing an intense emotional pain. On a continuing basis, other students at school tormented her on-line calling her an "Irish slut" and "a whore." Fearing she could no longer endure her peers' abusive taunts, Phoebe took her life. She was fifteen years old.

Friends described Tyler Clementi as a gentle, kind, and sensitive person who was an accomplished violinist at an early age. Tyler won a music scholarship to Rutgers University, and he was looking forward to his four years there and to a shining career. On September 22, 2010, however, that great potential ended when Tyler took his life by jumping off the George Washington Bridge. He was only 18 years old.

Tyler's roommate, Dharum Ravi and another Rutgers student Molly Wei, both 18 years of age, faced charges of invasion of privacy by secretly webcam recording and live streaming on the internet Tyler engaging in sexual activity in his room with another male student. Dharum tweeted to the over 150 of his followers: "I saw him making out with a dude." And then he tweeted, "Anyone with iChat, I dare you to video chat me between the hours of 9:30 and 12. Yes, it's happening again!"

Ravi was tried and convicted in 2012 on 15 counts, including invasion of privacy, tampering with evidence, and bias intimidation. The later charge is defined as a "hate-crime," which carries a penalty of up to 10 years imprisonment. Ravi served a total of 30 days in jail, and was sentenced to three years probation, ordered to complete 300 hours of community service and to attend counseling programs for cyber-bullying and education on tolerance for human differences. In addition, he must pay $10,000 to the probation department in increments of $300 per month, which will go to people targeted in hate crimes.

Michael P. Clough, Joanne K. Olson and Dale S. Niederhauser (Eds.), The Nature of Technology: Implications for Learning and Teaching, 269–290.

Ryan Patrick Halligan was born in Poughkeepsie, New York in 1990. His parents described him as a shy, sensitive, and affectionate young child with an infectious smile that early on drew people close. Before he entered school, his parents had concerns about his speech, language, and motor skills development, and from pre-school through fourth grade, they enrolled Ryan in special education services. The family moved to Essex Junction, Vermont, where, by the fifth grade, he encountered face-to-face bullying on a regular basis in his school. Rumors soon circulated throughout the school that Ryan was gay. By middle school, his classmates continually teased and harassed him for having a learning disability and for allegedly being gay. They soon extended their taunts against Ryan into cyberspace.

On October 7, 2003, feeling that he could no longer live with the constant and escalating abuse, Ryan Patrick Halligan took his life. He was 13 years old. Reports (Spero News, 2006) indicate that Ryan displayed many of the symptoms of youth targeted by cyberbullying: he spent long hours on his computer, and he was secretive regarding his interactions on social communication technologies. His parents saw him manifest a number of changes in his behavior: he increasingly lacked interest in engaging in social activities that included his peers, and he exhibited a pronounced change in his overall attitude, his appearance, and his habits.

John P. Halligan, Ryan's father, wrote: "I believe bullying through technology has the effect of accelerating and amplifying the hurt to levels that will probably result in a rise in teen suicide rates" (RyanPatrickHalligan.com). John established a web site in loving tribute to his son and as a clarion call to prevent what happened to Ryan from impacting the lives of any other young person. John expressed his hope:

> This site is dedicated to the memory of our son Ryan and for all young people suffering in silence from the pain of bullying and having thoughts of suicide. We hope young people become less ashamed to ask for help when feeling suicidal. We hope adults gain knowledge from our tragedy. As a society, we need to find better ways to help our young people through their most difficult growing years (RyanPatrickHalligan.org, retrieved April 19, 2009).

THE ADVENT OF CYBERBULLYING: THE IMPACT OF SOCIAL COMMUNICATION TECHNOLOGIES

Ross (2002) defines bullying as a form of terrorism since it involves an "unprovoked attack" upon another person or persons with the intention of causing harm. While bullying and harassment have long been problems for young people in our nation's schools at every level, advanced information and communication technologies bypass, filter, and thus mitigate social cues that act in ways to constrain bullying. At the same time, these information and communication technologies extend abusive and destructive bullying practices to virtually all aspects of a person's life. Thus, cyberbullying takes bullying to a more destructive and protracted level.

The American Psychological Association passed a resolution (2004) calling on educational, governmental, business, and funding agencies to address issues of face-to-face and cyberbullying. In the resolution, they particularly addressed acts of harassment "about race, ethnicity, religion, disability, sexual orientation, and gender identity" (p. 1). In addition, the resolution specifically emphasized the high rates of bullying around issues of sexual orientation and disability:

> WHEREAS children and youth with disabilities and children and youth who are lesbian, gay, or transgender, or who are perceived to be so may be at particularly high risk of being bullied by their peers (Dawkins, 1996; Hershberger & D'Augelli, 1995; Hunter, 1990; Nabuzka & Smith, 1993; Pilkington & D'Augelli, 1995; Rigby, 2003; Yude, Goodman, & McConachie, 1998).

Young people and even adults continue to endure schoolyard and workplace bullying and harassment. Bullying—also termed "face-to-face" ("f2f"), "real life" ("RL"), "traditional, "in-person," or "offline" bullying—involves deliberate and repeated aggressive and hostile behaviors by an individual or group of individuals intended to humiliate, harm, and/or control another individual or group of individuals of lesser power or social status (Nansel et al., 2001). Bullying is a specific type of aggression in which (1) the behavior is intended to harm or disturb, (2) the behavior occurs repeatedly over time, and (3) there is an imbalance of power, with a more powerful person or group attacking a less powerful one (Garret, 2003; Olweus, 1991; Olweus 1993; Orpinas & Horne, 2006; Ybarra & Mitchell, 2004). This asymmetry of power may be physical or psychological, and the aggressive behavior may be verbal (e.g., name calling, threats, taunting, malicious teasing), physical (e.g., hitting, kicking, spitting, pushing, taking personal belongings), or psychological (e.g., spreading rumors, engaging in social exclusion, extortion, or intimidation) (Farrington, 1993). In our era of advanced information and social communication technologies, however, a new variation has emerged, for we now live in the age of cyberbullying.

Cyberbullying occurs over information and social communication technologies such as Internet web sites, e-mail, chat rooms, mobile phone and text messaging, and instant messaging, and includes instances such as: (1) people sending hurtful, cruel, and oftentimes intimidating messages to others (e.g., "Flame Mail") designed to inflame, insight, or enrage; (2) "Hate Mail" (also known as "Cyberharassment"), which constitutes hate-inspired and oppressive harassment based on actual or perceived social identities in terms of race, ethnicity, religion, sex, gender, sexuality, physical and mental abilities, socioeconomic class, and others; (3) people stealing other peoples' screen names and sending inflammatory messages under those screen names to others; (4) anonymous postings of derogatory comments about another on web journals called "blogs" or on MySpace or Facebook; (5) young people creating online polling booths; (6) individuals taking pictures of others in gymnasium locker rooms with digital phone cameras and sending those pictures to others, or posting them on Internet web sites (a form of "sexting"); (7) people creating web sites with stories, cartoons, caricatures, pictures, or "jokes" ridiculing or mocking others; (8) posting material about a person involving private, sensitive, or embarrassing information, for example, "outing" a person's sexual identity to classmates and sometimes to the targets' parents or guardians; (9) sending intimidating or threatening messages (also known as "Cyberstalking"); (10) intentional interruption and harassment on gaming websites (so-called "Griefing") causing grief to website players; or (11) actions designed to isolate and exclude a person from online social communication technologies.

RATES OF CYBERBULLYING

Reports indicate that cyberbullying has increased exponentially as technologies have become more accessible and as new and advanced technologies continually emerge (Findelhor et al., 2000, 2006; Kaiser Family Foundation, 2005). A study conducted by the Pew Research Center (2010) found that 73% of U.S.-American wired teens use social networking websites, up from 55% in 2006 and 65% in 2008. In comparison 47% of online adults use social networking websites compared with 37% in 2008 (Lenhart, Purcell, Smith, & Zickuhr, 2010). As far back as 2003, the University of California-Los Angeles determined that overall Internet usage by young people was approximately 91% of 12- to 15-year olds and almost all teens (99%) ages 16 to 18 have access to and use the Internet. The Youth Internet Safety Surveys (YISS-1 & YISS-2) (Findelhor et al, 2000, 2006), led by Dr. David Finkelhor, Director of the Crimes against Children Research Center at the University of New Hampshire to assess Internet victimization of youth between the ages of 10 and 17, found that Internet usage within this population had grown enormously during the 1990s. According to Finkelhor et al., (2006), by 2006, an estimated 61% of ten-years-olds to an estimated 81% of 16-year-olds were Internet users on a regular basis. In addition, the Pew Internet and American Life Project (Lenhart et al., 2005), found that 87% of 12- to 17-year-old use the Internet.

An additional report, *Generation M: Media in the lives of 8–18 year-olds*, released by The Henry J. Kaiser Family Foundation found that already by 2005, home environments of the overwhelming majority of this country's youth were found to be "media saturated" (p. 10): 86% reported households with at least one computer, while 50% reported having three or more computers. Moreover, 64% reported home access to the Internet, and 60% reported having instant messaging programs. Nearly 75% of respondents reported living in homes containing at least three television sets. The study also reported that race was a factor in in-home media access: 80% of White youth, 61% of Black youth, and 67% of Hispanic youth have home access to the Internet, and 63% of White youth, 47% of Black youth, and 55% of Hispanic youth have access to instant messaging programs. The survey also found that by age, Internet usage ranged from 63% among 8- to 10-year-olds to about 80% among 11- to 18-year-olds. A parallel trend was reported for access to instant messaging. Fifty-one percent of all teenage Internet users, or approximately 11 million, are estimated to go online on a daily basis (Lenhart et al., 2005). Many young people spend large amounts of time online communicating with other young people. The Generation M (Kaiser Family Foundation, 2005) survey found that

> A combination of increased access to computers and the emergence of new, highly popular computer activities have resulted in more than a doubling of the amount of time U.S. kids spend with computers compared with the previous five years (p. 30).

The most popular online activities reported by youth in the Generation M report included visiting web sites and participating in chat rooms, sending and receiving instant messaging and email messages, and playing games online. Young people are also constructing personal websites of their own in greater numbers including MySpace and Facebook (Lenhart et al., 2005). The Pew survey also found that 45% of teenagers own cell phones, and 33% use it to text message: "Instant messaging has become the digital communication backbone of teens' daily lives..." with "roughly 32% of *all* teens" using "IM every single day" (p. iii, emphasis in original). In addition to sending text messages, they also share website links, music, photos, and video. Seventh grade was found to be the period of enormous growth in usage of all of these information and communication technologies:

> Going to Junior High seems to be the tipping point when many teens who were not previously online get connected. While about 60% of the 6th graders in our sample reported using the Internet, by 7th grade, it jumps to 82% who are online. From there, the percent of users in the teen population for each grade climbs steadily before topping out at 94% for eleventh and twelfth graders (Lenhart et al., 2005, p. v.).

The Cyberbullying Research Center found in their 2011 random sample study of 4,441 youth between the ages of 10 and 18, within the previous 30 days, approximately 43 percent reported experiencing at least one of several incidents that could be considered cyberbullying. For example, 13.7% reported receiving mean or hurtful on-line comments, 15.8% had someone post mean comments on their MySpace, 12.9% had on-line rumors spread about them, and 17% reported generally being cyberbullied two or more times during the prior month.

Adolescent girls were more likely to have experienced cyberbullying than boys (25.8% vs. 16%). The study also found that girls are more likely to spread rumors while boys more often posted hurtful pictures and videos. Research also showed that all races are evenly represented as both cyberbullying targets and as perpetrators. i-SAFE America, an Internet safety education foundation, conducted a nationwide survey of 1,566 students from grades four to eight to determine their experiences with cyberbullying (National i-SAFE Survey, 2004). The Survey found that 57% of students reported receiving hurtful or angry messages online with 13% saying it happens "quite often;" 44% have received mean or threatening e-mails; 43% admit sending mean or hurtful things to someone online, and 7% admit to doing it "quite often"; 45% have been threatened online with 5% saying it happens "quite often"; 42% reported being bullied online with 7% saying it happens "quite often."

Researchers conducting the two Youth Internet Safety Surveys (Finkelhor et al., 2000, 2006) reported significantly increased rates of behaviors attributable to online bullying and to online victimization in the intervening years between the two surveys with young people increasingly using information and communication technologies to threaten, embarrass, harass, and humiliate. One in 11 (9%) of survey participants reported being harassed online (Finkelhor et al, 2006) with almost one-third of the youth surveyed participating in behaviors attributable to online bullying. This is compared to 1 in 17 (6%) in YISS-1 (Finkelhor et al, 2000) harassed online with 12%-15% of those surveyed participating in behaviors attributable to online bullying.

Young people targeted for harassment varied somewhat by sex: 58% were female, and 42% male (Finkelhor et al, 2006). Youth targeted for harassment reported that 50% of the harassers were male, 28% were female, and in 21% of the cases, they did not know the sex of the perpetrator. In addition, the Pew (Lenhart et al, 2005) study, found that 21% of teens who use email, instant or text messaging, reported that after sending a message that was meant to remain private, it was then forwarded to others by the recipient.

GLSEN (Gay, Lesbian, and Straight Education Network), in its 2005 National School Climate Survey of 1,732 lesbian, gay, bisexual, and transgender students between the ages of 13 and 20, found that both face-to-face and online bullying "remain common in America's schools" (p. 4).

> [H]omophobic remarks were the most common type of biased language heard at school, with three-quarters of the students (75.4%) hearing remarks such as "faggot" or "dyke" frequently or often at school....[N]early a fifth (18.6%) of the survey respondents reported hearing homophobic remarks from their teachers or other school staff (p. 4).

Nearly two-thirds (64.3%) felt unsafe at school because of their sexual identity, while 40.7% felt unsafe because of how they expressed their gender. Cyberbullying was a significant problem as well. More than two-fifths (41.2%) of respondents reported receiving threatening or harassing text messages or email from other students.

In their subsequent 2009 National School Climate Survey of 7,261 middle and high school students, GLSEN reported that on the basis of sexual identity, 84.6% were verbally harassed, 40.1% physically harassed, 18.8% physically assaulted, and 61.1% generally felt unsafe on their campuses. In terms of gender identity and expression, 63.7% were verbally harassed, 27.2% physically harassed, 12.5% physically assaulted, and 39.9% generally felt unsafe. The report also showed increased levels of victimization of LGBT youth who were related to increased levels of depression and anxiety and decreased levels of self-esteem.

THE PSYCHOLOGY AND SOCIOLOGY OF CYBERSPACE: THE "FUNCTIONS" AND MEANINGS OF ABUSIVE ACTIONS

Psychology and sociology, as ever-expanding fields, have connected and redeployed their theories and concepts of human behavior to the virtual realities of cyberspace. What follows includes some of the theoretical foundations that may hold promise in addressing abuse of human-computer interactions.

The "Online Disinhibition Effect"

A number of similarities and differences exist between face-to-face (f2f) or real life (RL) bullying and cyberbullying. Both are about human relationships, power, and control, and actions can occur on numerous occasions. Also, both may involve what psychologists call the "leveling effect": people who cyberbully often do so to diminish others to inflate their own egos reflecting their insecurities. In addition, both do not simply involve those who abuse and those who are abused (the "dyadic view") but rather involve a number of "actors" or roles across the social/workplace/school environment (see e.g. Sutton & Smith, 1999).

The very nature of social communication technologies, however, establish the conditions that make it possible for users to perform and act in cyberspace in ways they would not ordinarily act in f2f interactions. Suler (2001) proposes a conceptual framework enumerating a number of basic psychological features when taken individually and in differing synergistic combinations within various online environments, help to explain how people experience themselves and others. Taken together, these elements explain what Suler (2001) terms as the "online disinhibition effect": the nature of social communication technologies combined with the relative anonymity of cyberspace create the conditions for users to experience fewer behavioral inhibitions than in RL f2f situations.

The online disinhibition effect can manifest itself in positive as well as negative ways. For example, through the anonymity of cyberspace, users can exhibit extraordinary acts of kindness or charity that they may have felt *inhibited* from expressing in RL. Suler describes this as "benign disinhibition." For example, an individual can initiate a kind or charitable action pulling in a seemingly endless number of others from distances great and near anonymously that if identified might make the benefactor or the recipient uncomfortable. On the other hand, this anonymity and altered psychological environment may allow users to communicate more objectionable needs and desires onto others, Suler's expression of "toxic disinhibition."

Cyberbullying is a particularly cowardly form of abuse. Social communication technologies permit people who engage in cyberabuse to hide in the anonymity of cyberspace. With anonymity, those who cyberabuse do not have to "own" their actions, and they often do not fear being punished. The technology can also shelter the user from tangible feedback about consequences of one's actions, which can result in minimized empathy or remorse for the target of the bullying (Media Awareness Network, 2000). In cyberspace, according to Suler, the user experiences reduced or filtered sensual input, often unable to see or hear the person or people on the other end: no facial expressions signaling emotional output, no ability to see or read body language and voice intonations. This is particularly evident in print e-communications: texting, instant messaging, SMA, website blogs, email, chat rooms, social media like Facebook and MySpace, and others. Even when employing audio and visual means of communication such as videoconferencing, podcasting, and internet phoning, much of the sensory input remains limited at best. Therefore, people who engage in cyberbullying can inflict pain without having to see the effects, which can result in a "deeper level of meanness" (Harmon, 2004). In addition, people who cyberbully can also communicate their hurtful messages to a wider audience (even thousands of people simultaneously near and far) with incredible speed.

Much of cybertime exists asynchronically (Turkle, 1995), that is, people often do not have to interact in real time, which can add to the disinhibition effect when one does not have to deal with the immediate reactions of others. Cyberspace also transcends distance by virtually shrinking space making geography irrelevant. This feature has advantages and disadvantages. It can bring people closer together, but for those with anti-social motives, the nature of social communication technologies can enable the user to abuse others not only next door, but also on the other side of the planet.

In addition, people can alter, change, emphasize different aspects of their personalities or identities, or totally reinvent themselves in cyberspace—they can show different *personae* (Latin for "that through which the sound comes" or the actor's mask) (Turkle, 1995). Human-computer interactions permit individuals to engage in masquerade and change into a virtual costume known as an "avatar." When communicating only with typed text in social communication technologies, the user has the option of being oneself, expressing only parts of one's identity, assuming imagined identities, or remaining completely anonymous—in some cases, being almost invisible as with the "lurker" (Suler, 2005, p. 2).

Computers embody one of postmodernism's important tenets by challenging, contesting, and ultimately destabilizing identities. Through computer-mediated interactions, individuals continually redeploy identities as fluid, changing, multifaceted, and non-essentialized. "In its virtual reality, we self-fashion and self-create" (Turkle, 1995, p. 180). This identity destabilization presents a number of possibilities, for it can allow individuals to relate in genuinely open and honest ways online about themselves that might be frightening offline to discuss in real life personal encounters. On the other hand, with anonymity, the individual can act out hostile or sadistic emotions by abusing others online.

Cyberspace can have an equalizing effect. People begin on a relatively level playing field. Suler refers to this as the "equalized status" that regardless of one's actual social status in terms of relative wealth, race, gender, sexual and gender identity and expression, physical and mental attributes, and many other social identities and characteristics, the nature of the technology aids in creating a virtual net democracy dependent largely on the technical skills of the user. Those of lesser social status or those who are the targets of abuse in RL can gain power and status, sometimes abusing others in cyberspace.

In a virtual sense, then, cyberspace communication can alter perceptions and one's state of consciousness by becoming a make-believe world, a dream-like experience, even a game to the user in which the rules of RL no longer apply. Social communication technologies, therefore, have created a kind of transitional space that becomes an extension of the user's intrapsychic world. Suler contends that sitting and gazing at a computer monitor for spans of time creates an experience that is somewhat surrealistic producing a dream-like state of consciousness. This state makes it quite attractive to the user, which may explain the forms of computer and cyberspace addictions.

In addition, cyberbullying is often even more invisible to adults than other forms of youthful bullying. In fact, i-SAFE (National I-SAFE, 2004) found that 58% of respondents would not or have not told their parents or other adults about negative experiences online. Young people fear not only that reporting instances of cyberbullying would break a perceived peer norm of silence, which might increase the attacks on themselves or result in further isolation from peers, but also, they fear that adults might take from them the technology for the expressed purpose of ending the cyberbullying. Social communication technologies have opened new windows to the world for young and old alike. While the targets of cyberbullying might in the short term gain from closing the windows by shutting down the technology, overall, by taking this action (or having this action taken to protect them), they close as well these social windows since virtually all of their peers remain connected.

Freud Peering from the Computer

In psychoanalytic parlance, the concept of "transference" (as introduced by Sigmund Freud, 1912) refers to an individual's unconscious redirection of feelings from one person to another. Kapelovitz (1987) defines transference as "the inappropriate repetition in the present of a relationship that was important in a person's childhood" (p. 66). Taking this concept into the realm of social communication technology interactions, Suler (2005) makes clear that though the technology is certainly not a past or current member of our human family, "... but rather we recreate in our relationship with the computer some aspect of how we related to our family members" (p. 2). Often, and primarily on an unconscious level, the very nature of the information and social communication technologies provide the (cyber)space for individuals to recreate and replay past relationships, and also to satisfy unmet, frozen, or thwarted needs from childhood. In addition, Suler discusses the notion of "erotic transference," which he makes clear does not consist of sexual feeling *per se* toward computers, "...but rather the perception of the computer as powerful, perhaps in ways similar to how parents are perceived as powerful" (p. 4). Related to cyberbullying, social communication technologies provide the users the means to act out unmet needs for power and control over others, or to transfer frozen needs for attention or acknowledgement not sufficiently satisfied within the family constellation.

Brain (n.d.) theorizes that perpetrators of computer-generated abuse, and in particular, the production and transmission of computer viruses, do so for a number of psychological motivations. Some may transmit viruses simply for the emotional "rush" or thrill, much the same way as would an individual who vandalizes or intentionally sets destructive fires. In addition, creating and transmitting a computer virus works much the same as an explosion for someone who finds joy in watching cars crash or bombs explode. Another reason is simply finding adventure in and claiming bragging rights for exploiting security holes in computer systems before someone else beats them to it.

The same emotional thrill and sense of adventure can result for perpetrators of cyberbullying, possibly to the same degree as those fashioning and disseminating explosive computer viruses. By posting a hurtful or threatening message through social communication technologies, the perpetrator places an emotional bomb on an unsuspecting potential victim.

Social communication technology users now have the potential to cyberbully any time and to any place. Home, therefore, is no longer a refuge from this abuse. Although perpetrators of cyberbullying often do so outside the parameters of the school grounds or workplace, it invariably affects the overall school and workplace climate and environment, and the individuals' educational or work performance, as well as their short- and long-term psychological states. Policies and legislation have not always caught up with cyberabuse, for it is often outside the legal reach of workplaces, schools, and school boards when it occurs external to the workplace site or school property.

Cyberbullying and face-to-face bullying are similar in that both are associated with higher rates of mental health problems. Low self-esteem, feelings of loneliness (Asher, Hymel, & Renshaw, 1984; Asher, Parkhurst, Hymel & Williams, 1990; Hymel, Rubin, Rowden & LeMare, 1990), depression (Rigby, 2003; Strauss, Forehand, Freme & Smith, 1984; Vosk, Forehand, Parker & Rickard, 1982), suicidal ideation, and anxiety disorders (Prinstein, Boergers, & Vernberg, 2001; Swearer, Grills, Haye, & Cary, 2004; Kim, Koh & Leventhal, 2005; Bond, Carlin, Thomas, Rubin & Patton, 2001; La Greca & Harrison, 2005) have been shown to be some of the most common mental health outcomes related to being the target of abusive behaviors. These targets may also experience increased school and workplace absenteeism (Owens, Shute, & Slee, 2000) and they often find it difficult to concentrate on schoolwork (McClure & Shirataki, 1989). Studies report that between 5 and 10% of students stayed at home to avoid being bullied (Rigby & Slee, 1999; Smith, Morita, Junger-Tas, Olweus, Catalano & Slee, 1999). Also, a study conducted by Education Statistics Services Institute (DeVoe, J., Kaffenberger, S., Chandler, K., 2005; Kosciw, J. & Diaz, E.,

2005) found that student targets of bullying were more likely to receive lower grades than their non-bullied counterparts. Targets of both direct and indirect bullying (e.g. actions such as social isolation and spreading rumors) were more likely to receive mostly Ds and Fs than those bullied either directly only or indirectly only.

According to GLSEN's (2009) national survey of 7,261 middle and high school lesbian, gay, bisexual, and transgender youth, within the past month prior to the survey, 29.1% missed a class at least once and 30% missed at least one day of school because of safety concerns. This is compared with the overall national average of 8.0% missing a class at least once and 6.7% missing at least one day of school for students of all social identities due to safety concerns. In addition, grade point averages of students more frequently harassed because of sexual or gender identity was almost one-half grade lower than students who were less often harassed: 2.7 vs. 3.1 respectively.

Related to those who have suffered bullying and cyberbullying, in some cases, targeted students have turned into the perpetrators of school violence. For example, this was the case in 75% of school shootings during the 1990s and early 2000s (Pescara-Kovach, 2006). In addition, students who were bullied at least once a week experienced poorer health. Garrett (2003) provides a comprehensive list of symptoms and disorders, including constant high levels of stress and anxiety; frequent viral infections, aches and pains in the joints and muscles with no obvious cause; also back pain with no obvious cause, which will not go away or respond to treatment; headaches and migraines; tiredness, exhaustion, constant fatigue; insomnia; poor concentration; and other symptoms. Face-to-face bullying and cyberbullying can also lead to suicidal ideation, attempts, and completion as noted earlier.

With all of this taken into account, it becomes clearer that cyberspace can also inhibit a user's sense of responsibility for actions online. Researchers (e.g., Staub, 1978 in Harrington, 1995) suggest that denial of responsibility (RD) can be seen as an enduring human trait measured along a wide continuum from high to low. Those low in RD tend to accept responsibility for their actions, while those closer to the high side of the scale tend to deny responsibility, tend not to be responsible for the well-being of others, and are likely not to follow societal or personal rules. BloomBecker (1990 in Harrington, 1995), who has investigated computer-related crimes, found that this denial of responsibility is a major factor leading to computer abuse. For example, BloomBecker profiled Robert Morris, a graduate student who lacked a sense of responsibility (high RD), though he was raised in a family where considerable attention focused on his moral development. Morris, who methodically infected a large number of computers with his Internet worm, when discovered and apprehended, rationalized his actions as being beneficial in that he contributed to the identification of weaknesses in the nation's computer networks and systems. He justified his actions as providing a valuable service.

In a study of cyberbullying (Blumenfeld and Cooper, 2010), a perpetrator, when identified and asked why he sent abusive messages to others online retorted, for example, "I was only telling the truth. She is ugly, and I felt she had to know it!"

(from unpublished transcripts). His rationalization—denial of responsibility—centers on offering the target of his abuse supposed needed and useful "information."

Social Dominance, Social Identity, and Social Rank Theories

Psychologists Sidanius and Pratto (2001) proposed their Social Dominance Theory, which posits that human societies are predisposed toward arrangements that maintain and perpetuate group-based hierarchical systems of social organization. In this connection, a number of researchers argue that the mere recognition of two groups into dichotomous social categories is sufficient for hostility—that is, group membership itself has profound effects on psychological functioning, irrespective of personality types and other individual differences. It is thought that the individual is transformed in group situations. People will show favoritism toward the in-group and hostility and discrimination toward the out-group even: 1) when group membership is random and anonymous, 2) in the absence of intergroup interaction, 3) where there is no history of explicit intergroup competition, enmity, conflict, or status concerns, and 4) where no self-interest is involved.

The so-called "minimal group[1]" studies demonstrate that group members will compete with perceived out-groups even when there is no rational reason to do so. This is acted out not only in face-to-face encounters, but also through cyberspace. Tajfel and Turner (1986) give a possible explanation for this: an individual's self-esteem is often connected to the position of their group (collective self-esteem) and that even within these "minimal groups," there is implicit competition for valued status. (According to Crocker and Luhtanen [1990], collective self-esteem is defined as the extent to which one evaluates group membership positively.)

The "minimal group" discriminatory factor was one of the conclusions stemming also from the Sherif et al. (1961) summer camp studies. Sherif and his colleagues (Sherif, 1964; Sherif et al., 1961; Sherif and Sherif, 1953) are principal proponents of the notion that conflict (emphasized by hostility, negative stereotyping, and aggression) arise over competition for scarce resources. Sherif looked at the "objective relationship" between groups: the relationship emphasized by competition and by cooperation between the groups. This classic study on the effects of competition and cooperation in intergroup relations was based on the theories of Morton Deutsch (1949).

Sherif and his colleagues (1961/1988 reissued) conducted their "Robbers Cave" study at a boys' summer camp over a two-week period. Several days after the twenty-four 11- and 12-year-olds arrived, researchers quasi-randomly divided them into two groups, the Rattlers and the Eagles in one study (the Bulldogs and the Red Devils in another version of the study), and placed them in competitive activities: football, tug-of-war, and cabin inspections. Hostility soon developed between the two groups culminating in name-calling, stereotyping, glorification of the in-group's achievements and denigration of the out-group's achievements, vandalism of one another's cabins, and a massive food fight at a camp picnic. When the presence of another group was definitely announced the Rattlers immediately wanted to challenge them, and to be the first to challenge (Sherif et al., 1961, p. 94).

Later in the camp session, researchers devised cooperative activities to determine whether this would improve relations between the groups. Counselors staged a number of "emergencies" such as having a camp vehicle break down and finding a split in the camp's water line, which required cooperation between members of the Rattlers and Eagles. Researchers discovered that the introduction of a goal that members of both groups worked toward cooperatively significantly reduced tensions and conflict between the groups—hostility between groups declined substantially, the boys made friends with members of the other group, and they even began to work alongside one another spontaneously.

Researchers concluded that maladjusted, neurotic, or unstable psychological tendencies or behaviors were *not* necessary for the development of intergroup conflict and hostility. What was required, however, was an "objective" or "functional" relationship of competition (a perceived opposition of real vested interests), giving rise to a degree of ethnocentrism and prejudice emphasized by negative stereotyping, negative perceptions, and hostility, along with a high level of in-group solidarity and cohesion, feelings of in-group superiority, and justification for negative opinions of the out-group (Sherif and Sherif, 1969). Conflict, discrimination, and negative stereotyping come about when there were either limited resources or a goal in which only one group of two or more could attain.

Researchers replicated results in a number of independent studies in several other countries using a wide variety of subjects—female and male, old and young (e.g., Brewer, 1979; Tajfel, 1982). Results vary, however, by country reflecting different cultural norms. For example, Wetherell (1982) concluded that while all three groups in their study (children of European, Samoan, and Maori origin) clearly showed in-group favoritism, the latter two showed somewhat less than the first by repeatedly choosing to share with the perceived out-group. Wetherell explained this by the value Polynesian societies place on generosity to others as a marker of high status.

Tajfel's and Turner's overall "Social Identity Theory" attempts to explain the minimal group findings by proposing a three-pronged psychological process (social categorization, social identification, and social comparison) linking a group member's need for a positive self-image to discriminatory intergroup behaviors.

Social Categorization: Bruner (1956) states that "the main function of categorization is to reduce the complex object world to a more simple and manageable structure" (in Taylor & Doria, 1981, p. 83). This categorization process in the formation of social groupings is the same process associated with the construction and maintenance of stereotypes.

Social Identification: Tajfel defines "social identification" as the knowledge that one belongs to a group, along with the emotional and value significance attached to that membership. For Sumner (1906), the categorization of individuals into distinct ethnic groupings originated in the first human's struggles (and competition) to meet their basic needs. Social Identity theories insist, however, that the simple fact of belonging to one group over another, the mere subdivision or categorization of persons into in-groups and out-groups—even when issues of competition for scarce resources and incompatible group goals are absent—is enough to trigger ethnocentric (xenophobic, discriminatory) attitudes favoring the in-group (Allen & Wilder, 1975; Billig and Tajfel, 1973; Tajfel, Billig, Bundy, & Flament, 1971). A major premise in social identify theory, as proposed by Tajfel, is that social identities themselves create and maintain attitudinal and behavioral discriminations favoring the in-group (Tajfel, 1978, 1982; Tajfel and Turner, 1986). The stronger are the individuals' identification with their in-group, the greater is the tendency to perceive out-group members as undifferentiated members of another social category, and to perceive oneself and other in-group members as different or dissimilar from the out-group (i.e., the "out-group homogeneity effect"). This in turn provides the basis for stereotyping out-groups and out-group members.

Social Comparison: Social Comparison Theory states that identity is organized and maintained through intergroup comparison. It is the process by which individuals will pursue a positive self-identity by comparing one's sense of self with the relevant out-group, and in the process clarifying and crystallizing one's self identify. Therefore, for the individual to feel positive about membership in a social group, they must first feel positive about that social group. Group theorists, such as Festinger (1954), argue that "individuals are attracted to groups in which the members have opinions similar to their own so that they can evaluate their own opinions with precision." In this process, group formation is enhanced.

Fried (1961) proposed that the degree and intensity of conflict differed depending on the organization of the society. He differentiated between three levels of social organization: 1) Egalitarian Societies, that lack rank statuses; 2) Ranked Societies that have status differentials but not differentiated access to strategic resources of the society; and 3) Stratified Societies that have status differentials giving different access to strategic resources. He proposed that the intensity and severity of conflict increases from one level to the next. At the final level, however, the economic imperative for conflict becomes dominant because of the subgroup differentiation between those who have access to strategic resources and those who lack such access.

By deploying these social identity theories, we can begin to understand why perpetrators of f2f bullying and cyberbullying may select specific targets for malevolence over others. The mere fact that they comprise different social groupings, even though very superficial differentiations actually exist, could be enough to create, at least in the perpetrators' minds, sufficient reason for competition. This, along with the addition of the nature of social communication technologies, creates conditions for and often rationalizations cyberbullying.

In this connection, "Social Rank Theory," as used by Hawker and Boulton (2001), proposes that aggressive individuals actually hold a higher rank, power, or status within a social group. Therefore, aggressive behavior, and abuse in particular, may be reinforced, and it provides those who engage in aggressive behaviors a sense of belonging. Hawker and Boulton contend that aggression toward others serves a number of functions. First, it establishes and maintains a social hierarchy within a given group (an "in-group"), and second, it maintains distinctions between members of the in-group from members of other groups ("out-groups").

In addition, Teräshjo and Salmivalli (2003) contend that aggressive individuals fulfill the social "function" of establishing and reinforcing social norms. They found that students often justify abusive behaviors by blaming the targets of their attacks, and emphasizing that they somehow deserve the peer aggression or that they in some way deviate from the established peer social norms. This can be considered as a form of "ruthless socialization."

Experiences of being cyberabused, regardless of whether it is physical, relational, or verbal, appear to have devastating short- and long-term effects on youth as well as adults with consequences far beyond mere embarrassment.

Social Learning Theory

In social learning theory (sometimes referred to as "social cognitive theory)," Bandura (1986) proposes that individuals learn by observing others. Salancik and Pfeffer (1978) found that an individual's values, attitudes, and behaviors are greatly impacted by co-workers and peers. Even when individuals judge a particular behavior or actions to be morally wrong, the organizational environment—that is, the perceived attitudes and behaviors of peers or co-workers—can severely "neutralize" their previously held moral judgments. They then often take on the actions consistent with the perceived organizational climate (e.g., Vitell and Grove, 1987), especially individuals who are particularly susceptible to social influences, what Synder (1979) refers to as those high in "self-monitoring" who rely to a great extent on cues from social interactions to shape appropriate attitudes and behaviors. In this sense, then, behavior is not always an indication of beliefs or values, for an individual may take on actions in accordance with perceived accepted organizational or peer actions, even when those actions run counter to the individual's ethical judgment.

These findings have implications for cyberbullying as well. For instance, in the case of what has come to be known as online polling booths, young people create "booths" for large numbers of their peers, for example, to rate girls and boys as the "hottest," "ugliest," "most boring," "biggest dyke," or "wimpiest fag" in the school. Someone who might not ordinarily engage in such abusive behaviors can be caught up in the storm of social pressure to "contribute" their vote in order to fit in and also to avoid becoming a target of such abuse if they had not joined in the perceived group norms of engagement.

In addition to the nature of social communication technologies, which alters psychological conditions for users from that of usual interface interactions, we need to attempt to explain some of the possible reasons why perpetrators of cyberbullying target and victimize certain individuals and groups over others.

Bullying must not simply be seen as a "youth problem," but must be viewed as resulting from larger psychological and societal issues. Institutional bullying and harassment do not exist within a vacuum, but rather reflect and actually reproduce the messages and actions stemming from the social environment. I refer to this as "the social ecology of bullying and harassment" (see Starobin and Blumenfeld, 2013). Ecology can be defined as the relationships between organisms and their environment. We must, therefore, investigate the larger sociological and psychological environment for us to determine, understand, and if necessary, institute procedures to change our institutional environments.

Looking through the lenses of Social Rank Theory and Social Learning Theory, perpetrators of cyberbullying may need or desire higher social status within their social environments. Young people learn the social rules by others around them: family, peers, teachers, the media, social and political leaders, celebrities, and others. From them, they learn what and who are valued and scorned in their societies. This helps to explain some of the reasons why users of social communication technologies more often select certain individuals from specific social groups for cyberbullying than others.

According to the Federal Bureau of Investigation (2010), of the reported 6,624 single bias incidents, the motivation for 47.3% was a racial bias on the part of the perpetrators, 20.0% were religious biases, 19.3% comprised sexual orientation biases, and 12.8% comprised an ethnicity/national origin bias. Disability biases accounted for 0.6 percent of single-bias incidents (FBI, 2010).

People selected Phoebe Prince for cyberbullying as "a foreigner" from Ireland, Tyler Clementi for being gay, and Ryan Patrick Halligan for being perceived as gay and having a learning disability. As the song "You've Got To Be Carefully Taught" from the Rogers and Hammerstein (1947) Broadway musical makes clear:

> You've got to be taught to hate and fear. You've got to be taught from year to year. It's got to be drummed in your dear little ear. You've got to be carefully taught.
>
> You've got to be taught to be afraid of people whose eyes are oddly made, And people whose skin is a diff'rent shade. You've got to be carefully taught….

Young people in the schools and employees in the workplace learn the social norms and rules, while social communication technologies supply the nature and the means to initiate and maintain their "ruthless socialization."

A STRATEGY FOR CHANGE

Social Norms Theory

A number of strategies have been suggested to counter perceived social support for unethical actions both in real life and in cyberspace. One in particular of these theoretical foundations has come to be known as "Social Norms Theory." First suggested by Perkins and Berkowitz (1986), Social Norms Theory is based on the premise that behavior is often influenced by erroneous perceptions of how other members of a social group think and act. What an individual believes others think and do (in social norms theory called a "perceived norm") and what in fact are others' real attitudes and actions (an "actual norm") are often at odds. The distance between a perceived and an actual norm is referred to as "misperception." For example, Perkins and Berkowitz (1986) found that college students often overestimated the extent to which their peers supported unhealthy drinking behaviors, and that these misperceptions predicted how individuals drank.

Social Norms Theory involves interventions that are intended to correct misperceived social norms. A critical element in this approach is to correct misperceptions of norms by focusing on the positive and healthy attitudes and behaviors of the majority in an attempt to increase it. This element should be developed in consort with the use of information regarding these positive norms to direct interventions with abusers. Fabiano (1999) enumerates six stages in the social norms intervention process: 1) assessment to collect data; 2) selection of the normative message; 3) testing the message with the target group; 4) selecting the normative delivery strategy; 5) determining the "dosage" (amount, form) of the message; and 6) evaluation of the effectiveness of the message.

Focusing on peer influences, social norms interventions have shown promise, especially when combined with other strategies—for example, with detailed policy changes—in addressing issues related to changing unhealthy patterns of alcohol consumption and the use of tobacco, prevention of sexual assault, improvement of overall academic climate in an educational institution, and reducing discriminatory behaviors.

Social Norms Theory can be an effective strategy in the reduction of abusive behavior in real life and in cyberspace. In one study (Salmivalli et al, 1996), researchers found that between 80 to 90% of young people expressed aversion to bullying behavior and disapproved of people who bully others, though this proportion decreased somewhat during adolescence. The same study showed, however, that merely 10 to 20% of those surveyed actively intervened on behalf of those who were victimized by the bullying behavior of a peer or peers. In addition, Bigsby (2002) examined perceptions of bullying behavior in an elementary school and found that students and their parents overestimated (misperceived) the degree and amount of bullying behavior that occurred. This indicates that while bullying behaviors—and aggression in general—may be (mis)perceived as being an accepted norm by a significant number of people in a given environment, in reality, the vast majority find these behaviors distasteful at best. Social Norms Theory in many contexts has proven effective in empowering those that oppose an unhealthy or abusive behavior, as well as empowering "by-standers" who are aware of negative behaviors, but who feel powerless to intervene.

In addition, while it is not the intention here to give a comprehensive narrative on how to reduce and ultimately eliminate instances of cyberbullying, which social technologies provide in the public schools and communities—for what might work effectively in one area or school might not function in another—some foundational guidelines for educators can be considered.

Assessment: Hold public hearings, and/or conduct interviews, or distribute research surveys in your school, community, and/or your state/province to access the level and forms of face-to-face and cyberbullying.

Policies: State Departments of Education, school districts, and individual schools are encouraged to develop policies protecting students, staff, and faculty from bullying and cyberbullying.

Personnel Trainings: Schools are encouraged to offer on-going and comprehensive training to school personnel in violence prevention, suicide prevention, and specifically issues of bullying and cyberbullying. These training sessions will address issues of reporting as well.

Counseling: Schools and communities are encouraged to provide affirming school- and community-based counseling for the targets as well as the perpetrators of bullying and cyberbullying and their families.

Information in Libraries: School and community libraries are encouraged to develop and maintain up-to-date and age-appropriate collections of books, videos, CDs, DVDs, journals, magazines, posters, internet websites, and other information on issues that address bullying, and the nature of social communications technologies and cyberbullying prevention.

Curriculum & School Programs: It is imperative that school administrators and curricular specialists develop and incorporate accurate, up-to-date, on-going, comprehensive, and age-appropriate educational units across age groups and academic disciplines in school courses, programs, assemblies, and in school and community newspapers related to the nature of social communication technologies and how they can promote detrimental and inappropriate actions, including cyberbullying.

Teacher Certification: College and university teacher education programs are encouraged to include detailed lessons on the nature of social communication technologies and how they can provide users the space for cyberbullying and other toxic cyberactions.

In general, a holistic approach is the best approach to take. This means that all areas of the school as well as the local community need to come and work together to address the problem of bullying in all its forms, including on-line.

CONCLUDING QUESTIONS

Today, the world is undergoing a technological revolution to match and surpass the industrial revolution of previous decades and centuries, the likes of which still have unforeseen benefits and opportunities, challenges and consequences. Though there is no going back, we must consider and address a number of critical questions as we proceed. Sherry Turkle (1995) provides three questions worth exploring regarding the impact of social communication technologies:

- What will social communication technologies do to our commitment to other people?

- Will social communication technologies satisfy our needs for connection and social participation, or will they further undermine fragile relationships?
- What kind of responsibility and accountability will we assume for our virtual actions? (p. 178).

How we answer these and other important questions will have implications for this and future generations. Drawing students' attention to these questions, human thinking and behavior, the nature of technology, and the ways technology impacts human thinking and action all have an important role to play.

NOTE

[1] This is a research methodology used in social psychology for investigating the minimal conditions essential for discrimination to occur between groups. Researchers employing this approach have shown that even random or superficial differences between groups (e.g. the color of their hats) can activate a propensity to favor one's own group and disfavor or act against others.

REFERENCES

Allen, V. L., & Wilder, D. A. (1975). Categorization, belief similarity, and group discrimination. *Journal of Personality and Social Psychology, 32*. 971–977.

American Psychological Association. (2004, July). APA resolution on bullying among children and youth.

Asher, S., Hymel, S., & Renshaw, P. D. (1984). Loneliness in children. *Child Development, 55*, 1456–1464.

Asher, S. R., Parkhurst, J. T., Hymel, S., & Williams, G. A.(1990). Peer rejection and loneliness in childhood. In S. R. Asher & J. D. Coie (Eds.), *Peer rejection in childhood.* (pp. 253–273). Cambridge, UK: Cambridge University Press.

Bandura, A. (1986). *Social foundations of thought and action.* Englewood Cliffs, NJ: Prentice-Hall.

Bigsby, M. J. (2002). Seeing eye to eye? Comparing students' and parents' perceptions of bullying behavior. *School Social Work Journal, 27*(1), 37–57.

Billig, M., & Tajfel, H. (1973). Social categorization and similarity of intergroup behavior. *European Journal of Social Psychology, 3*. 27–52.

BloomBecker, B. (1990). *Spectacular computer crimes*, Homewood, IL: Dow Jones-Irwin.

Blumenfeld, W. J., & Cooper, R. (2010). LGBT and allied youth responses to cyberbullying: Policy implications. *International Journal of Critical Pedagogy, 3*(1), 114–133.

Bond, L., Carlin, J. B., Thomas, L., Rubin, K., & Patton, G. (2001). Does bullying cause emotional problems? A prospective study of young teenagers. *British Medical Journal, 323*, 480–484.

Brain, M. (n.d.). How computer viruses work. Howstuffworks. Retrieved November 4, 2009 from http://www.howstuffworks.com/virus.htm/printable.

Brewer, M. B. (1979). In-group bias in the minimal intergroup situation: A cognitive-motivational analysis. *Psychological Bulletin*, 307–324.

Bruner, E. M. (1956). Primary group experiences and the process of acculturation. *American Anthropologist, 58*, 605–623

Crocker, J., & Luhtanen, R. (1990), Collective self-esteem and ingroup bias. *Journal of Personality and Social Psychology, 58*,60–7.

Dawkins, J. (1996). Bullying, physical disability and the pediatric patient. *DevelopmentalMedicine and Child Neurology, 38*, 603–612.

DeVoe, J., Kaffenberger, S., Chandler, K. (2005) *Student reports of bullying. School crime supplement to the National Crime Victimization Survey*. Education Statistics Services Institute.

Fabiano, P. (1999). Learning lessons and asking questions about college social norms campaigns.

Presentation given at the Second National Conference on the Social Norms Model: Science Based Prevention, Big Sky, MT, July 29.

Farrington, D. P. (1993). Understanding and preventing bullying. In M. Tonry & N. Morris (Eds.) Crime and Justice, Vol 17, (pp. 381–458). Chicago: University of Chicago Press.

Federal Bureau of Investigation. (2010). *Hate crimes statistics, 2010.* (http://www.fbi.gov/news/pressrel/press-releases/fbi-releases-2010-hate-crime-statistics) Retrieved, August 25, 2012.

Festinger, L. (1954). *A theory of social comparison process.* Human Relations, *7*, 117–140.

Finkelhor, D., Mitchell, K., Wolak, J. (2006). *Online victimization of youth: Five years later.* (YISS-2). Alexandria, VA: National Center for Missing & Exploited Children.

Finkelhor, D., Mitchell, K., Wolak, J. (2000). *Online victimization: A report on the nation's youth.* (YISS-1). Alexandria, VA: National Center for Missing & Exploited Children.

Fried, M. H. (1961). Warfare, military organization, and the evolution of society. *Anthropologica, 3.* 134–147.

Freud, S. (1912). *The dynamics of transference. The standard edition of the complete psychological works of Sigmund Freud, Volume XII* (1911–1913): The case of Schreber, papers on technique and other works. London: The Hogarth Press and the Institute of Psychoanalysis.

Garrett, A. (2003) *Bullying in American schools. Causes, preventions, interventions.* Jefferson: Mcfarland & Company, Inc., Publishers.

GLSEN (2005). National School Climate Survey. New York: Gay, Lesbian, and Straight Education Network.

Harrington, S. J. (1995). The anomaly of other directedness when normally ethical I.S. personnel are unethical. *Special interest group on computer personnel research annual conference proceedings of the 1994 computer personnel research conference on reinventing I.S.: Managing information technology in changing organizations*, Alexandria, VA. pp 35–43.

Harrington, S. (1994). The impact of codes of ethics on information systems personnel. Special interest group on computer personnel research annual conference proceedings of the 1994 computer personnel research conference on reinventing I.S.: Managing information technology in changing organizations, Alexandria, VA. pp 199–207.

Hawker, D. S. J., & Boulton, M. J. (2001). Subtypes of peer harassment and their correlates: A social dominance perspective. In J. Juvonen & S. Graham (Eds.), *Peer harassment in school: The plight of the vulnerable and victimized.* (pp. 378–397). New York: Guilford Press.

Hershberger, L. & D'Augelli, A. (1995). The impact of victimization on the mental health andsuicidality of lesbian, gay, and bisexual youths. *Developmental Psychology, 31*, 65–74.

Hunter, J. (1990). Violence against lesbian and gay male youths. *Journal of Interpersonal Violence, 5*, 295–300.

Hymel, S., Rubin, K. H., Rowden, L., & LaMare, L. (1990). Children's peer relationships: Longitudinal prediction of internalizing and externalizing problems from middle to late childhood. *Child Development, 61*. 2004–2021.

Kaiser Family Foundation (2005). *Generation M: Media in the lives of 8–18 years-olds.* (http://www.kff.org/entmedia/7251.cfm). Retrieved March 28, 2007

Kapelovitz, L. H. (1987). *To love and to work/A demonstration and discussion of psychotherapy.* Lanham, MD: Jason Aronson Inc.

Kim Y-S, Koh Y-J, Leventhal B. L. (2005). School bullying and suicidal risk in Korean middle school students. *Pediatrics, 115*, 357–363.

La Greca, A. M., & Harrison, H. W. (2005). Adolescent peer relations, friendships, and romantic relationships: Do they predict social anxiety and depression. *Journal of Clinical Child and Adolescent Psychology, 34*, 49–61.

Lenhart, A., Purcell, K., Smith, A., & Zickuhr, K. (2010). Social media & mobile internet use among teens and young adults. Pew Internet and American Life Project.

Lenhart, A., Madden, M., & Hitlin, P. (2005). Teens and Technology: Youth are Leading the Transition to a Fully Wired and Mobile Nation. Washington D.C.: Pew Internet & American Life Project.

McClure, M. & Shirataki, S. (1989). Child psychiatry in Japan. *Journal of the American Academy of Child Adolescent Psychiatry, 28*, 488–492.

Media Awareness Network. (2000). Young Canadians in a wired world: Parents and youth focus groups in Toronto and Montreal. Environics Research Group.

Nabuzka, O., & Smith, P. (1993). Sociometric status and social behaviour of children with and without learning difficulties. *Journal of Child Psychology and Psychiatry, 34,* 1435–1448.

Nansel, T. R., Overpeck, M., Pilla, R. S., Ruan, W. J., Simons-Morton, B., & Scheidt, P. (2001). Bullying behaviors among U.S. youth: Prevalence and association with psychosocial adjustment. *JAMA: Journal of the American Medical Association, 285*(16), 2094-2100.

National i-SAFE Survey (2004, June 28). National I-SAFE survey finds over half of students are being harassed online. (www.isafe.org).

Olweus, D. (1993). *Bullying at school: What we know and what we can do.* Cambridge: Blackwell.

Olweus, D. (1991). Bully/victim problems among schoolchildren: Basic facts and effects of a school based intervention program. In D. J. Pepler & K. H. Rubin (Eds.), *The development and treatment of childhood aggression* (pp. 411–448). Hillsdale, NJ: Erlbaum.

Orpinas, P., & Horne, A. M. (2006). Bullying prevention: Creating a positive school climate and developing school competence. Washington, DC: American Psychological Association.

Owens, L, Shute, R., Slee, P. (2000). "Guess what I just heard!" Indirect aggression among teenage girls in Australia. *Aggressive Behavior, 26*(67), 67–83.

Perkins, H. & Berkowitz, A. (1986). Perceiving the community norms of alcohol use among students: Some research implication for campus alcohol education programming. *International Journal of the Addictions, 21*(9/10), 961–976.

Pescara-Kovach, L. (2006). Using multimedia to reduce bullying and victimization in third-grade urban schools. Professional School Counseling. http://www.thefreelibrary.com/Using+multimedia+to+reduce+bullying+and+victimizatio n+in+third-grade...-a0157032927. Retrieved on 3/20/2007.

Pilkington, N. & D'Augelli, A. (1995). Victimization of lesbian, gay, and bisexual youth in community settings. *Journal of Community Psychology, 23,* 34–56.

Prinstein, M. J., Boergers, J., & Vernberg, E. M. (2001). Overt and relational aggression in adolescents: Social-psychological functioning of aggressors and victims. *Journal of Clinical Child Psychology, 30,* 477–489.

Rigby, K. (2003). *New perspectives on bullying.* London: Jessica Kingsley.

Rigby, K., & Slee, P. T. (1999). Australia. In P. K. Smith.. *The nature of school bullying.* London, Routledge, pp. 324–339.

Salancik, G., & Pfeffer, J. (1978). A social information processing approach to job attitudes and task design, *Administrative Science Quarterly, 23.* 224–253.

Salmivalli, C., Lagerspetz, K., Björkqvist, K., Österman, K., & Kaukiainen, A. (1996). Bullying as a group process: Participant roles and their relations to social status within the group, *Aggressive Behavior, 22,* 1–5.

Sherif, M., Harvey, O. J., White, B. J., Hood, W. R., & Sherif, C. W. (1961). *Intergroup conflict and cooperation: The robbers' cave experiment.* Norman: University of Oklahoma Press.

Sherif, M., & Sherif, C. W. (1953). *Groups in harmony and tension.* New York: Harper

Sidanius, J., & Pratto, F. (2001). Social dominance: An intergroup theory of social hierarchy and oppression. Cambridge, UK: Cambridge University Press. Smith, P. K., Morita, Y., Junger-Tas, J., Olweus, D., Catalano, R., & Slee, P. (Eds.). (1999). *The nature of school bullying: A cross-national perspective.* London: Routledge. Brothers.

Smith, P. K., Morita, Y., Junger-Tas, J., Olweus, D., Catalano, R., & Slee, P. (Eds.). (1999). *The nature of school bullying: A cross-national perspective.* London: Routledge.

Spero News. (2006, August 17). 1 of 3 teens are victims of cyber bullying. http://www.speroforum.com/site/article.asp?idCategory=31&idsub=129&id=4494&t=1 of 3 teens are victims. (Retrieved March 13, 2007.)

Starobin, S. S., & Blumenfeld, W. J. (2013). Utilization of social ecology of bullying and social reproduction frameworks to interrogate workplace bullying at community colleges. In J. Lester (Ed.), *Workplace bullying in higher education.* New York: Routledge.

Staub, E. (1978). *Positive social behavior and morality—Volume I: Social and personal influences.* NY: Academic Press.

Strauss, C. C., Forehand, R., Frame, C., y Smith, K (1984). Characteristics of children with extreme scores on the Children's Depression Inventory. *Journal of Child Clinical Psychology, 13*, 227–231.

Suler, J. (2005). The basic psychological features of cyberspace: Elements of a cyberpsychology model. http://users.rider.edu/~suler/psycyber/basicfeat.html, Retrieved May 4, 2012.

Suler, J. (2001). Psychology of cyberspace—The online disinhibition effect. http://www.rider.edu/~suler/psycyber/disinhibit.html. Retrieved May 22, 2004.

Sutton, J., & Smith, P. K. (1999). Bullying as a group process: An adaptation of the participant role approach. *Aggressive Behavior, 25*, 97–111.

Swearer, S. M., Grills, A. E., Haye, K. M., & Cary, P. T. (2004). Internalizing problems in students involved in bullying and victimization: Implications for intervention. In D. L. Espelage & S. M. Swearer, (Eds.). *Bullying in American schools: A social-ecological perspective on prevention and intervention*. (pp. 63–83). Mahwah, NJ: Lawrence Erlbaum Associates.

Tajfel, H. (1982). *Social identity and intergroup relations*. Cambridge, UK: Cambridge University Press.

Tajfel, H. (1978). *Social psychology of minorities*. Cambridge: Cambridge University Press.

Tajfel, H., & Turner, J. C. (1986). The social identity theory of intergroup behavior. In S. Worchel & W. G. Austin (Eds.), *Psychology of intergroup relations* (2nd Edition). Chicago: Nelson-Hall.

Tajfel, H., Billig, M. G., Bundy, R. F., Flament, C. (1971). Social categorization and intergroup behavior. *European Journal of Social Psychology, 1*. 149–177.

Taylor, D. M., & Doria, J. R. (1981). Self-serving and group-serving bias in attribution. *The Journal of Social Psychology 113*, 201–211.

Teräshjo, T., & Salmivalli, C. (2003). "She is not actually bullied." The discourse of harassment in student groups. *Aggressive Behavior, 29*. 134–154.

Turkle, S. (1995). *Life on the screen: Identity in the age of the Internet*. New York: Simon & Schuster.

UCLA Internet Report, 2003, February. *UCLA Internet report: Surveying the digital future—year three*. (http://www.freep.com/money/tech/mwend17_20031117.htm). Retrieved July 23, 2004.

Vitell, S. J. & Grove, S. J. (1987). Marketing ethics and the techniques of neutralization. *Journal of Business Ethics, 6*, 433–438.

Vosk, B., Forehand, R., Parker, J. B., & Rickard, K. (1982). A multi-method comparison of popular and unpopular children. *Developmental Psychology, 18*, 571–575.

Ybarra, M., Mitchell, K. (2004). Youth engaging in online harassment: Associations with caregiver-child relationships, Internet use, and personal characteristics. *Journal of Adolescence, 27*(3), 319–336.

Yude, C., Goodman, R., & McConachie, H. (1998). Peer problems of children with hemiphegia in mainstream primary schools. *Journal of Child Psychology & Psychiatry & Allied Disciplines, 39*, 533–54.

Warren J. Blumenfeld
School of Education,
Iowa State University

SECTION III

TEACHER EDUCATION AND THE NATURE OF TECHNOLOGY

CHAPTER 15

BENJAMIN C. HERMAN

CONVERGENCE OF POSTMAN AND VYGOTSKY PERSPECTIVES REGARDING CONTEMPORARY MEDIA'S IMPACT ON LEARNING AND TEACHING

How can a society that exists on instant mashed potatoes, packaged cake mixes, frozen dinners, and instant cameras teach patience to its young?

– Paul Sweeney

INTRODUCTION

The nature of technology (NOT) addresses issues such as what technology is, how and why technology is developed, the unequal beneficial and malevolent consequences that accompany any technology, the way technology influences how to think and act, and the relationship between society and technology. However, the general public's thinking about technology mostly entails examples of technology and their uses, not deeply knowing about technology and its nature. The International Technology Association's (ITEA, 2007) Standards for Technological Literacy (STL), notes: "Everyone recognizes that such things as computers, aircraft, and genetically engineered plants are examples of technology, but for most people the understanding of technology goes no deeper" (p. 22). Unfortunately, technology education is, with rare exceptions, directed primarily, perhaps solely, at training students how to use technology while ignoring educating students about the NOT. Niederhauser & Lindstrom (2006), after reviewing 716 teachers' narratives about their technology instructional practices, stressed that the social, ethical, and human issues inherent in technology must be given much higher profiles in classrooms.

The STLs themselves are partly to blame due to their relatively vague nature regarding the NOT. For instance, ITEA's Standards (2007, p. 22) represent the NOT as: "what technology is, its general core concepts, and the relationships among various technologies and between technology and other areas of human endeavor." Despite linking the NOT with "areas of human endeavor", an emphasis on how technology is situated within a social, cultural, and environmental context, they omit

Michael P. Clough, Joanne K. Olson and Dale S. Niederhauser (Eds.), The Nature of Technology: Implications for Learning and Teaching, 293–328.

much of what is fundamental for understanding technology and its nature (e.g. see Postman, 1985; 1995). For instance, the STLs (2007, p. 62) most relevant to the NOT advocate 9–12 grade students learn that:

1. Changes caused by the use of technology can range from gradual to rapid and from subtle to obvious.
2. Making decisions about the use of technology involves weighing the trade-offs between the positive and negative effects.
3. Ethical considerations are important in the development, selection, and use of technologies.
4. The transfer of a technology from one society to another can cause cultural, social, economic, and political changes affecting both societies to varying degrees.

While these STL are important and do touch on some NOT issues (e.g. how technology has altered people's perceptions), text accompanying the STL tends to frame technology use and its consequences in the context of deliberate human choice, rather than making evident the inherent messages and biases of technologies that impact human choice, thought, and action. This perspective is evident in the STL (p. 30) which states:

> Technology is intricately woven into the fabric of human activity and is influenced by human capabilities, cultural values, public policies, and environmental constraints. Students need to recognize these influences and how their integration affects technological development.

A more accurate and fair statement would be:

> Technology is intricately woven into the fabric of human activity and is influenced by and *influences* human capabilities, cultural values, public policies, and environmental constraints. Students need to recognize these influences and how their integration affects technological development *and is also determined by technological development.*

To illustrate how technology influences, and in some ways determines, human thinking and action without our awareness, this chapter focuses on the nature of media technology (particularly television and the internet) and how it impacts how people think, and the way they conceptualize learning, teaching and schooling. The chapter ends with a description of a course designed to educate preservice and inservice secondary science teachers about the nature of media technology, the implications for teaching and learning, and how to educate their students about the nature of media and the NOT more broadly.

THE NATURE OF TELEVISION AND INTERNET MEDIA

Media can be thought of as any agent that stores and delivers information. Television, radio, news services, books, magazines, articles, billboards, texting, the internet and road signs are common examples of media. Different media have different biases; characteristics that determine the form (e.g. context, scope, duration, use of imagery and symbols, language characteristics) of that medium's messages. Through the form of its messages, each medium, conveys and imposes upon the media consumer its own metaphor of what the world is like. That is, different media provide different templates for our thoughts, which can also be viewed as metaphors or "conceptual symbols[1]". Furthermore, the form of a particular medium also models for its consumers how to view learning and communicating. Therefore, if a particular medium is primarily used by a culture to transfer ideas, then that medium will define through its languages and symbols that culture's knowledge and views about learning and communicating ideas.

The assertions in the previous paragraph remain true even if different media intend to convey ideas about the same topic. For instance, because of the disparate nature of their language and symbol expression capabilities, Facebook posts about one's immersion in nature and Thoreau's book *Walden* (1854) convey very different metaphors of what being immersed in nature is like and how to think and communicate about being immersed in nature. The former was developed in the contemporary culture defined by television and internet media. The latter is much older and was developed in a culture defined by print media. The former medium encourages short messages and perhaps some images, which superficially convey one's experiences in nature. The latter medium devotes numerous pages of sequenced and illustrative prose to fully articulate Thoreau's experiences in nature and the impact of those experiences. The significance of Facebook posts about one's immersion in nature is marginalized because it exists within a media space that competes with other Facebook posts for the media consumer's attention and will either be quickly deleted or more likely be buried within a sea of other posts. The latter provides a platform devoted solely to Thoreau's message, which is frozen in text. The former medium calls for little deep thought about the extrinsic and intrinsic qualities gained by being in nature. Instead, those viewing the Facebook post are left with a fleeting thought about nature which, like most Facebook posts, is insignificant and soon forgotten. The latter requires investment from the reader to deeply contemplate Thoreau's experiences and connect them to their own experiences in nature. The former provides a model for the media consumer that significant experiences in nature should be communicated about through a few trivial statements and images that are disconnected from other life experiences. The latter provides a model for the media consumer that significant experiences in nature should be communicated through rich logical descriptions that connect to other aspects of life. That a comparison is being made between messages from Thoreau and a general Facebook poster matters little. Consider, for instance, the following: To what extent would Thoreau's writings and the meaning of those writings retain form and significance if he instead posted them on Facebook? Furthermore, to what extent would Thoreau's book *Walden* (1854) be perceived as significant and meaningful if it was published in today's culture which is defined by television and internet media?

Television and the internet are strikingly similar in terms of how they convey information and work against conceptual thought and meaningful discourse (Olson & Clough, 2001). However, many people wrongly maintain that meaningful discourse can and does occur on television and the internet, and that these media help people think about serious issues. This is because much of what we have learned and know about serious issues and how to communicate about them is garnered from news casts, documentaries, posts, and oratories that appear on television and the internet. With a click of the remote or the mouse, one can access a multitude of seemingly serious subjects. By having so much information at our fingertips we feel that we are more "informed" than ever about substantive and compelling content. However, herein lays the irony. These media deliveries are potentially harmful to our culture because they: 1) are the primary source of public information about seemingly credible, substantive, and compelling issues; and 2) are perceived as a credible model of how to think and engage in discourse about these issues. Therefore, through internet and television consumption people *feel* as if they are deeply informed and know how to effectively learn and communicate about serious issues. The very form and nature of television media models for and teaches people that learning and communicating about serious matters should be entertaining, immediate, restricted to trivialities, limited in context, and occur in short bursts of information. In other words, what has been modeled for the public is exactly the opposite of what is required for meaningful discourse and learning about important ideas and issues - both crucial for participating in and preserving our democratic society.

What about the nature of television and the internet may contribute to the demise of key components in a judicious democratic society (e.g. the free exchange of important ideas through critical discourse)? The form of television and internet media is limited to conveying short and abrupt segments of trivial information through superficial and entertaining imagery. Furthermore, image based information through television and internet media need not follow a logical sequence, is largely independent of previous and subsequent information, and quickly expires. However, the most dangerous qualities of television and internet media are that these media often convey their scattered, discontinuous, and entertaining trivialities in a contrived manner thus *appearing* to be compelling, informative, and delivered in serious contexts. Media consumers can wrongly perceive they actually have important information, deeply understand that information, and can effectively think, act, and communicate about important matters. Media consumers are apt to think that learning, knowing, and communicating about difficult and complex ideas is akin to the game Trivial Pursuit – entertaining factoids devoid of context and meaning. Because the American public's primarily source of news increasingly comes from the internet and television, the ability for our public to engage in meaningful discourse and decision making about complex issues that truly matter may very well be in peril.

An illustrative example is how the form of television and the internet have impacted political discourse. During the 1992 U.S. presidential election, the

Washington Post presented research from two independent studies noting that the average sound bite length of presidential candidates' statements on the network evening news had dramatically shortened from thirty-four uninterrupted second in the 1968 elections to nine uninterrupted seconds in the 1988 elections. While thirty-four seconds may seem long compared to nine seconds, it pales in comparison to the uninterrupted hours Lincoln and Douglas debated political philosophies in the mid-1800's (Postman, 1985). This research sparked great concern among academicians, politicians, and journalists. For instance, Michael Dukakis was quoted as saying "If you couldn't say it in less than 10 seconds, it wasn't heard because it wasn't aired" (Fehrman, 2011). Reacting to this research, CBS Corporation (CBS) instituted a policy for the 1992 elections that required sound bites of presidential candidates to be at least 30 uninterrupted seconds. Noting a marketing opportunity, CBS dubbed the policy as a "public service" (Fehrman, 2011). However, CBS's efforts failed because the political candidates had great difficulty engaging in discourse for 30 continuous seconds. Perhaps, as Montagne (NPR, 2011) noted, they "had learned to keep their thoughts very, very, short." Television returned to airing short sound bites and the internet followed suit. Current television sound bites of politicians are just over eight seconds and politicians are advised by consultants how to win arguments using short unsubstantial quips. Also, politicians use the internet to reach the public through You Tube and Twitter. Interestingly, as Montagne (NPR, 2011) points out "tweets are limited to 140 characters - words that would take about nine seconds to say." Whether citizens, conditioned by television and internet media to receive their political "discourse" in short entertaining sound bites, could today follow political discourse of any length is highly suspect. As Postman (1985, p. 7) noted, "You cannot do political philosophy [i.e. meaningful discourse about serious issues] on television. Its form works against the content."

This example is one of many (e.g. news casts about climate change, infomercials about weight loss pills, documentaries about the conflicts in the Middle East, the current economic crisis, etc.) that illustrate how the form of television and internet media limits serious in depth discourse and encourages a limited, trivial, and dangerous education. That is, television's and the internet's form is one of quick shifting imagery that does not encourage or permit time for arduous reflection. Camera shots and verbal explanations do not linger for more than a few seconds before the viewer is provided another stimulating jolt of insignificant information shrouded in entertainment. Again, as Postman (1985) noted, the most serious outcome of all this is that the media consumer comes to think (to expect!) that learning and communication about serious matters should be entertaining, require little time and effort, be restricted to trivialities, be limited in context, and occur in short bursts of information. This is dangerous when an entire culture (e.g. the American public) consumes television and internet media as their primary sources of information about how to think, communicate, and take action about serious issues.

Adatto (1990) notes that as sound bite lengths decreased from 1968 to 1988, visual images of U.S. candidates, absent of their orations, increased by more than

300%. Also, television increasingly forced U.S. presidential campaigns to become grounds for entertainment rather than for meaningful civil discourse of major issues such as foreign policy, the environment, and education. U.S. presidential campaigns are increasingly "orchestrated" to appear on the news (Adatto, 1990), and the television commercial is now the "fundamental metaphor for political discourse" (Postman, 1985, p. 126). From 1968 to 1988 the number of excerpts from candidates' commercials broadcast by the evening news during campaigns had increased from two to 125 (Adatto, 1990). As former U.S. President Ronald Reagan noted, "Politics is just like show business" (Postman, 1985, p. 125).

Adatto (1990) describes the ascension of television imagery use and its conflicted attempt to avoid revealing the guilefully contrived nature of that imagery:

> The language of political reporting was filled with accounts of staging and backdrops, camera angles and scripts, sound bites and spin control, photo opportunities and media gurus. So attentive was television news to the way campaigns constructed images for television that political reporters began to sound more like theater critics, reporting more on the stagecraft than the substance of politics. (p. 20)

Attempting to *appear* thoughtful while conveying messages through imagery, television downplays meaningful and serious discourse about politics. For instance, Adatto notes that former U.S. President Bush's Labor Day campaign appearance at Disneyland was an image rich event covered as a "performance for television":

> 'In the war of Labor Day visuals,' CBS's Bob Schieffer reported, 'George Bush pulled out the heavy artillery. A Disneyland back drop and lots of pictures with the Disney gang.' When Bruce Morton covered Dukakis riding in a tank, the story was the image. 'In the trade of politics, it's called a visual,' said Morton. 'The idea is pictures are symbols that tell the voter important things about the candidate. If your candidate is seen as weak on defense, put him in a tank.' (Adatto, 1990, p. 20)

The aforementioned examples reveal how the very form of television and internet media frames and thus hinders, deep learning and discourse about truly important issues. First, the images are numerous, quickly shifting, over stimulating, entertaining, and distract the media consumer from actually focusing on whether the important issues were addressed. But this is not perceived as negative by the television and internet consumer. Rather, to remain favorable to the television and the internet consumer these media must simultaneously entertain while *appearing* to address important issues. Second, any narrative accompanying the images does not focus on the serious issues being addressed. They focus on what demands the most attention....the image itself. Therefore, little room is left for an informing, logically substantiated, and substantive message.

However, because the image based messages are associated with a serious context and *appear* to address important issues, the media consumer feels as if they have learned something compelling and substantive about those issues. More worrisome, the media consumer also takes away the overarching message regarding how to effectively think, communicate, and act about those issues. This is what makes this context and form of television and internet media the most dangerous for society. That is, again, they imply meaningful learning and communication should be entertaining, immediate, restricted to trivialities and short bursts of information, and requires little to no arduous discourse on even the most serious matters.

COMMUNICATION AND ITS IMPLICATIONS FOR LEARNING

Because television and the internet are the dominant media forms in our culture, they exert a significant impact on the form of other media and the kind of learning that occurs via those media. *USA Today* with its relatively short articles, news snippets, and extensive graphics exemplifies how newspapers have been impacted by television and the internet. The manner we communicate with one another through Facebook is further indication of television's and the internet's influence. Texting and twittering, also reflect the influence of television and the internet, but in a more subtle manner: communication more and more reflects the short attention span that television and the internet promote. This section presents how the media people use impacts what they learn and powerfully conveys what learning and communication look like.

How People Learn through Social Contexts

> "Consciousness is reflected in a word as the sun in a drop of water. A word relates to consciousness as a living cell relates to a whole organism, as an atom relates to the universe. A word is a microcosm of human consciousness." –Lev Vygotsky (1986, p. 256)

Vygotsky's social constructivism accounts for how learning is subject to the media present within the social contexts people experience. Common to other constructivist frameworks (e.g., Piaget, 1952; 1973a & b; 1977; 1978), Vygotsky's social constructivism frames learning as an active process where people conceptually engage already held knowledge and beliefs to make sense of the world around them. However, Vygotsky's sociocultural approach to understanding learning moves beyond other frameworks and establishes that learning and development through "human activities take place in cultural contexts, are mediated by language and other symbols, and can be best understood when investigated in their historical development" (John-Stiener & Mahn, 1996, p. 191).

Conceptual development in social contexts through semiotic mechanisms. John-Stiener and Mahn (1996, p. 192), citing Vygotsky, claim that people mentally develop through the "transformation of socially shared activities into internalized process." Vygotsky advocated that the social contexts an individual experiences will determine their conceptual development, self-reliance, and higher order thinking:

> Every function in the cultural development of the child comes on the stage twice, in two respects; first in the social, later in the psychological, first in relations between people as an interpsychological category, afterwards within the child as an intrapsychological category…All higher psychological functions are internalized relationships of the social kind, and constitute the social structure of personality. (Vygotsky, 1981, as cited in Valsiner, 1987, p. 67)

Within any social context are symbols (i.e. semiotic mechanisms) that carry culturally and socially agreed upon meanings and play a key role in moving thought development from an "interpsychological category" to one that is an "intrapsychological category". In other words, semiotic mechanisms provide a medium for bridging social and individual cognitive function, where socially created and accepted symbols become internalized as personalized conceptual symbols. Language and symbolic symbols (Vygotsky, 1981), and computers, calendars, and even paint brushes (John-Stiener & Mahn, 1996, p. 193) are all examples of semiotic mechanisms used in "sociocultural discourse [and]….central to the appropriation of knowledge through representational activity by the developing individual." Semiotic mechanisms carry socially agreed upon meaning and directives about how to think and act.

Words, thought and language. Vygotsky (1986, p. 255; and in Howe, 1996, p. 41) emphasized how words and language serve as semiotic mechanisms that play a crucial role in thought and conceptual development. Language and thought share a dynamic interplay where language is a vehicle for thought's inception and thought gives language meaning and life (Howe, 1996). Written and spoken language (What Vygotsky refers to as *communicative speech* or *external speech*), like other semiotic mechanisms, serve as a medium in social contexts for purposefully bridging social function and individual thought. Vygotsky noted how semiotic mechanisms conveyed in social contexts impact the development, mediation, and expression of verbal thought *within* individuals (i.e. inner speech) which helps create our perceptions of what the world is like. For instance, consider the following words and phrases: "*stop sign*"; "*green traffic light*"; "*serious thought*"; "*science teaching*"; "*Einstein*"; "*television*"; "*internet*"; and "*books*". Each of these words and phrases summon a set or, perhaps a network, of conceptual symbols through inner speech. Furthermore, different conceptual symbols will be prompted with different combinations of words. For instance, the words: "*green traffic light*" together prompt different conceptual metaphors than the words: green; traffic; and light in isolation.

Vygotsky (1986) described the pathway through which external speech is translated into personal conceptual symbols and vice versa. He specifically focused on the role inner speech plays as gatekeeper in this pathway (see also Figure 1).

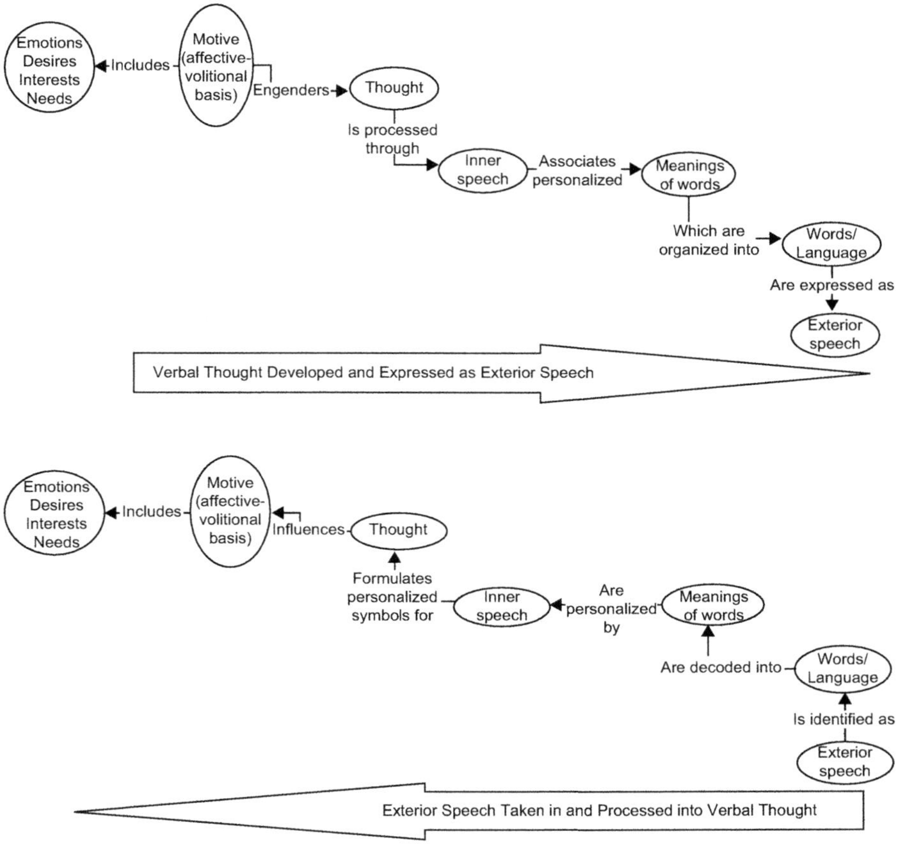

Figure 1. Vygotsky's pathways of verbal thought to exterior speech and vice versa.

Inner speech is for one's self; external speech is for others (p. 225)....Absence of vocalization per se is only a consequence of the specific character of inner speech which is neither an antecedent of external speech nor its reproduction in memory, but is, in a sense the opposite of external speech. The latter is the turning of thoughts into words, their materialization and objectification. With inner speech the process is reversed, going from outside to inside. Overt speech sublimates into thoughts (p. 226)....[I]nner speech serves as a draft not only in written speech but also in oral speech (p. 243)...In reality, the development of verbal thought takes the opposite course: from the motive that engenders a thought to the shaping of a thoughts, first in inner speech, then in meaning of words, and finally in words.

Figure 1 is a very simplified illustration of Vygotsky's ideas. The transition from inner speech to external speech is not a simple translation of language. Rather, inner speech works as a "filter" by processing one's thoughts into personalized conceptual symbols or meanings that are organized and transmitted as words, language, and other expressions through semiotic mechanisms. In reverse, inner speech serves again as a "filter" of sorts for bridging language and thought through engaging one's own personalized meanings or conceptual symbols. Vygotsky (1986, p. 249) describes the nature of inner speech below:

> Inner speech is to a large extent thinking in pure meanings. It is a dynamic, shifting, unstable thing, fluttering between word and thought, the two more or less unstable, more or less delineated components of verbal thought.

Moreover, inner speech is saturated with meaning, contextualized, predicated, joined in complex ideas, determines the extent an idea is contemplated and beyond the comprehension of anyone but the beholder. In other words, inner speech is saturated by the highly personalized conceptual symbols that are accumulated through personal experience.

Internalization. Vygotsky's concept of internalization explains how experiences, past and present, in various social contexts influence the development and engagement of personal conceptual symbols through inner speech. Through inner speech people draw from previous and current experiences in social contexts to construct and internalize personal conceptual symbols and meanings about those experiences. Therefore, the social contexts and their accompanying semiotic mechanisms are no longer something that must be concretely experienced in order for the individual to develop a relevant thought. Rather, the individual may employ inner speech to draw upon their conceptual symbols about those previous experiences, particularly when thinking about relatable contexts.

Zone of proximal development. Vygotsky (1978, p. 86) refers to the zone of proximal development as "the distance between an individual's actual development through independent problem solving and the individual's potential development via problem solving under adult guidance and or in collaboration with more capable peers." What this means is that the social context we experience impacts our ability to understand ideas. These social contexts play a significant role in determining the nature and scope of our inner speech, conceptual symbols, and internal verbal thoughts. John-Steiner and Mahn (1996, p. 198), citing several sociocultural theorists (Cole & Engstrom, 1993; Chang-Wells & Wells, 1993; John-Steiner, Panofsky, & Smith, 1994; Rogoff, 1994), describe learning within the zone of proximal development as a consequence of participating in a community of practice, and that the learning in a community is shared, interactive, and contextual. Citing Brown (1992) and Brown *et al.*, (1993), they also note that the agents affecting learning within the zone of proximal development include not only people, but also tools (i.e. semiotic mechanisms) within the social context — such as books, videos, and computers — that carry symbols of learning and knowing.

Spontaneous and nonspontaneous concepts. One of the more profound ideas proposed by Vygotsky was that of the interplay between spontaneous concepts learned outside of school and nonspontaneous (or in Vygotsky's terms " scientific") concepts learned in school. Spontaneous concepts are learned in social contexts devoid of purposeful instruction, and tend to be presented and learned in a scattered and unsystematic fashion with little networking to other concepts. Alternatively, nonspontaneous concepts are learned in social contexts that are focused on the explicit teaching and learning of systematically organized and networked concepts (e.g. school). Vygotsky (1986) argues that these two types of concepts and their development are interdependent:

> We believe that the two process-the development of spontaneous and nonspontaneous concepts are related and constantly influence each other. They are parts of a single process: the development of concept formation which is affected by varying external and internal conditions but is essentially a unitary process, not a conflict of antagonistic, mutually exclusive forms of thinking. (p. 157)

Thus, how an individual learns and communicates in one social context will transfer to and impact how they learn and communicate in another. This is because learners retain and use the inner speech and conceptual symbols they constructed and internalized through previous social contexts.

Summary. The socio-constructivist framework accounts for what types of thinking and communicating an individual and a community of individuals can accomplish. To summarize:

1. Conceptual development occurs in social contexts.
2. Semiotic mechanisms (e.g. language, television, books, etc.) experienced in social contexts provide a medium for socially created and accepted meanings to become internalized as personalized conceptual symbols and meanings for the developing individual.
3. These conceptual symbols are like internalized metaphors engaged through "inner speech" and play a significant role in how people perceive, contemplate, and communicate about the world.
4. The social context in which we learn governs the breadth and depth to which we can understand and communicate ideas. All learning occurs within a "zone of proximal development" that partially accounts for the extent people can conceptually develop autonomously and with assistance from teachers, peers, or learning tools. The characteristics of one's inner speech and conceptual symbols are determined by the quality of the interactions between the entities within the

zone of proximal development.

5. How learning and communicating occurs in one social context will transfer to and affect, through the expression of peoples' inner speech and conceptual symbols, how learning and communicating occurs in another. Therefore, the quality of one's learning and communicating in one social context will determine the quality of their learning and communicating in another social context.

This means that the similarities and differences between the social contexts individuals experience impact what is learned and how communication and learning are viewed. More specifically, the characteristics of an individual's inner speech and conceptual symbols are shaped by the semiotic mechanisms and social contexts they experience. Therefore, the extent that individuals' inner speech and conceptual symbols are congruent depends partially on the extent that they share common experiences within similar social contexts. For instance, proposing the phrase "*atomic bomb*" would invoke a very different set of conceptual symbols among scientists who worked on the Manhattan Project compared to secondary school age children learning about the atomic bomb. Furthermore, the scientists and children would share within their respective groups more similar inner speech and conceptual symbols about what an atomic bomb is, and how to think and communicate about an atomic bomb. The role of social contexts and semiotic mechanisms are crucial for determining what people know, how people know, and how people value learning and communicating.

CONVERGENCE OF VYGOTSKY'S AND MEDIA CRITIC NEIL POSTMAN'S IDEAS

Media scholar and cultural critic, Neil Postman, drawing on the work of media theorist Marshall McLuhan and others, made many contributions that reflect Vygotsky's core argument. For instance, in *Amusing Ourselves to Death* (Postman, 1985) he wrote:

> Each medium, like language itself, makes possible a unique mode of discourse by providing a new orientation for thought, for expression, for sensibility….. the forms of our media, including the symbols through which they permit conversation …are rather like metaphors, working by unobtrusive but powerful implication to enforce their special definitions of reality. Whether we are experiencing the world through the lens of speech or the printed word or the television camera, our media-metaphors classify the world for us, sequence it, frame it, enlarge it, reduce it, color it, argue a case for what the world is like. (p. 10)

The media Postman speaks of is synonymous to the semiotic mechanisms found in social contexts that Vygotsky and other learning theorists note play a crucial role in conceptual development. Postman's metaphors are akin to socially created conceptual symbols described earlier. Therefore, from both Postman's and Vygotsky's perspective, media serve as semiotic mechanisms that transmit symbols embedded with socially created and accepted meanings. These socially created symbols become internalized as personal conceptual symbols and meanings for the individual to engage through inner speech that assist them in understanding the world in which they live. Furthermore, the characteristics of these personalized conceptual symbols and inner speech are influenced by the scope and kind of media with which the individual interacts — Vygotsky's zone of proximal development. Again, Postman likens this to the creation of our media-metaphors. From a Vygotsky framework, through interaction with these semiotic mechanisms results the creation of our inner speech, conceptual symbols, and internal verbal thought. Both Postman and Vygotsky emphasize how media (i.e. semiotic mechanisms) influence the construction of our inner conceptual symbols (i.e. metaphors) and through the resulting inner speech "classify the world for us, sequence it, frame it, enlarge it, reduce it, color it, argue a case for what the world is like" (Postman, 1985, p. 10). Thus, the media we encounter influence what we know, how we know, and how we think learning and communicating about ideas should occur.

THE IMPACT OF TELEVISION AND THE INTERNET ON MEANINGFUL DISCOURSE, LEARNING AND SCHOOLING

The primary media people today consume is not lengthy discourse or scholarly books, but rather the internet and television. The Kaiser Family Foundation's (Rideout, Foehr, & Roberts, 2010) most recent findings about the media use of children include:

1. 8–18 year-olds devote an average of 7 hours and 38 minutes to using media across a typical day (more than 53 hours a week). Because they often use more than one medium at a time, they actually pack a total of 10 hours and 45 minutes of media content into those 7½ hours.
2. From 2004 to 2009 the amount of time spent watching regularly-scheduled TV actually declined, by 25 minutes a day. However, the many new ways to watch television programs (e.g., the Internet, cell phones, and iPods) resulted in an *increase* in total average daily television program consumption from 3 hours 51 minutes to 4 hours 29 minutes.
3. Popular new activities like social networking also contribute to increased media use. Top online activities include social networking (22 minutes/day), playing games (17 minutes/day), and visiting video sites such as YouTube (15 minutes/day). 74% of 7th-12th graders say they have a profile on a social networking site.
4. Time children spent daily with every medium other than movies and print increased over the past five years – 47 minute increase for music/audio, 38 minute increase for TV content, 27 minute increase for computers, and 24 minute increase for video games. TV remains the dominant type of media content consumed, at 4 hours 29 minutes per day, followed by music/audio at 2 hours 31 minutes, computers at 1 hour 29mimnutes, video games at 1 hour 13 minutes, print media

at 38 minutes, and movies at 25 minutes.

5. High levels of media used simultaneously also contribute to the amount of media young people consume each day. Approximately 40% of 7th-12th graders say they use another medium "most" of the time they're listening to music (43%), using a computer (40%), or watching TV (39%).
6. Texting. 7th-12th graders report spending an average of 1 hour 35 minutes a day sending or receiving texts. (Time spent texting was *not* counted as media use in this study.)

The *USA Today* (Marcus, 2011), tacitly acknowledging that meaningful discourse cannot compete with new media, reported on the above study in a brief 350 word front page article (The six points above contain just over 320 words). The article noted that minority children use media devices an average of 13 hours a day, vs. Caucasian children who use media devices an average of 8.5 hours a day. The article provided short statements from three seemingly qualified experts, two who merely restated in a few words that American youth spend far too much time with media and miss important face-to-face interactions. None of the statements provided adequate insight why children blindly and uncritically overuse media. Whether the experts did or did not provide more extensive insight cannot be ascertained, but even if they did provide detailed and scholarly responses, the *USA Today* media format could not have fit those types of responses into a 350 word article.

The brevity of reporting on a serious matter is symptomatic of the influence television and internet media have on other media and society. Postman (1995, p.192) wrote that "A new technology usually makes war against an old technology. It completes with it for time, attention, money, prestige, and a "worldview". The aforementioned USA Today article is just an illustrative example of how television and the internet has made war against other contemporary media, and how newspapers more and more are conforming to the template set forth by television and internet media. A lengthy and cognitively demanding newspaper article that seriously addressed children's media usage would likely be met with apathy or resistance by today's media consumer. Ironically, on the same newspaper page containing the above article, more space and attention was devoted to promoting the very media that claim so much of our children's time: the new Nintendo Wii U's ability to bridge console and portable games; how to view news photos of the day on your smart phone; reference to an article about which recent Snow White film rendition is better; where to find the TV listings; and an image of an hurried girl on a smart phone with survey results about how impatient customers respond to telephone customer service representatives.

The most serious consequence of all this is that those having grown accustomed to such media and the cognitive biases they promote now largely expect meaningful discourse and learning experiences to be brief, undemanding and at least somewhat entertaining. This has ominous consequences for meaningful discourse, learning, teaching and schooling.

Expert Knowledge and Discourse as Modeled by Television and the Internet

Television and the internet confound important differences between novice and expert knowledge, and this influences the general public's understanding of and attitude toward expert knowledge and meaningful discourse. Bransford, Brown, & Cocking (2000, p. 31) note several characteristics of experts:

1. Experts notice features and meaningful patterns of information that are not noticed by novices.
2. Experts have acquired a great deal of content knowledge that is organized in ways that reflect a deep understanding of their subject matter.
3. Experts' knowledge cannot be reduced to sets of isolated facts or propositions but, instead, reflects the context of applicability: that is, the knowledge is "conditionalized" on a set of circumstances.

The very format of television and internet discourse blurs novice and expert knowledge. Because these media are image based, imagery often trumps the accuracy, importance and wise application of knowledge. As McLuhan (1964) noted, "The medium is the message." That is why, in this age of television and internet dominance, political campaigns and even politics itself is more about visible imagery and sound bites rather than accuracy of information. Unaware of the imagery bias of television and much of the internet, media consumers often develop a false sense that they are truly informed about serious issues, when they really have only sound bites. Furthermore, they are rendered inept with making value judgments about who truly are experts with expert knowledge, and come to expect that the template for expert knowledge and meaningful discourse should resemble that of novices. Bluntly, television and internet media marginalizes expert knowledge and what being an expert entails.

The Purposes of Schools as Modeled by Television and the Internet

How learning and communicating occur in one social context will transfer to and impact how they occur in another (Vygotsky, 1986). Not surprisingly, because television and the internet are the dominant media in contemporary society, they influence individuals' conceptual symbols regarding the purposes of schooling. For example, Postman (1995) argues that the American public has come to accept that schools exist to serve the purposes of economic utility, consumerism, and technology. These views of schooling are encouraged through what people learn from television and the internet. Because of this, the noble intent of schooling, which is to educate people "how to make a life" (see Postman, 1995, p. x) in a democracy, and contribute to that democracy, is eroded.

Consumerism and economic utility feed off one another. Postman (1995 p.33) states "the latter postulates that you are what you do for a living; the former that you are what you accumulate." By the time children in the U.S. reach eighteen years of age, they have viewed an estimated 500,000 television commercials. Adults are also bombarded with messages that a successful life is measured by holding a high profile well-paying career and vast material possessions. Not surprisingly, U.S. policymakers and the general public now primarily speak and expect of schools these purposes. Consequently, schools become job training, focused on skills rather than education, obedience rather than civil education.

Postman (1995, p. 38) claims that "nowhere do you find more unexamined enthusiasm for Technology than among educators." Much of this exuberance for technology in education is because educators confuse educating with information delivery, often in a fun and entertaining manner. Furthermore, television and the internet eschew the extensive human interaction, critical discourse, and wrestling with ideas that is necessary for understanding and applying complex ideas. Postman (1985, p. 143) notes that:

> Whereas a classroom is a place of social interaction, the space in front of the television is a private preserve. Whereas in a classroom, one may ask a teacher questions, one can ask nothing of a television screen. Whereas school is centered upon the development of language, television demands attention to images. Whereas attending school is a legal requirement, watching television is an act of choice. Whereas in school, one fails to attend to the teacher at the risk of punishment, no penalties exist for failing to attend to the television screen. Whereas to behave oneself in school means to observe rules of public decorum, television watching requires no such observances, has no concept of public decorum. Whereas in a classroom, fun is never more than a means to an end, on television it is the end itself.

Students learn from television, the internet, and educators' efforts to gamify schooling that learning should be fun and entertaining, or at the very least, not cognitively and emotionally challenging.

Summary

That people uncritically consume vast amounts television and internet media has significant implications for discourse, learning and schooling. Contemporary media brings access to unfathomable amounts of information, and at the same time confuses information with education. Vygotsky's "nonspontaneous concepts" or what Bransford *et al.,* (2000) characterize as "expert knowledge" demands far more than information access. Furthermore, television, the internet, and other contemporary media convey to users that meaningful and empathic intrapersonal and interpersonal communication can be achieved primarily if not solely through fragmented and abrupt distant messages. Finally, what has been well established regarding learning and teaching appears stale and dull compared to the ever distracting and entertaining media of television and the internet. Educators are equally caught up in the biases of contemporary media and their influence on thinking and action, including thinking about learning, the purposes of schooling, and teaching. Helping educators understand the nature of technology, accurately identify and critically examine the biases of any technology, and make appropriate choices based on that analysis is crucial for schooling and society.

A COURSE FOCUSED ON DEVELOPING MEDIA LITERATE TEACHERS

Media literacy is too often conceptualized as merely how to use media. A more comprehensive and important aspect of media literacy is understanding the nature of media, including, but not limited to, the kinds of issues raised in this chapter (Postman, 1985 & 1995; McLuhan, 1964; and others). Serious attention to this aspect of media literacy in K-postsecondary schools would create a population having a perspective of media and technology useful to civil life (Postman, 1998). Lacking this education, Postman warns that the reigns of civil life will be surrendered to those controlling the media. However, few teachers are themselves media literate in the sense presented in this chapter, nor are they prepared to promote meaningful media literacy among their students.

Here I describe my effort to make media literacy a major component of a dual listed *Communication Skills in the Science Classroom/Reading and Communication in Science Education* course I teach. The overarching goal of the course is to prepare teachers who understand the nature of media and its implications for learning and teaching. Specifically, the course is directed at helping teachers understand:

1. the biases inherent in all technology and media;
2. the anticipated and unanticipated impact of various media;
3. the pros and cons of various technologies and media, and how those are unequally distributed;.
4. the messages and biases that technology and media possess about knowledge, learning, and communicating about ideas;
5. how media impact learning and knowing about science subject matter and the nature of science;
6. how technological change and changes in media consumption changes the way people view learning, teaching, and schooling; and
7. how different technologies and media impact the way people learn, communicate about, and engage in socioscientific decision making.

Examples to Initiate Interest and Concern Regarding the Impact of Contemporary Media

Students entering the course, with rare exception, view technology in a favorable light. Any concerns they have with technology in general, and media more specifically, has to do with how it is used. They have not considered how technology can and does change the way we think and act — how it may use us — without our awareness. This is unsurprising given the pervasive messages in our society that technology is undeniably good, and any ill impact is merely a result of how we use it. For instance, students are aware of the inordinate amount of time people spend watching television and surfing the internet, but again, those are seen simply as choices. Thus, instruction in the course employs a conceptual change approach (e.g. Appleton, 1993; Posner, Strike, Hewson, & Gertzog, 1982). For instance, examples that are familiar to the students are used to promote dissatisfaction with their tightly held and potentially emotionally charged views toward television and the internet. These examples also provide a base to scaffold to more abstract instruction that promotes media literacy.

Analysis of media: Beck vs. Ratigan. During the second class session I show students videos of the Glenn Beck program and the Dylan Ratigan program. The hosts of the two shows serve as pundits for opposing political sides, and during the chosen programs use their shows to direct a one-way scientific sounding monologue at each other about climate change. Before beginning the videos, students are told that the pundits were selected because they are engaged in unilateral attacks on each other's episodes about climate change — not because of any political agenda. To provide context, brief descriptions of each pundit's show appear below:

> The Glenn Beck Show: Aired 02/10/2010
>
> Retrieved from: http://www.youtube.com/watch?v=ghzcr2zYaMI
>
> Transcript available at: http://www.glennbeck.com/content/articles/article/198/36153/
>
> In this video, Glenn Beck denounces climate change and climate change advocates. The clip begins with Beck satirically claiming Al Gore and Robert F. Kennedy Jr., both who acknowledge global warming is occurring, have not been seen since arctic like weather pummeled Washington D.C. the weeks prior to the airing of this program. Beck is insinuating that the brutal winter demonstrates global warming is a farce and global warming advocates are in hiding so they will not have to admit this. Shortly thereafter, Beck exclaims that not every global warming nut is hiding out during the blizzards. Simultaneously, Beck begins to show a forty second clip of Dylan Ratigan. In the clip, Ratigan makes the claim that the blizzards in Washington D.C. are extreme weather events that climate scientists predicted are a consequence of climate change. Ratigan proceeds to state that warmer air holds more moisture, and when subjected to colder winter temperatures create more snow. While this brief explanation is occurring, Beck is visible in a smaller window exhibiting sarcastic non-verbal responses. The clip of Ratigan ends, Beck resumes control of the entire screen, and opens his unidirectional "debate" by saying, "It does not take a genius to see through the more snow is proof of global warming claim." Beck then runs over to a chalk board and argues if this claim is true

then the traditional thermometer is wrong. On the chalk board, Beck designs a new thermometer where the top curves down to the bottom bulb. Concurrently, Beck sarcastically claims that according to the logic presented by Ratigan, the hotter it gets, the colder it becomes. Beck then quickly attempts to use this mock lesson to support a short quip about how every claim made by scientists and the government about global climate change are false. After devoting less than four minutes to this one way engagement, Beck quickly and smoothly transitions to the next target: President Obama and the economic recession.

The Dylan Ratigan Show: Aired 02/11/2010

Retrieved from: http://www.msnbc.msn.com/id/21134540/vp/35354694#35354694

The next video shows Dylan Ratigan responding to Glenn Beck's attacks. Ratigan begins by making reference to Beck's and other opposing pundits' remarks through entertaining metaphors such as "storms of misinformation". Ratigan then turns his attention to the earlier episode of Beck's show in an eye for an eye fashion. Similar to Beck, Ratigan shows a brief clip of Beck's episode aired the previous day (described above). The clip lasts just 45 seconds, interspersed with commentary by Ratigan and also contains much of the footage of Ratigan portrayed on Beck's show. Ratigan follows Beck's approach further by using a chalk board and scientific jargon to superficially demonstrate how Beck misrepresented the "climate change theory". In his defense, Ratigan makes clear that he does not believe the snow storms are proof of climate change, and he does not believe the storms are not proof of climate change. The lecture lasts three minutes and fifteen seconds and concludes with Ratigan taking a moderate political position on climate change and an invitation for Beck to discuss further issues more directly. Ratigan then welcomes his guest Bill Nye the Science Guy, whom Ratigan claims is afraid of nothing and a great student of Carl Sagan. Nye, who built his career performing on entertaining television science programs, is asked by Ratigan to provide commentary about "politics and social theories basically bastardizing science to prove anything they believe whether it is climate change or anything else". Nye, in response to this poorly worded directive, stammers through an approximately four minute explanation on the matter.

After viewing each clip, I ask students to avoid each pundit's political affiliation and obvious differences regarding the veracity of global climate change, and instead focus on the manner that both programs convey ideas. The intent of this discussion is to illustrate how television and internet media often refer to important socio-scientific issues in what can *appear* as discursive and informative debates between experts. However, the ideas expressed by both programs are largely trivial, filled with misconceptions, shrouded in entertainment, and presented by individuals lacking a credible understanding regarding the complexity of global climate change and weather. But most viewers are in no position to critique the claims and can easily think they have received and understand credible ideas regarding a complex scientific and socio-scientific issue. Further discussion points regarding the video clips include:

1. Both pundits use classroom props (e.g. chalkboard) and scientific jargon to make their messages appear legitimate. While doing so, they either superficially or erroneously portray scientific content (e.g. the relationship of temperature and climate change) and the nature of science (NOS) (e.g. belief is credible grounds in science, science ideas are proven).
2. Both create what appears to be a debate through using unchallenged monologue to respond to selected clips from each other's shows.
3. Both rely on approaches that are entertaining and convey trivialities, but ignore authentic complexity in the issue. For instance, little attention and no depth were given to the topics addressed before transitioning to a new topic and context. Visual imagery and illustrations were used to sensationalize points. Multiple deliveries of image rich media occurred simultaneously while the issue of climate change was being addressed. For instance, fragmented news lines about unrelated stories flash at the bottom of the telecasts. Additionally, references to other media outlets (e.g. the internet) where Beck's messages can be consumed also appear in the lower right corner of the screen, and Ratigan's show has the major market indices at the top of the screen.

During this discussion, I ask many questions that draw students' attention to the issues above, how television and internet media influences what people know, how they know what they know, and the ways they value learning and communicating ideas. For instance:

1. In what ways are the exchanges between the pundits an intellectual debate and in what ways are they not an intellectual exchange of ideas? How might the exchange occur differently if the two pundits were debating in person? How would the debate be different if they were discussing the issues in front of us rather than through the television medium? How does their method of engaging each other's position on climate change impact the intellectual rigor of the debate?
2. To what extent do the clips aim to inform? To what extent do they aim to entertain? With the multiple bits of information occurring simultaneously (e.g. scrolling news lines, stock price updates flashed periodically), to what extent is the viewer able to meaningfully attend to and critically think about the pundit's claims? To what extent would television and internet media that portrayed a lengthy, in-depth and serious discourse (without the associated entertaining diversions) about subjects such as climate change be popular? How well does television's and the internet's form work for cognitively demanding discourse about complex issues?

3. To what extent do the pundits deeply address the climate change issue? In what ways does the use of scientific jargon and props make the pundits *appear* to engage in conceptually challenging discourse? Neither one of these individuals are experts in climate science. In what ways does the conveyance of this information by these two individuals marginalize the conception of what it means to be an expert or hold expert knowledge?
4. People often walk away from broadcasts such as these feeling informed about a scientific topic. In what ways are they really informed and understand the topics addressed? What message do media such as this convey about what it means to know, learn, and communicate about difficult issues and key ideas embedded in those issues?
5. How does mass broadcasting of programs such as these impact our society's ability to engage in socioscientific decision making about subjects such as climate change? What would truly scholarly discourse about this subject look like?

Analysis of media: Magazine covers. Several other examples of science portrayals in the media (e.g. movies and magazine articles) are subsequently used to help students understand how media shapes what people know, how they know what they know, and how they value learning and communicating ideas. For instance, I have students compare the May 20, 1996 *Time* magazine cover and the advertising poster for the movie Twister.[2] Specifically, I make evident how both images depict a large tornado swooping through a farm setting with tornado researchers in the forefront and the word "Twister(s)" splashed colorfully across. Media critic Mark Crispin Miller noted that the similarities between the two are not coincidental (Postman, 1998). Time Warner owns both companies and used *Time* magazine to subtly promote its movie, *Twister*. The apparent similarities are problematic because Time Magazine is perceived as a credible and informative media outlet, whereas the movie Twister is intended to entertain. However, by appearing extrinsically similar, both media sources imply they also have similar intrinsic qualities. That is, their informational contexts and messages about knowledge delivery are meshed for the media consumer.

Students are attended to how media examples such as this blur the line between entertainment and education, for example, how the credibility attributed to *Time Magazine* as a source of valid information lends credibility to a movie like *Twister* as a similarly valid source of information. Consequently, media consumers may identify entertaining media (e.g. a movie that employs science-like themes) as a way of becoming informed about scientific knowledge and its societal ramifications. The danger is that media consumers will also feel like they have learned something compelling and credible, and come to expect all forms of learning to be entertaining. For example, while teaching science in 2004, the movie *The Day After Tomorrow* was released. I was surprised at the number of my students and their parents who perceived that the film accurately portrayed abrupt and catastrophic global warming.

Developing Teachers' Media Literacy

The *Communication Skills in the Science Classroom/Reading and Communication in Science Education* course seeks to help educators understand that different forms of media convey different messages about what effective learning and communication of ideas entails. Students in the course explore how even when different media forms intend to convey similar messages about an identical topic, different ideas may be expressed about the topic and what it means to effectively communicate about that topic. This is because the different media forms favor different semiotic mechanisms (e.g. text, sound, rapid imagery) about that topic, which influences what and how media consumers think. The very nature and form of a medium and its constituent semiotic mechanisms, impact what inner speech and conceptual symbols (or in Postman's (1985, p. 10) broad term "metaphors") individuals and a culture will construct.

These ideas are difficult to grasp, and the contemporary media examples and associated questions provided above are designed to provide authentic cases that raise issues, promote deeper thinking about these issues, and provide a more concrete foundation as students read literature regarding the NOT. Drawing from these experiences, I engage students in several concept mapping activities and discussions that address how different media result in disparate messages, even if the content of the messages remains constant. Figure 2 illustrates a class generated concept map conveying the form of communication favored by the printing press and television/internet media, and what these favored forms convey about communication and learning.

Discussions that occur during concept map generation center on the following overarching ideas:

1. Unequal negative and positive trade-offs that result from TV and the internet replacing printed text as the dominant media consumed by our society.
2. The printing press promoted textual communication which exerts a bias toward logical expression and progression of ideas — that is, expression of ideas in text is sequence dependent, persists over time, builds upon old knowledge, and lays the foundation for new knowledge.
3. Television and internet media exerts a bias toward imagery — that is, expression of ideas via images that are not restricted to a logical sequence, quickly expire, and are largely independent of previous and subsequent information.
4. The nature of written discourse requires its consumer to engage in higher order thinking, cognitively hold multifaceted arguments, and internally sequence and decipher those arguments in a logical fashion. Furthermore, the very focused and logical structure of meaningful written discourse models for the media consumer that learning, knowing, and communicating about abstract ideas requires lengthy uninterrupted concentration and arduous thinking.
5. The nature of television and internet media is predisposed toward more superficial and disassociated ideas that require less concentrated attention and thought.

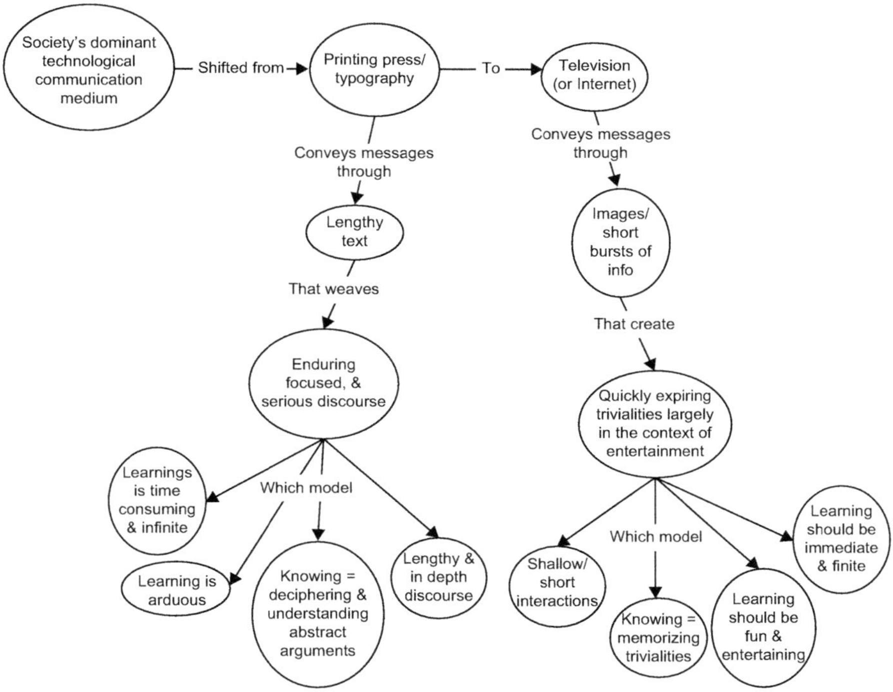

Figure 2. Class-generated concept map comparing the biases of the printing press to television/internet.

However, through quickly shifting imagery (often aggrandized), television and internet media easily conveys shallow ideas as thoughtful, compelling, and effectively communicated. This models for the media consumer that learning, knowing, and communicating about abstract ideas entails minimal cognitive effort, time, and dialogue – visual presentation trumps conceptual substance.

Classroom discussions also focus on how each media emphasizes, models and promotes different types of intrapersonal (e.g. inner speech that engages personalized conceptual symbols) and interpersonal communication, and how intrapersonal communication and interpersonal communication are entangled, interdependent, and reflect and reinforce one another's qualities. Key ideas addressed are:

- Written discourse demands the media consumer engage in lengthy and cognitively demanding internal dialogue (i.e. Intrapersonal communication) about the ideas presented in text — a concept commonly referred to in class discussions as "wrestling with an abstract argument". Engagement in lengthy and cognitively demanding intrapersonal dialogue is requisite for deeply discussing complex ideas with others (i.e. Interpersonal communication). Written discourse requires the media consumer follow logical language sequences and forms appropriate for communicating complex ideas. The logical sequence of language and extended text also models for media consumers how to interpersonally communicate ideas that retain logic and meaning, which models for others how to effectively communicate complex ideas.

- Dialogue on television and the internet lack continuity because of interruptions (e.g. commercials, multiple shifting images). Thus, these media promote among consumers short disassociated bursts of internal dialogue about information, and convey interpersonal communication in a trite, unilateral and superficial way, even when dialogue occurs among experts. This models and teaches that interpersonal communication need not and perhaps should not be lengthy and complex, even when the issues being discussed are.

Addressing the Importance of Social Context for Teaching and Learning

When students sufficiently understand the concepts outlined above, they are introduced to the importance of social context in which learning takes place. Students read literature regarding social constructivism (e.g. Vygotsky, 1986; Howe, 1996; John-Steiner & Mahn, 1996) and I draw their attention to the following ideas (described in detail earlier in this chapter).

1. Conceptual development occurs in social contexts.
2. Semiotic mechanisms (e.g. language, television, books, etc.) experienced in social contexts provide a medium for socially created and accepted meanings to become internalized as personalized conceptual symbols and meanings.
3. These conceptual symbols are like internalized metaphors engaged through "inner speech" and are used by learners in their perception of and communication about the world.
4. The social contexts in which we learn govern the breadth and depth to which we can understand and communicate about ideas. All learning occurs within a "zone of proximal development" which partially accounts for the extent people can conceptually develop autonomously and with assistance from teachers, peers, or learning tools. The characteristics of one's inner speech and conceptual symbols are determined by the quality of the interactions between the entities within the zone of proximal development.
5. How learning and communicating occurs in one social context will transfer to and affect, through the expression of peoples' inner speech and conceptual symbols, how learning and communicating occurs in other contexts. Therefore, the quality of one's learning and communicating in one social context will impact the quality of their learning and communicating in other social contexts.

I have students complete two activities followed by discussions that assist them in understanding the readings about the concepts outlined above and how television and internet media impacts teaching and learning. The first activity and discussion focuses on ideas 1- 3 and helps students learn how the social contexts and semiotic mechanisms people experience shape their inner speech and conceptual symbols. Moreover, students learn that aspects of our inner speech and conceptual symbols are often similar because both are based upon and triggered by semiotic mechanisms that carry socially agreed upon meanings. While the first example describes a specific video, any media that provides disparate audio and visual sensory perceptions which trigger common prior conceptions will work.

The second activity and accompanying discussions build upon ideas addressed during the first activity. Focusing on ideas 1–5 above, the second activity and accompanying discussions address the importance of the nature of social contexts and semiotic mechanisms, particularly language, in determining the effectiveness of teaching and learning.

Through these experiences, discussions and follow-up readings, students learn that language and other semiotic mechanisms are the vehicles for thought transmission. Students also come to understand that each social context experienced is like a learning ecosystem, each with a unique combination of entities (i.e. people and semiotic mechanisms) that the learner interacts with to generate knowledge. How the learner interacts with those entities (e.g. types of language and media used) forms what Vygotsky refers to as the learner's "zone of proximal development" and shapes the learner's knowledge, thinking, and values about learning and communicating ideas. Thus, how ideas are learned and communicated about in one social context will affect learners' expectations about how ideas should be learned and communicated in another social context.

The Man with the Golden Voice (Portnoy, & Lawrence, 2010)

Retrieved from: http://abcnews.go.com/Entertainment/homeless-man-golden-radio-voice-ohio/story?id=12540522

In 2010–2011, videos of Ted Williams, a homeless man in Columbus, Ohio, permeated the internet. In the video Williams appears disheveled and unkempt, much like one would expect a homeless man to appear. However, to panhandle money Williams mocked a radio announcer's voice that was not at all reflective of his appearance. Rather, Williams sounded like a hip and confident oldies radio announcer.

I initiate this activity by instructing students close their eyes and pay attention to the mental images they develop while listening to the clip of Williams. Once the clip is played, I have students draw their mental images on white boards. The majority of students tend to draw similar images of a stereotypical oldies radio announcer. A few students who have seen the clip before on television and the internet draw a likeness of Williams. I quietly tell these students to not mention why their image looks like Williams.

Students then display their whiteboard drawings at the front of the room and discuss why the images possess similarities and differences. I first focus on students' recorded mental images that closely resemble a stereotypical oldies radio announcer to help them understand that the semiotic mechanisms they experienced in previous social contexts influenced them to develop and embed within their inner speech socially agreed upon conceptual symbols of 1) what a radio announcer is; 2) how a radio announcer communicates and uses language; and 3) how to think and communicate about a radio announcer. I also help students understand that their inner speech will initiate and potentially modify these conceptual symbols whenever they have subsequent experiences associated with radio announcers. Because most of their experiences with radio announcers are largely similar, their conceptual symbols and inner speech about radio announcers will also be largely similar. To help students engage in and begin to understand these concepts I ask questions such as the following:

1. How do you account for your images appearing like a stereotypical radio announcer?
2. Rather than use the word "stereotypical" what are some reasons why the images you drew might be thought of as "socially agreed upon symbols" of what an oldies' radio announcer looks like?
3. How would these images look different if the context of the radio announcement was switched from oldies to heavy metal?
4. What role did "inner speech" play in your development of conceptual symbols about what a radio announcer represents?
5. What were the characteristics of your inner speech?
6. What role do prior experiences and the social settings of those experiences play in your developing these conceptual symbols and inner speech?
7. Given what we have discussed, how do we account for each of you possessing similar conceptual symbols and inner speech about the meaning behind something like a stoplight?

Once we have satisfactorily accounted for students' drawings of an oldies radio announcer, I turn their attention to the students' drawings that look like Williams. I ask a similar set of questions to scaffold students to the same idea that previous experiences prompted particular students' inner speech to engage conceptual symbols that are drastically different than those representing stereotypical radio announcers. Furthermore, the fact that while these students envisioned and drew Williams, they simultaneously recognized Williams as an aberration from the stereotypically held image of oldies radio announcers. That is, based on their prior experiences of hearing and seeing Williams, through solely listening to the clip they were able to manifest context specific conceptual symbols that resemble Williams. To help students understand these concepts I ask questions such as:

1. How do you account for the fact that these images look drastically different than those that look like a stereotypical announcer?
2. What role do you think prior experiences played in developing these conceptual symbols and inner speech about a radio announcer?

3. In what ways did you cross-compare conceptual symbols of Williams and what a radio announcer should look like when listening to the clip? How did prior experiences facilitate this cross-comparing?
4. Why does this image of a disheveled old homeless man not fit the common societal image of an oldies radio announcer?
5. What would have to occur for the disheveled old homeless man to become the conceptual symbol that most of our culture holds as what an oldies radio announcer looks like?

At the conclusion of this discussion I have students view and listen to the video of Williams. The students who have not previously experienced this video are quite shocked when they find that Williams' appearance does not at all match what they conceptualized based solely on the audio clip. In fact, many of the students' initially reject the video's authenticity. This is an opportunity to help students further understand that we carry inner speech and conceptual symbols that are determined by the social contexts we have previously experienced. Our conceptual symbols and inner speech are based largely on socially agreed upon semiotic mechanisms, and significantly impact the development of our "world view". Because students hold socially agreed upon conceptions from their experiences regarding the appearance of radio announcers, their initial reaction to the video of Williams is shock and disbelief.

I then address how the video media changed the nature of the message the media consumer receives, and therefore impacts the conceptual symbols the media consumer materializes through their inner speech. For instance, students note that when they were only experiencing the audio clip of Williams, they were better able to focus on his message. However, once they also watched Williams they found focusing on his message far more difficult, even though they had previously heard it. This important point is later brought up when the teachers discuss television and internet media's impact on our ability to focus on the ideas they convey. To help students understand these concepts I ask questions such as the following:

1. How did your ability to focus on William's message change when it was transmitted on an image focused media vs. a media strictly devoted to sound?
2. How do you account for the reduced ability to focus on and think about William's message?
3. In what ways does "adding" a layer of imagery on top of William's articulation result in sensory overload?

Activity 3. The Turbo Encabulator

Retrieved from: *http://www.youtube.com/watch?v=rLDgQg6bq7o*

http://www.youtube.com/watch?v=2fjcJp_Nwvk&feature=related

http://www.youtube.com/watch?v=MXW0bx_Ooq4&feature=related

This activity centers on three spoof videos that address a fictional piece of technology called the "turbo encabulator". Students are directed to watch the three videos, take notes, and then describe what the turbo encabulator is to the class. Each of the three videos presents content about the turbo encabulator in fictitious language that is intended to humorously parody the sometimes confusing languages of science and engineering. Although the three videos present similarly confusing dialogue, their explanations about the turbo encabulator are increasingly detailed and accompanied with concrete representations.

Students find they have learned very little and struggle in describing this particular technology. Initially, teachers claim their struggles are merely due to the language used in the videos. However, I push students further by reminding them of Vygotsky's (1978) "zone of proximal development" and then make the following points salient:

1. Regardless how many times we watch the three videos, we may never understand or be able to effectively communicate about the turbo encabulator.
2. The social context and semiotic mechanisms, particularly language, used to convey information about the turbo encabulator was essentially beyond our zone of proximal development (ZPD).
3. The reason our comprehending the turbo encabulator exceeds our ZPD is because, in part, a critical mass of the semiotic mechanisms used in the videos were incomprehensible by the inner speech and conceptual symbols we hold and employ. That is, most of the language and symbols conveyed about the turbo encabulator were undecipherable by our inner speech and a mismatch to the conceptual symbols we have previously developed and hold through experiences in other social contexts. Therefore, we could not engage in meaningful inner speech about the turbo encabulator.

Significantly, viewers are attended to the humorous nature of the turbo encabulator video because it parodies the conceptual symbols and inner speech people often develop via schooling, television and the internet about science and engineering language. This point is emphasized so that students understand that the social context they create and the semiotic mechanisms they use when teaching science will impact what and how their students learn. Teachers must understand the need to use semiotic mechanisms and language that are within their students' zone of proximal development. However, teachers must also be aware of how students' learning in social contexts outside of school impacts their attitudes toward and expectations of teaching and learning in the classroom – often in ways that are detrimental to their education.

Educating Teachers about Contemporary Media's Impact on Meaningful School Discourse and Learning

Now grasping the implications of social constructivist framework for teaching and learning, I direct students to contemplate how the learning contexts of television, the internet, and other contemporary media impact teaching and learning in schools. Because television and the internet are the dominant social contexts in which children interact, they have a disproportionately large influence on what children learn and their expectations for learning and communicating about ideas. To prime these discussions, I have students review how television and internet media model communicating, wrestling with, and learning about complex ideas. I draw from previous classroom experiences to make apparent that what students have tacitly learned about conducting inner speech stems from consuming television and internet media, and is reflected in the students' expectations regarding learning and interpersonal communication skills. That is, students can easily come to expect learning interactions to be abbreviated, center primarily on retrieval of information, conceptually undemanding, and entertaining. Teachers can just as easily come to expect that learning tasks should reflect those characteristics and require little meaningful and extended teacher-student dialogue. To assist students draw from previous coursework and understand these ideas, I pose questions such as the following:

1. In what ways are the semiotic mechanisms present in schools similar and different from those of television and the internet?
2. How do the communication practices in the Beck and Ratigan videos, internet chats and twitter posts model for students what effectively communicate about ideas entails?
3. Consider students' high rate of television and internet media consumption. In what ways does this model for them and shape the extent they deeply contemplate ideas through inner speech?
4. To what extent does learning from television and internet media shape students' expectations of how learning should occur in formal school contexts?
5. To what extent would a typical student's zone of proximal development be optimally identified and engaged through television or the internet?

Teachers readily share their new found understanding about how television and internet media teaches students their expectations for school teaching and learning. Furthermore, they recognize that Postman's ideas and Vygotsky's framework are highly convergent. For instance, one teacher's e-mail reaction to the sequence of course instruction was as follows:

> Well I just read the Vygotsky reading, and holy good scaffolding (for lack of better words)! This reading is very well placed in the sequence of concepts we are learning in this class. In many ways it reflects Vygotsky's ideas on how learning is a zigzag process, moving back and forth between concepts that the reader may be familiar with (ideas we have read from Postman and discussed in class) to more "systematic" ideas that may be more abstract. Many of the ideas Vygotsky proposes are more readily understood from the Postman framework. So, it is critical to have had those readings first.

Another student's understanding was conveyed in the following e-mail:

1. Thanks for the overview/review. It is insightful and helpful particularly how you described that: We think in metaphors. They are like conceptual symbols.
2. Metaphors are the organ of perception. Through metaphors we see the world as

one thing or another.

3. Thought is transmitted through language. This is directly related to the learning theories through the examples that illustrate the connections you are trying to highlight.

   You said: "Language is the vehicle through which metaphors are transmitted. If this is the case, then the medium also determines the characteristics of the language (vehicle) and the metaphors that are used in the social context currently experienced, and in other social contexts experienced at other times." Thus, the very foundation for an argument of why our traditional 50 minute classes which encourage a limited student attention span are favored is because students are conditioned to have a 30–50 minute attention span for one topic because TV programs are of a similar length. This is an example of the medium determining characteristics of language and students' expectations for learning in other social contexts.

   Furthermore, technology has shaped the epistemology of our culture. The way we know things and how we acquire knowledge is utterly determined by our means of communication. For instance, a student who is thoroughly engaged (and even scarier, one who is not fully engaged) often thinks that by watching a 30–60 minute show about Africa, he or she knows all there is to know about Africa. This is because the student may perceive that the technology and those delivering the message about Africa are authorities on Africa. Therefore, the concepts portrayed/communicated through this medium are "the most important/ all that is important" on the subject of Africa. Consequently, people perceive that by watching a show one can "know" and be an expert about a topic. The kicker then is that "our schools are a symptom of our culture." There is an underlying expectation/demand for our classes to mirror our technology (television and the internet), whether it is the best means of communication or not.

Other students express their understanding of these concepts via observations of how television and the internet have influenced their own students' views of school teaching and learning. For instance, they have witnessed students often balking at teaching and learning that requires serious lengthy cognitive effort which would result in the construction of what Vygotsky calls nonspontaneous concepts. They further note that many students in their classrooms see no need for serious and prolonged contemplation and discourse about ideas because of how television and internet media models contemplating and communicating about ideas.

Students' understanding is also illustrated in their concept maps that outline the types of social discourse, learning, and the purposes of schooling which are promoted through television and internet media (see Figures 3 and 4). Interestingly, students often convey that their concept maps will never be fully complete because deep learning about serious issues and abstract ideas is a continual process shaped by both interpersonal and intrapersonal dialogue.

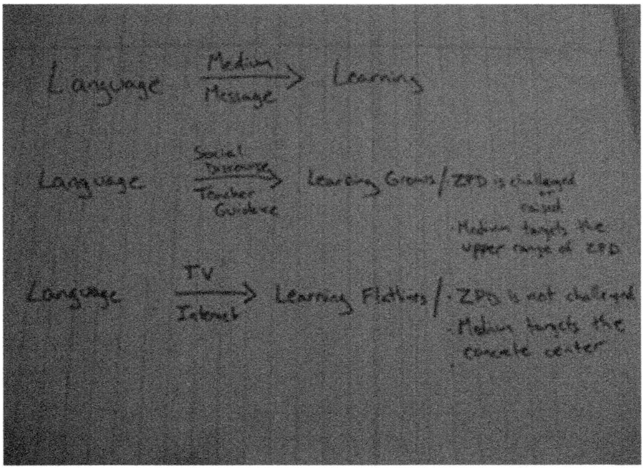

Figure 3. Student's comparison about how language and learning differs between effective teacher guidance and television and internet media.

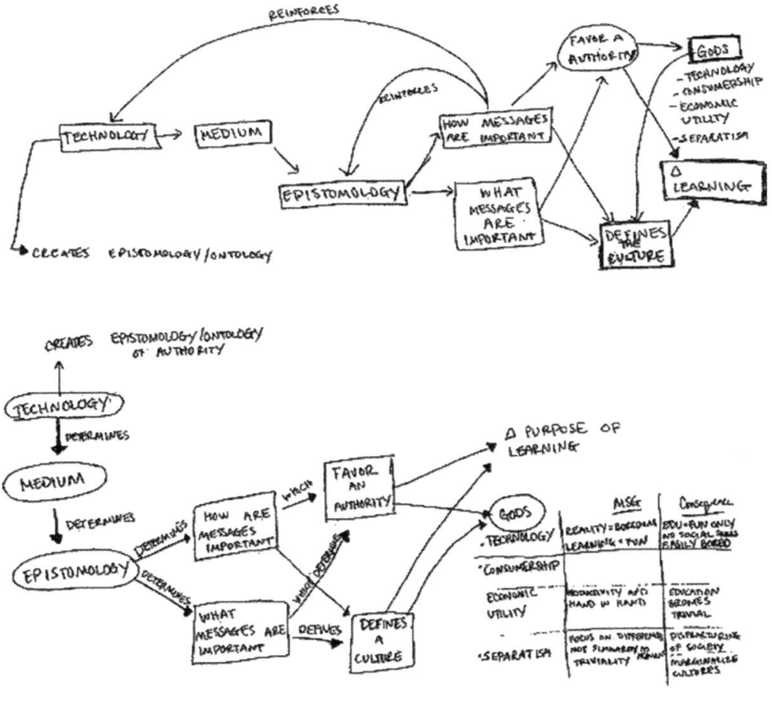

Figure 4. Student generated concept map illustrating how media shapes epistemology and our culture's perceptions about the purposes of schools.

Of course, the teachers are also expected to demonstrate their understanding about the nature of media through summative classroom assignments. For instance, the teachers analyze the extent various teaching strategies can be effectively implemented in the classroom through their respective media. More specifically, the teachers analyze how the media required for these strategies influence what students know, how students think, and students' values about learning and communicating ideas. For instance one group analyzed the Technology-Enhanced Formative Assessment (TEFA; Beatty & Gerace, 2009) strategy that employs clickers in the classroom and arrived at the following conclusion:

> [A]nother down side of using clickers with multiple choice questions is that it limits students' need for critical thinking; instead it promotes "trivializing knowledge". Since the answer choice is limited to approximately four choices per question, students tend to guess. Therefore, thinking through the idea and connecting their previous knowledge to the current material is no longer necessary since the students can just guess the right answer. Clickers in particular have a limited number of letters and numbers on the apparatus itself. The numbers on the keyboard run zero through nine, and letters are "A though J". This type of response will limit the students' opportunity to progress their language and communication skills because they are not writing out the answers. Due to limited characters on the clicker response system, students are only able to choose the answer to the question by entering the appropriate character for that question. Free-response versions of the clickers exist but are essentially twitter machines that limit the number of characters the students are allowed to use. Therefore, they limit the language that they have to choose from (I'm sure you can divulge everything you need to express about conceptual change in 140 characters [Note: this was written with a tone of sarcasm]).
>
> Compared to an open-ended discussion question that can be offered by teachers when communicating without clickers, a multiple choice question format does not encourage students' self-talk (i.e. inner speech) and critical thinking while trying to answer the question. When using such technology, students tend to respond to the question by simply choosing the answer provided below. If the answer is not known by the individual, they do not continue to wrestle with the idea. Instead, they make a guess about which response will be best suited as the right answer. After the choice has been made, some are no longer concerned about it……
>
> The use of technology, such as clickers, can limit student-teacher interactions. Once the teacher has this alternative way to communicate with the students, the interaction between them tends to decrease. Instead of using the clicker system to enhance the dynamics of the classroom, the multiple-choice question on the classroom projector becomes the primary mode of interaction between the teacher and student. Although, during the discussion portion of the teaching strategy, the teacher has an opportunity to develop interaction with his students,

some teachers may lessen their effort to do so due to the simplicity and time saved when using comprehension measurements through TEFA.

The value that my students find in the course is also conveyed in their asking how they can educate their own students to become media literate. I use this opportunity to draw students' attention to the structure of my course, the concrete experiences that provided a foundation for rich discussions, the kinds of questions I asked along with other teacher behaviors I used to create extensive intra and interpersonal dialogue, and the placement of readings and tasks that together promoted an understanding of complex ideas such as:

1. What conceptual learning entails and the value of this type of learning.
2. The characteristics of school teaching that promote conceptual learning.
3. How television and internet media covertly teaches people what they know; how they think; and their values about learning and communicating ideas.

Creating a social context conducive for conceptual learning entails complex teacher decision-making and many teacher behaviors that in concert create a highly discursive and mentally engaging intellectual climate (Clough, Berg & Olson, 2009). Teachers must model for and overtly teach students what meaningful and deep learning entails, and point out particular teacher decisions and teacher behaviors that engage learners in meaningful intrapersonal and interpersonal discourse about abstract ideas (Kruse, Wilcox & Herman, 2010). For instance, I model questions that can be asked of students of most any age:

1. When I ask an open ended question and wait expectantly, how does that impact the extent that you think and the way you think?
2. In what ways is your learning impacted if I make you discuss the question in small groups rather than if I immediately provided the answer?
3. What are some reasons why we engage in discussions about concepts, rather than just presenting them on digital presentations?
4. In what ways is your inner speech about the concepts affected when you discuss them in groups? How does that impact how you learn something?

I also point out that media examples like those used in class abound. Selecting examples related to the discipline specific content being taught, along with discussion like those modeled in class, will assist K-college students to become media literate while at the same time conceptually engaging more deeply in the discipline specific content being taught. Thus, promoting media literacy need not detract from, and actually may enhance, content instruction.

SUMMARY

The most prevalent media sources that many consume and learn from are television and internet media. Thus, they are a pervasive and powerful influence on what people know, how they think, their values about learning and communicating ideas, and their actions. The messages promoted by television and internet media, which are internalized and communicated by those in our society, are antithetical to much of what research has for decades made clear is necessary for deep and meaningful learning of complex ideas.

The values about learning and communication set forth by television and internet media are readily infused into schools because they are internalized, held, and promoted by education stakeholders (e.g. students, parents, teachers, administrators, and policy makers). Therefore, school teaching, learning, and discourse can often resemble how television and internet media communicate information. This causes schools and our culture to fall short in preparing the youth to develop into well-adjusted individuals who can effectively and contemplatively interact within and contribute to a democratic public. Rather, schools have become something to be tolerated rather than revered, and their purposes serve as Postman (1995) describes: economic utility, consumership, and technological worship.

Unfortunately, because television and internet media satiates people with pleasure and appears to meaningfully inform, most of the general public is oblivious to the effects that television and internet media have on conceptual learning and meaningful discourse. Because schools are symptomatic of how a culture thinks and acts, and are an agent of cultural change, teachers must come to understand the nature of media, particularly television and internet media, through their teacher education programs. Teachers must also learn how to teach the nature of media to their students. Without this aspect of media literacy being taught to students, the very cultural fabric of our society may very well be in peril.

NOTES

1 The term "conceptual symbols" is used throughout the chapter. It refers to a mental metaphors used in a person's inner speech (addressed later) to identify and demarcate things (e.g. objects, historical events, traditions).

2 To view the *Time* magazine cover go to: http://www.time.com/time/covers/0,16641,19960520,00.html. To view Twister movie posters go to: https://www.google.com/search?q=twister+movie+poster&hl=en&tbo=u&qscrl=1&rlz=1T4ADFA_enUS489US489&tbm=isch&source=univ&sa=X&ei=WQUdUcTBLoS69QS6p4HoBQ&ved=0CDAQsAQ&biw=1311&bih=534

REFERENCES

Adatto, K. (1990). The incredible shrinking sound bite. *The New Republic. 202* (22) pp. 20–23.

Appleton, K. (1993), Using theory to guide practice: Teaching science from a constructivist perspective, School Science and Mathematics *93*(5), 269–274.

Beatty, I. D. & Gerace, W. J. (2009). Technology-enhanced formative assessment: A research-based pedagogy for teaching science with classroom response technology. Journal of Science Education and Technology, *18*(2), 146–162.

Beck, G. (2010, February 10). *Glenn Beck: Global Warming Colder Winters [video file]. Retrieved from:* http://www.youtube.com/watch?v=ghzcr2zYaMI;

Bransford, J. D.,Brown, A. L. & Cocking, R. R. (2000). *How people learn: Brain, mind, experience, and school.* Washington: National Academy of Press.

Brown, A. L. (1992). Design experiments: Theoretical and methodological challenges in creating complex interventions in classroom settings. *Journal of the Learning Sciences, 2*(2), 141–178.

Brown, A. L., Ash, D., Rutherford, M., Nakagawa, K., Gordon, A., & Campione, J. C. (1993). Distributed expertise in the classroom. In G. Solomon (Ed.), *Distributed cognitions: Psychological and educational considerations* (pp. 188–228). New York: Cambridge University Press.

Chang-Wells, G. L. M., & Wells, G. (1993). Dynamics of discourse: Literacy and the construction of knowledge. In E. A. Forman, N. Minick, & C. A. Stone (Eds.), *Contexts for Learning: Sociocultural Dynamics in Children's Development* (pp. 58–90). New York: Oxford University Press.

Clough, M. P., Berg C. A. & Olson J. K. (2009). Promoting effective science teacher education and science teaching: A framework for teacher decision-making. *International Journal of Science and Mathematics Education, 7*(4) 821–847.

Cole, M. & Engstrom, Y. (1993). A cultural-historical approach to distributed cognition. In G. Salomon (Ed.), *Distributed Cognitions: Psychological and Educational Considerations* (pp. 1–46). New York: Cambridge University Press.

Fehrman, C. (2011). The incredible shrinking sound bite. *The Boston Globe*. Retrieved 08/13/2011 from: http://www.boston.com/bostonglobe/ideas/articles/2011/01/02/the_incredible_shrinking_sound_bite/

Howe, A. C. (1996) Development of science concepts within a Vygotskyian framework. *Science Education, 80*(1), 35–51.

International Technology Association (ITEA), (2007). *Standards for technological literacy: Content for the study of technology* (3rd ed); Reston, Virginia: ITEA.

John-Stiener, V. & Mahn, H. (1996). Sociocultural approaches to learning and development: A Vygotskian framework. *Educational Psychologist. 31*(3/4), 191–206.

John-Steiner, V., Panofsky, C. P., & Smith, L. W. (1994). *Sociocultural approaches to language and literacy: An interactionist perspective.* New York: Cambridge University Press.

Kruse, J. W., Wilcox, J., & Herman, B.C. (2010). *Learning dispositions: Explicit instruction concerning the nature of learning and assessing students' views on learning.* Presented at the 2010 Association for Science Teacher Education Conference in Sacramento. CA, Jan 14, 2010.

Marcus, M. B. (2011). Minority kids use media more. *USA Today*. Published June 8th, 2011.

McLuhan, M. (1964). *Understanding the media: The extensions of man.* Massachusetts: MIT Press

Montagne, R. (2011). The incredible shrinking sound bite. *National Public Radio*. Retrieved 06/15/2011 from: http://www.npr.org/2011/01/05/132671410/Congressional-Sound-Bites

Niederhauser, D. S. & Lindstrom, D. L. (2006). Addressing the NETS for students through constructivist technology use in K-12 classrooms. *Journal of Educational Computing Research. 34*(1) 91–128.

Olson, J. K. & Clough, M. P. (2001). Technology's tendency to undermine serious study: A cautionary note. *The Clearing House, 75*(1), 8–13.

Piaget, J. (1952). *The origins of intelligence in children.* M. Cook, trans. New York: International Universities Press.

Piaget, J. (1973a). *The child and reality: Problems of genetic psychology.* New York: Grossman.

Piaget, J. (1973b). *The language and thought of the child.* London: Routledge and Kegan Paul.

Piaget, J. (1977). *The grasp of consciousness.* London: Routledge and Kegan Paul.

Piaget, J. (1978). *Success and understanding.* Cambridge, MA: Harvard University Press.

Portnoy, S. & Lawrence, C. (2010). *Homeless man with a golden radio voice: 'I'll announce for food'* Retrieved 02/05/2011 from: http://abcnews.go.com/Entertainment/homeless-man-golden-radio-voice-ohio/story?id=12540522

Posner, G. J., Strike, K. A., Hewson, P. W. & Gertzog, W. A. (1982). Accommodation of a scientific conception: Toward a theory of conceptual change. *Science Education 66*(2), 211–227.

Postman, N. (1985). *Amusing ourselves to death: Public discourse in the age of show business.* Penguin: New York.

Postman, N. (1995). *The end of education: Redefining the value of school.* Vintage: New York.

Postman, N. (1998). Technology & Society. C-SPAN, Presented at Calvin College, January 12, 1998. *Retrived from http://www.c-spanvideo.org/program/TechnologyandSo.*

Ratigan, D. (2010, February 11). *The Dylan Ratigan Show* [video file]. Retrieved from: http://www.

msnbc.msn.com/id/21134540/vp/35354694#35354694
Rideout, V. J., Foehr, U. G. & Roberts, D. F. (2010). Generation M^2: Media in the lives of 8 to 18 year olds. *The Henry J. Kaiser Family Foundation.* Retrieved from: http://www.kff.org/entmedia/entmedia012010nr.cfm
Rogoff, B. (1994). Developing understanding of the idea of communities of learners. *Mind, Culture, and Activity*, 1, 209–229.
Sweeney, P. (2012). Retrieved July 17, 2012. http://thinkexist.com/quotes/paul_sweeney/
The Turbo Encabulator Retrieved 02/05/2011 from: http://www.youtube.com/watch?v=rLDgQg6bq7o January 15, 2011.
The Turbo Encabulator Retrieved 02/05/2011 from:http://www.youtube.com/watch?v=2fjcJp_Nwvk&feature=related January 15, 2011.
The Turbo Encabulator Retrieved 02/05/2011 from: http://www.youtube.com/watch?v=MXW0bx_Ooq4&feature=related January 15, 2011.
Thoreau, H. D. (1854). *Walden; Or, life in the woods.* Boston: Ticknor and Fields.
Valsiner, J. (1987). *Culture and the development of children's action: A cultural-historical theory of development.* Chichester, UK: Wiley.
Vygotsky, L. S. (1978). *Mind in society: The development of higher psychological processes.* (M. Cole, V. John-Steiner, S. Scribner, & E. Souberman, Eds.) Cambridge, MA: Harvard University Press.
Vygotsky, L. S. (1981). The instrumental method in psychology. In J. V. Wertsch (Ed.), *The concept of activity in Soviet psychology* (pp. 134–188). Armonk, NY:Sharpe.
Vygotsky, L. S. (1986). *Thought and language*. A. Kozulin, (Ed). Cambridge, MA: MIT Press.

Benjamin C. Herman
Department of Secondary Education,
University of South Florida

CHAPTER 16

HEATHER TILLBERG-WEBB & JOHANNES STROBEL

IDEOLOGIES IN THE CONCEPTUALIZATION AND USE OF EDUCATIONAL TECHNOLOGY

Using Huxley, Orwell and Forster to Inform a Humanizing Framework for Educational Technology Practice

INTRODUCTION AND THEORETICAL CONTEXT

> "Information … now … comes indiscriminately, directed at no one in particular, disconnected from usefulness; we are glutted with information, drowning in information, have no control over it, don't know what to do with it."
>
> – Postman

The concern about information overload predicted by Postman is that in the information society that exists in the 21st century, humans are left without agency against a barrage of relentless information. Postman's arguments, positioned alongside the technological dystopian literature of Orwell, Huxley, and Forster, provide lessons applicable to educational technology practice. *1984* (Orwell, 1949), *Brave New World* (Huxley, 1992) and *The Machine Stops* (Forster, 1909), demonstrate the resulting cultural contexts when technology is allowed to develop hand-in-hand with a totalitarian socio-political culture that attempts to suppress and control human life, its impulses, desires and hopes. In this way, all these works invoke Mumford's view of authoritarian technics (1964), which undermine democratic ways of life. Accordingly, a key responsibility of modern education must be to help learners manage, apply, evaluate, and synthesize information for effective decision-making and for creatively contributing back to the world of information.

Though education espouses the free flow of information, we must critique the technological structures that allow for an increasingly rapid generation and exchange of information to ensure that the development of human capital is paramount. Postman's point of view is that our "glut" of information (1993), mediated by "technological" thinking, provides increased opportunities to distract humans from issues of true value. Postman also posits that Huxley's vision of the future, where people unquestioningly love their gadgets and entertainment, removing their capacity to think independently,

Michael P. Clough, Joanne K. Olson and Dale S. Niederhauser (Eds.), The Nature of Technology: Implications for Learning and Teaching, 329–344.

is an oppressive future more dangerous and likely than Orwell's fear-driven society (1986). Forster's Machine is a fitting middle ground where everyone is dependent on the technology, and the ubiquitous technology mediates interpersonal relationships in a way that prevents useful and applied thought.

Through an analysis of these works and their messages about educational theory and philosophy, we construct a vision of educational technology philosophy and practice that interweaves a humanizing framework with the tensions of navigating the socio-historical context in which one operates, the development of critical thinking, independent thinking skills and an individual's own sense of motivation and affective engagement in fulfilling an individual's sense of agency vis-a–vis technology.

Though all three authors portray technological dystopias, Huxley's *Brave New World* has its denizens entertained into stupor, while Orwell's inhabitants of *1984* are bullied into obedience, and Forster's citizens of the Machine are an interesting mix of both—both indoctrinated with fear of what lies outside the Machine, and entertained by meaningless lectures of disconnected knowledge. Viewing the socio-cultural historical context of educational technology through the lenses of technological and social determinism within these works, paints the background milieu for the importance of human agency within a technological society. In the case of communication technologies, each technology tool is an extension of the "public sphere," (Habermas, 1989) an important tool of government and institutions, and carries with it an embodiment of structured human relations, particularly in the case of information communication technology (ICT), which structures and restructures interpersonal communication, collaboration, and interaction. Then, analyzing the challenges of cultivating independent thinking in light of theories of linguistic and media determinism illuminates the importance of cultivating empathetic, independent thinkers as part of a democratic society. An individual is raised and immersed in a cultural context and with particular patterns of communication and interaction with technology; but each individual also brings his/her own motivation and aspirations. The juxtaposition of these philosophies provides a toolkit for those concerned that education in the information age might start to look like any of the dystopianist views presented by these authors.

This chapter expands on an earlier version of a critical and humanizing framework of educational technology (Strobel & Tillberg-Webb, 2008), where we explored the relationship of technology and education solely from the perspective of ideologies toward technology and on a paper for *Techne* (Tillberg-Webb & Strobel, 2011) focused solely on the technological philosophies evident in the technological dystopian works inherent in these works of Huxley, Orwell, and Forster. Each of the works of Orwell, Huxley, and Forster describes characteristics of the educational system inherent in the technological dystopias therein. As we've established in previous work (Tillberg-Webb & Strobel, 2011), hardware and software computer-based technology should be seen as just one form of technology – though certainly a driving and motivating force in the existing focus on technology in education. There are other technologies and forces influencing communication, interaction and behaviors in a community. Other technologies in education include "standard lessons, curricula, and assessment schemes that make up the technical reality in schools" (Charles, 2004). Technology in this chapter is defined as the whole of the systems and processes that structure how we organize and implement these tools, in other words, the cultural context of our technological adoption (Pacey, 2000). The integration of technological tools in education implies an entire culture of practice surrounding that tool. Teaching someone to simply use a tool is not a neutral act (Winner, 1980; Kranzberg 1986), as the tool itself is not neutral. With each technology tool in the classroom, certain practices are supported through the technology design, and other practices and interactions are either purposely or unintentionally omitted. Furthermore, the socio-cultural environment surrounding "contemporary technology is not neutral but favors specific ends and obstructs others" (Feenberg, 2004).

Because these dystopian literary works contextualize education as an important technology of control within totalitarian systems, choosing this analytical lens necessarily focuses our attention to the role of our technologies as we envision an educational system that balances technological culture with humanistic focus. By tying together the educational themes identified in these works, particularly where technological systems are used to oppress humans, we amplify the importance of human-centered approaches in education.

Huxley, Orwell, and Forster's novels illuminate aspects of the educational system in place in each dystopia. The educational model in each novel provides the central point of control, and this should give us pause. Educators might like to view their role as helping to empower learners to become technology savvy, independent-thinkers and involved, ethical and productive citizens. Teaching about technology as a content add-on to everything else that is taught in the curriculum or utilizing technology solely as a tool to achieve learning gains fall short in recognizing the much larger space technology is already occupying and reduces technology to a separate entity.

Central to the argument that technologies have created humans that are driven by fear (Orwell and Forster) or driven by the pursuit of pleasure (Huxley), is that humans in these worlds are not permitted to exercise their own agency. In fact, each book has a character or two who are driven by a quest that their purpose in life must transcend beyond the world to which they have been limited: Bernard and the Savage in *Brave New World*; Winston and Julia in *1984*; Kuno in *The Machine Stops*. In this way, society in these texts has not been completely successful in any of these totalitarian worlds at completely suppressing the human spirit for curiosity, a thirst for "free will," and a suspicion that there is more to life than what they have been shown. In all protagonists of the three novels, the rebellion against their world starts with a cognitive realization of the discrepancy produced by the system and their own experienced life. In this facet of our humanizing framework, we will also start by focusing on the cognitive aspects of learning.

SOCIO-CULTURAL CONTEXT AND TECHNOLOGICAL DETERMINISM

> "How we have advanced, thanks to the Machine." – Everyone, The Machine Stops

In *The Machine Stops* (Forster, 1909), people are portrayed as viewing their existence and progress in relation to "The Machine," a central artificial intelligence device that controls all the actions of the citizenry. All citizens acknowledge the machine's greatness by exclaiming this mantra, "How we have advanced thanks to the machine." This vision of the future through the lens of technological determinism portrays a world completely at the whim of the progression and disintegration of technology, and has implications for human agency when confronted with technological change.

In this perspective, the technological world view is the fundamental force of social change, and technology is seen as the prime "mover" in a historical context (Chandler, 1995) or through analyses which view progress through the lens of technological change (Pacey, 2000). Blacker labels technological determinism as a "substantive" view of technology, rooted historically in Ellul's conceptualization of *la technique* as "an all-embracing and evil power that has come to enslave all our endeavors, from art and politics to education" (1994). Technological tools require a systematic way of thinking, an understanding and logic that then starts to inform our worldview. This logic does not require a value shift on the part of tool users, but it provides another experience that continues to shape ways of thinking and being in the world.

The historical and cultural patterns that inform our educational practices are important to examine. In educational technology practice, our inherent ideological biases guide us to integrate technology tools and new processes without considering long-term ramifications or the inherent assumptions the technologies with their embedded practices carry with them. Likewise, in the dystopian literature presented in this chapter, social leaders would prefer that the citizens of their societies did not consider the patterns of the past or the present. Activity theory will be used as a mechanism for holistically exploring the historical context set forth in the literary writings we used.

Agency is a central theme in technological determinism, as humans are viewed as having little to no agency in a world run by technologies (Smith & Marx, 1994; Winner, 2004). The intentional stance (Dennett, 1987) is an example of the ways in which we betray our own agency to computers, and grant them more power than they actually have, for example when we say, "the computer won't let me do that." In educational settings, how an educator orients the learning with the technology; whether the technology is perceived as something to be mastered versus something that constrains activities will send important messages to learners about one's agency when dealing with technology.

Social Determinism. An antidote to technological determinism is a social deterministic perspective. From a social deterministic point of view, humans make calculated and responsible decisions about the most effective ways in which to use new technologies, and do not just view technologies, such as the computer, as a monolithic force (Feenberg, 2001). As a converse perspective to technological determinism, the social determinist lens portrays technology as a product of social change. If technological tools are in the focus at all, then what matters is not the technologies themselves, but the larger processes, the social, political, cultural and economic systems in which those technologies are embedded (MacKenzie & Wajcman, 1985). Whereas technological determinism identifies autonomy as an attribute of technological innovation, and society has "to catch up" with the next technological innovation; in social determinism, society and individual players within are seen as autonomous and thus humans drive the development of technology to serve the need and goals of society.

TECHNOLOGICAL AND SOCIAL DETERMINISM AND HISTORICAL PERSPECTIVE

"History is bunk. History, he repeated, is bunk." -Brave New World

Applying a historical perspective to the discussion of technological and social determinism reveals complex tensions between humans and technology, rather than a binary relationship. Huxley recognized the importance of understanding history and controlling the historical message. In a *Brave New World*, by declaring history to be bunk, Mustapha Mond, the World Controller in *Brave New World* can remove the context of the past, and imprison the citizenry within their present. For this reason, the citizens of *Brave New World* are conditioned to despise books, a fear that is reinforced from a very young age.

History need not refer to just a geopolitical sense of history. Systems have patterns, which may be replicated over time. A method for examining events in a socio-historical content can be the use of activity theory to analyze systems. Employed in educational settings, activity theory can be a helpful framework for effecting change while respecting historical perspectives and cultural patterns. Recent work by Yamagata-Lynch has examined activity theory as a guiding framework for participants in recognizing the system and patterns around K-12 community partnerships (L. C. Yamagata-Lynch & Smaldino, 2007) and learning reflections (L. Yamagata-Lynch, Smaldino, & Click, 2008). The activity theory framework includes: mediating artifacts, activity members, their roles/the distribution of labor, and rules of the community, as well as proposed goals and actual outcomes. By embedding inquiry in a systemic framework and exploration of both successful and unsuccessful past practices, educators can examine how current implementation of new technologies mirrors these longitudinal patterns. Activity theory combined with praxis can provide educators with a helpful approach that examines artifacts, community, roles, and goals.

An important dimension of activity theory is the concept of expansive visibilization. To understand a phenomenon, one must capture snapshots of the activity system repeatedly over time, to start to understand the historicity of systems, meaning how they "take shape and become transformed over lengthy periods of time." (Engestrom, 1999, pp. 136–37). Historicity grounds educational technology adoption within the larger picture of previous innovations and their ensuing impact. This historical aspect is not a uni-dimensional time-line that we all share, but includes several dimensions: (a) the history of the theoretical ideas and tools that shaped the activity (system); (b) the activity system including the students and the instructors who bring their own history, as Freire asserts: "through their continuing praxis, men and women simultaneously create history and become historical beings" (1999); (c) the activities and objects present in the local context have a history; and (d) the interaction between the different layers of historicity, which is different from the histories of individuals in a community or the community or the objects, but also include the tensions and contradictions that naturally occur between these different layers (Engeström, 1999).

Linguistic Determinism

> "… the whole aim of Newspeak is to narrow the range of thought?…The whole climate of thought will be different. In fact, there will be no thought, as we understand it now. Orthodoxy means not thinking—not needing to think. Orthodoxy is unconsciousness." Syme, 1984

In the above quote from *1984*, mandated limitation of language through "Newspeak" is central to the control of thought. In linguistic determinism, language is the critical factor that shapes thinking. According to the Sapir-Whorf hypothesis (for a description of the evolution of Sapir-Whorf, see Koerner, 1992), one's native language can be seen as framing our ability to make sense of the world, which gives new question to the concepts of agency, freedom, and responsibility (Cameron, 1999). From this viewpoint, our thoughts are constrained by the available means of expressing them. Thus the development of Newspeak in *1984* is a means for shaping, constraining and controlling thought.

Language, like technology, is a complex system developed by humans to mediate relationships with other humans. In Orwell, Huxley, and Forster's view of a totalitarian technology-controlled future, the power of the structure of language is an important theme. Observations about how language can be used to control people politically and to limit the scope of what they will think about aligns with deterministic philosophy. With vocabulary limited in a language, such as it is in *1984*; with language disconnected from experience, as we see in *The Machine Speaks*; and with language disconnected from literature, in *Brave New World*- the ability to fully express oneself, describe experience, and create allusions to broader cultural metaphors is stymied.

Linguistic Determinism and Independent Thinking

> "…old men in the bad old days used to renounce, retire, take to religion, spend their time reading, thinking – thinking!" - Brave New World

> It is intolerable to us that an erroneous thought [one that is not consistent with those in power] should exist anywhere in the world, however secret and powerless it may be. - 1984

In *Brave New World* and *1984*, thinking is the precursor to questioning authority, and thus a dangerous activity. Interestingly though, in *The Machine Stops*, the only activity people engage in is thinking, but that thinking is disconnected from activity or application. Huxley, Orwell, and Forster all recognize the power that thoughts have in informing language, which in turn impacts one's sense of independent thought and action in the world- making this an important area of emphasis for educators.

Independent thinking can be conceptualized as a combination of Bloom's (1984) categories of analysis, synthesis, and evaluation; Jonassen's (2000) critical thinking and creative thinking; Gagné's (1985) attention to both general problem-solving strategies and domain-specific problem-solving as important; Paul's (1993) Thirty-Five Dimensions of Critical Thinking, encompassing both affective and cognitive strategies; and Reeves (2006) conative domain. The cognitive macro-skills include comparing analogous situations, evaluating the credibility of sources of information, reading critically, questioning deeply, generating or assessing solutions, and making interdisciplinary connections. Paul's cognitive micro-skills include distinguishing relevant from irrelevant facts and exploring implications and consequences.

In educational settings, the shortcoming of technology integration has been a focus on adding equipment without structuring pedagogy to promote the development of critical thinking. If there is a focus only on rote instruction, our schools sometimes look like a dystopian setting of students engaged in "drill and kill" exercises instead of authentic problem-solving. Warschauer's qualitative two-site comparison (2000) identified that affluent schools asked students to engage in meaningful learning activities with technology whereas their counterparts at a nearby less affluent school were relegated to rote, skills-based exercise. Hester (2002) surveyed 323 K-12 teachers and reported similar results: Students at high socioeconomic status (SES) schools had teachers who reported integrating more critical thinking activities with computer use compared to their counterparts at low SES schools.

Just as mastering a culture depends on fluency with its language system, navigating technological tools and systems must be mastered if one wishes to realize agency and autonomy in an information society. As such the acquisition of basic writing skills with chalk or pen and the use of e-mail, chat or other technological means are crucial for the communication and interaction with ones' surrounding environment" (Taylor, 2007).

Media Determinism

> "I see something like you in this plate, but I do not see you. I hear something like you through this telephone, but I do not hear you. That is why I want you to come."- Kuno, The Machine Stops

In *The Machine Stops*, the protagonist's son, Kuno, expressed to his mother that he wanted to see her face-to-face to talk, instead of just having a mediated discussion through "The Machine," as he expressed in the above excerpt. "The Machine" mediated the image and sound, and it was the message of control and structure embodied in that medium that Kuno wished to resist. Media determinism is another variety of determinism, identifying media as the key technology that structures our thoughts and actions as humans. McLuhan (1964) and Hall (1966), key theorists in media determinism, rooted their work in the hard determinism of the Sapir-Whorf hypothesis that frames linguistic determinism. McLuhan envisioned that electronic communication would qualitatively shape and influence society (Carey, 1981). Postman's view of media in *Technopoly* ascribes to media a sort of anthropomorphizing, where "when media make war against each other, it is a case of world views in collision (Postman, 1993, p. 16).

Media, in this view, makes possible and cultivate human interactions through the way communication is structured (DiMaggio, Hargittai, Neuman, & Robinson, 2001).

The media for Huxley, Orwell, and Forster figure prominently into the control mechanisms used by society. For Orwell, this is done through a limitation of media: "The television screen in Orwell's *1984* is on constantly, cannot be turned off, and has only one channel. Media are thus primarily here an apparatus of surveillance and terror rather than indoctrination" (Kellner, 1990). In contrast in *Brave New World*, the media is the "feelies," which implies a tactile rather than verbal-visual experience. The experience is driven by a tactile pleasure, with no concern for cognitive or affective engagement. In *The Machine Stops*, there are endless distractions in the form of bells and "speaking-tubes," which prevent "isolation." In each instance, the importance of remaining focused in technological worlds of interruption from media is a prescient interpretation of the world we now live in. Accordingly, we can take the lessons from these fictional settings to note that education necessarily needs to be focused on learner engagement, but engagement cannot just be limited to whether or not an experience is enjoyable for a student or will result in student satisfaction.

Media Determinism and Conditioning

> "We condition the masses to hate the country…but simultaneously we condition them to love all country sports. At the same time, we see that all country sports shall entail the use of elaborate apparatus. So they consume manufactured articles as well as transport. Hence the electric shocks." – Brave New World

In *Brave New World*, the educational practice of hypnopaedia serves as tool by those in power to regulate the other citizenry. People are taught to consume products and activities without using any independent thought. The focus on fun and pleasure above all other human pursuits is what motivates Postman's concern in *Amusing*

Ourselves to Death. Hypnopaedia, we learn from the Director in *Brave New World*, was ineffective in its early experimentation in conveying intellectual knowledge without an applied context: "You can't learn a science unless you know what it's about." However, for moral education, which the Director tells us "must never be rational," it is extremely effective. For twelve years, children listen to nightly repetitive broadcasts that condition their thoughts and attitudes towards themselves, others, products, and cultural norms.

Although behaviorism is the historical grounding of educational technology, it has fallen out of favor in the current literature of educational technology in favor of epistemologies such as constructivism and theories of learning, which emphasize agency on the one hand by focusing on the meaning-making process, and postulate the conditioning through the environment on the other. Sometimes in our struggle to create meaningful, contextualized learning, we forget the power of reward and punishment. At the same time, overuse of reward structures and the gamification movement (Kapp, 2012) potentially reduces learning to entertainment. Entertaining students rather than engaging them plays into a Huxlean fear of mindless distractions undermining free will. Educators need to be reflective and purposeful in integrating policies that use these powerful features without reflecting on the approach.

Media Determinism and Questioning Authority

> All our science is just a cookery book, with an orthodox theory of cooking that nobody's allowed to question, and a list of recipes that mustn't be added to except by special permission from the head cook." - World Controller Mustapha Mond, *Brave New World*

Questioning is inherent to a critical perspective, where one is encouraged to question "why," rather than simply "how" (Hlynka, 1994). Central to all of these stories — *Brave New World, 1984, The Machine Stops* — is a fear of questioning. The consequences of challenging authority vary in each work, but in all of these worlds requires questioning the technological processes of the society. But the implication that voicing dissent is dangerous is consistent. Research analyzing stories of how teachers integrate technology into their classrooms corroborates that the most neglected parts of the National Educational Technology Standards Students (NETS*S) standards are those aimed at a critical evaluation of the role of technology on society and on ethics of work with computers like critically evaluating the information found on web sites (Niederhauser, Lindstrom, & Strobel, 2007)

Questioning or being critical of technology does not mean refusing to use technology tools, but does require a critical reflection on its effect on many different aspects, like constructing of meaning, verifying and justifying evidence and interpersonal communication. For Kuno in *the Machine Stops*, questioning the technology that literally kept everyone in the society alive, represented a breach of trust with the societal controls he was experiencing.

Everywhere we remain dependent to technology and technological processes, whether we passionately affirm or deny it. "Critical reflection on technology use need not be equated with a "luddite" or "laggard" (Rogers, 1995) attitude towards technologies, but the pejorative connotation of these terms referring to those who choose not to adopt new technologies reveals our cultural bias toward those who voice discomfort with or resist adopting new technologies (Davis, 2003). Typically, instead of critiquing our use of technological tools, educational technologists have not asked "if" we should use a tool, but how we can convince people to use it (Moore, 2005). If we fail to recognize and question the ways in which we let technology shape all aspects of culture in terms of language and its forms; interactions with friends, family, colleagues, and strangers; and use of recreational time, we are not only failing to question technology, we are blind to its sway over our lives.

Critical Theory and Human Agency

War is peace, slavery is freedom, ignorance is strength. - 1984

Critical social theory offers a perspective that allows for the questioning and critique of the socio-political and cultural structures that support various phenomena, with a particular focus on power structures. Huxley's society in *Brave New World* understood the power of allowing the questioning of authority, thus it was forbidden. The critical theorists traditionally viewed media as a powerful tool for stymieing human agency. So argued Adorno and Horkheimer, for example, in their interpretation of "culture" that high culture had no place for recorded media (Poster, 1995) as their message became enmeshed with a twinge of antipathy towards these "lower" forms of communication that have been introduced.

Habermas (1984), in contrast, "developed a theory of culture or symbolic interaction which locates the point of critique in the "lifeworld," the egalitarian space of the everyday rather than in the elite moment of high culture or in the philosophical labor of "negative dialectics" (Poster, 1995). Habermas' concept of the public sphere and the technology that mediates communication developed in an understanding of a mediating space between individuals and their immediate community and public resources such as the state. "What Habermas called the 'bourgeois public sphere' consisted of social spaces where individuals gathered to discuss their common public affairs and to organize against arbitrary and oppressive forms of social and public power" (Kellner, 2004, p.5). In a manner that helps us make more sense of the Habermasian public sphere from our modern context, Kellner suggests that "rather than conceiving one liberal or democratic public sphere, it is more productive to theorize a multiplicity of public spheres, sometimes overlapping but also conflicting" (2004, p.7).

From the educational technology literature, Blacker (1994) identified critical theory in technology as being an instrumentalist perspective, where we often view technology as "merely tools that human beings use in order to achieve the purposes we assign to them" (p. 3). Because of our instrumentalist tendencies towards technology adoption (seeing only the function of the tool rather than the entire system), our current uses of technology are "overwhelmingly driven by morally suspect motives," according to Blacker (1994). This ties in with Habermas' fear that media and technology might eventually become a tool of the state, "eroding the difference between state and civil society, between the public and private sphere... transforming citizens into consumers, dedicating themselves more to passivity and private concerns than to issues of the common good and democratic participation" (Kellner, 2004, p. 5). We allow this same passivity to play out in education contexts: Hlynka suggests that the traditional role of instructional designers, and by our extension educational technologists, is to implement strategies to achieve goals, not to question the goals themselves (1989, p. 185). Educators using critical theory as a framework may question the goals of instructional contexts.

Affective Engagement and Agency

> There will be no loyalty, except loyalty towards the Party. There will be no love, except the love of Big Brother. There will be no laughter, except the laugh of triumph over a defeated enemy. There will be no art, no literature, no science....There will be no curiosity, no enjoyment of the process of life. All competing pleasures will be destroyed. -1984

In *1984*, the destruction of human emotion is central to the ability of the state to control its citizenry. Given the interconnectedness between technology and our cultural values and the role of education as a systemic process with inherently expressed cultural values (Ogbu, 1982), attention to the affective component is a key aspect of a holistic approach to educational technology. A core affective focus can be on empathy, which can be seen as affective response or cognitive, as "a learned ability to grasp the world from someone else's point of view" (McTighe, 1998). Empathy has been identified as a necessary component of humane education (Aspy, 1975; Dixon & Morse, 1961; Olden, 1953). "Empathy is the capacity of the subject instinctively and intuitively to feel as the [other] does...that enables the subject instantaneously to sense the [other's] apparent emotions" (Olden, 1953, p. 112–3).

In *1984*, we observe a drastic result of lack of empathy when it comes to public hangings of "criminals":

> "Why can't we go and see the hanging? roared the boy in his huge voice. "Want to see the hanging! Want to see the hanging! Chanted the little girl. - 1984

This scene depicts children begging their mother to see the public hanging of the Eurasian prisoners, guilty of war crimes. The taking of other's lives as a public spectacle, with the humans being objects for entertainment is typical of the dehumanization we see in *1984*. In *Brave New World,* also with a divisive class structure, children are conditioned from a young age to see those who are not like themselves as something less than human. Everyone in society adheres to his or her assigned social status and role, which contributes to the perceived harmony of the society. Members of each caste in society do not interact with each other, or necessarily think of each other as human, but instead define themselves by their breeding of Epsilon, Alpha, Beta. The inability of Lenina in *Brave New World*, for example, to conceptualize the extent to which her own conditioning has shaped her limited view of other "classes" of people, makes it impossible for her to ever see others' perspective. Reeves (2006) advocates for attention to the conative domain in education, and defines the conative as "the will, desire, drive, level of effort, mental energy, intention, striving, and self-determination to actually perform at the highest standards" (p. 4). This type of affective engagement is critical for an individual to develop the skills to fulfil goals or aspirations.

Authentic Experiences and Agency

> "This is the first bit of useful information I have ever got from a lecture, and I expect it will be the last." – Kuno, The Machine Stops

Kuno, our disillusioned inhabitant of *The Machine Stops* pushed the boundaries of the constraints of the machine-controlled society. He expressed his desire to visit the surface of the earth to his mother's disdain. He did not request the proper permissions to do so, and snuck out of the machine-regulated world, finding himself for the first time in the natural world and needing to recall information about the patterns of the sun from a lecture. Though ideas proliferate and lectures are the modus operandi of most human interaction in *The Machine's* world, all of the knowledge is abstract and disconnected from experience, thus being useless except as a form of distraction. Likewise, in educational technology practice, the pedagogical focus is on creating opportunities for application of knowledge and construction of knowledge. Only as learners develop the skills of self-regulation and developing the ability to construct questions and seek out new knowledge will information become useful.

Implications for Humanizing Educational Technology Practice

> "Our world is not the same as Othello's world. You can't make flivvers without steel-and you can't make tragedies without social instability. The world's stable now. People are happy; they get what they want, and they never want what they can't get...And if anything should go wrong, there's soma. … You've got to choose between happiness and what people used to call high art." World Controller Mustapha Mond, Chapter 16, pg. 220, Brave New World

The three technological dystopian works addressed in this chapter present technology as integrated with a totalitarian political structure; and yet ignore a vision of how information and technology might fit with a pluralistic, postmodern society. Likewise, education's use of technology in the 21st century must be contextualized in a more nuanced socio-cultural perspective than represented by Orwell, Huxley, Forster, or Postman. The fragmentation of a cultural norm that was associated with modernism, whereby the structure of authority was very clear, creates a need to help develop people who are critically engaged, self-aware and self-regulated, true to their own values, and who also can understand and respect diverse perspectives.

For students in schools, economic realities drive the quest for education as the "mode of production" has shifted from industrial and manufacturing to "information." As a society considering what it means to be an educated citizen, we have a responsibility to adopt an intentionality around our use of technology in education that requires students to view themselves as humans with agency, capable of parsing through information to make informed decisions, but also capable of assessing and evaluating one's own values and choosing cultural practices with technology that align with their own best interest. This requires attention to the affective and conative domains of learning as well as a critical reflection on the cultural worldviews that shape our thinking, which our education system—driven by measurable behavioral objectives has shed away from. The commercial sphere, marketers, are well aware, however, of the power of messaging to tap into affective behaviors and result in forming children's values (Thorson, Chi, & Leavitt, 1992). While the fear of a disorganized glut of information on the part of Postman is very real, management of this information is the new currency, which the students of today must learn to navigate, organize, synthesize, and manage, so the substantive 'glut' (prescribing values) is replaced by a formal 'glut' (prescribing the process). The problems our society will face in environmental, political, and social arenas will require competent problem-solvers who understand that there is not necessarily a technological solution for every social problem and who understand their own impact in shaping the world in which they live.

Postman serves as example for the allure of blaming "technology" for our direction in technology integration. However, in doing so, we must also examine the role of education and pedagogy as among the arsenal of structures supporting the technological tools in their implementation. As educators, we have an important role in defining and modeling appropriate interaction and collaboration that amplifies our capabilities in a positive way.

CONCLUSION - TRANSLATION BETWEEN IDEOLOGY AND PRACTICE

> The only tyrant I accept in this world is the "still small voice" within. And even though I have to face the prospect of a minority of one, I humbly believe I have the courage to be in such a hopeless minority. –Gandhi

Our technology and tools are not outside of our cultural milieu, they are intertwined with how we live, eat, communicate, interact, collaborate and create new knowledge. Though the relationship between technology and culture is reciprocal, human need and curiosities should drive our creation and adoption of new technologies (for a further description of this "human-centered" approach see Strobel and Tillberg-Webb, 2008). Humans make decisions about what to create, how to implement tools and processes, and thus create systems that structure and restructure the way we live our lives. Our educational system is one of the most prominent technologies for indoctrinating or cultivating human capital. In the academic literature of educational technology, however, a technologically deterministic perspective often creeps in. Fictional works provide us with an opportunity explore educational culture in these other worlds, and compare and contrast our own experience.

One must always work as a part of one's socio-cultural and historical context, so simply accepting or rejecting the technological innovations of our time is too simplistic a solution. A robust educational system will critically evaluate the role of our instructional technologies and the impact they have on the human experience. Attention to not just the present, but also the past that brought us to our political, economic and social systems, is essential to an informed citizenry. Within the larger public sphere, cultivating a student's ability to act independently within the world is an essential part of a democratic and egalitarian system. Attending to the development of problem-solving and decision-making skills, as well as enhancing innate motivation and ethical values are part of a holistic educational experience. These messages are explicit in these dystopian works, where we see the ramifications of societies that suppress and oppress citizenry through technological mechanisms. Though the complexity of engaging in discussion about ideology and educational practice may be daunting, it is a necessary step in developing a society with members who are also critically reflective of their own understanding of the cultural implications of their views of technology.

REFERENCES

Aspy, D. N. (1975). Helping teachers discover empathy. *Humanist Educator, 14*(2), 56–62.

Blacker, D. (1994). Philosophy of technology and education: *An invitation to inquiry*. Philosophy of Education Retrieved Accessed March 28, 2004 from http://www.ed.uiuc.edu/EPS/PES-Yearbook/94_docs/BLACKER.HTM.

Bloom, B. (1984). *Taxonomy of educational objectives* Boston, MA: Allyn and Bacon.

Cameron, D. (1999). Language: linguistic relativity: Benjamin Lee Whorf and the return of the repressed. *Critical Quarterly, 41*(2), 153–156.

Carey, J. W. (1981). McLuhan and Mumford: the roots of modern media analysis. *The Journal of Communication, 31*(3), 162–178.

Chandler, D. (1995). *Technological or media determinism*. Retrieved June 17, 2007, from http://www.bos.org.yu/cepit/idrustvo/st/TechorMediaDeterminism.pdf.

Charles, M. T. (2004). Where are we going as we leave no child behind: La Technique and Postman, Papert, and Palmer – Part 1. Interface: *The Journal of Education*, Community and Values. *4*(1).

Davis, W. (2003). Interpreting a meaning of technology. *Unpublished Dissertation*, University of Virginia, Charlottesville, VA.

Dennett, D. (1987). The intentional stance. Cambridge: *MIT Press*.

DiMaggio, P., Hargittai, E., Neuman, W. R., & Robinson, J. P. (2001). Social implications of the internet. *Annual Review of Sociology Annual, 27*, 307–336.

Dixon, W. R., & Morse, W. C. (1961). *Prediction of teaching performance: empathetic potential*. Journal of Teacher Education, 12.

Engeström, Y. (1999). Expansive visibilization of work: An activity-theoretical perspective. *Computer Supported Cooperative Work, 8*, 63–93.

Feenberg, A. (2001). Whither educational technology? *International Journal of Technology and Design Education, 11*(1), 83–91.

Feenberg, A. (2004). *Critical theory of technology*. Retrieved November 9, 2008, 2008, from http://www-rohan.sdsu.edu/faculty/feenberg/ctt.htm.

Forster, E. F. (1909). The machine stops. *CreateSpace*.

Freire, P. (1999). *Pedagogy of the oppressed*. New York: The Continuum Publishing Company.

Gandhi, M. (1922). Speech at Pardoli Taluka conference. January 29, 1922. *In The Collected Works of Mahatma Gandhi*. http://www.gandhiserve.org/cwmg/VOL026.PDF, p. 260.

Gagne, R. (1985). *The conditions of learning (4th ed.)*. New York: Holt, Rinehart & Winston, Inc.

Habermas, J. (1984). *The theory of communicative action* (Vol. One). Boston: Beacon Press.

Habermas, J. (1989). *Structural transformation of the public sphere*. Cambridge, MA: MIT Press.

Hall, E. T. (1966). *The hidden dimension: man's use of space in public and private*. London: Bodley Head.

Hester, J. (2002). The influence of select variables on the instructional use of computers in Shelby County public schools. *Unpublished dissertation*, The University of Memphis, Memphis, TN.

Hlynka, D. (1989). Applying semiotic theory to educational technology. *Paper presented at the Association for Educational Communications and Technology*, Dallas, TX.

Hlynka, D. (1994). Glossary of Terms. *Educational Technology, 34*(2), 14–15.

Huxley, A. (1992). *Brave new world*. NY: HarperPerennial.

Jonassen, D. H. (2000). Computers as mindtools for schools: *Engaging critical thinking*. Upper Saddle River, NJ: Merrill/Prentice Hall.

Kapp, K. (2012). *The gamification of learning and instruction: game-based methods and strategies for training and education*. New York: Pfeiffer.

Kellner, D. (1990). From 1984 to one-dimensional man: critical reflections on Orwell and Marcuse'. *Current Perspectives in Social Theory, 10*, 223–252.

Kellner, D. (2004). Habermas, the public sphere, and democracy: A critical intervention. Retrieved December 2, 2004, 2004, from http://www.gseis.ucla.edu/faculty/kellner/essays/habermaspublicspheredemocracy.pdf

Koerner, E. F. K. (1992). The Sapir-Whorf hypothesis: a preliminary history and a bibliographical essay. *Journal of Linguistic Anthropology, 2*(2), 173–198.

Kranzberg, M. (1986). Technology and History:" Kranzberg's Laws." *Technology and Culture, 27*(3), 544–560.

MacKenzie, D., & Wajcman, J. (Eds.). (1985). *The social shaping of technology: how the refrigator got its hum. Milton Keynes:* Open University Press.

McLuhan, M. (1964). *Understanding media: the extensions of man*. New York: Mentor.

McTighe, G. (1998). Understanding by design. Alexandria, VA: *Association for Supervision and Curriculum Development*.

Moore, S. (2005). *The social impact of a profession: an analysis of factors influencing ethics and the teaching of social responsibility in educational technology programs*. Unpublished Dissertation, University of Northern Colorado, Greeley, CO.

Mumford, L. (1964). *Authoritarian and democratic technics*. Technology and Culture 5:1.

Niederhauser, D., Lindstrom, D., & Strobel, J. (2007). Addressing the NETS*S in K-12 Classrooms: implication for teacher education. *Journal of the Technology and Teacher Education. 15*(4), 483–512.

Ogbu, J. U. (1982). Cultural discontinuities and schooling. *Anthropology & Education Quarterly, 13*(4), 290–307.

Olden, C. (1953). On adult empathy with children. *Psychoanalytic Study of the Child*, 8, 111–126

Orwell, G. (1949). 1984. New York, Penguin Press.

Pacey, A. (2000). *The culture of technology*. Cambridge, MA: The MIT Press.

Paul, R. (1993). *Critical thinking*. Rohnert Park, CA: Foundation for Critical Thinking

Poster, M. (1995). *The second media age*. Cambridge, MA: Polity Press.

Postman, N. (1990). *Informing ourselves to death*. Retrieved December 1, 2004, from http://interact.uoregon.edu/MediaLit/mlr/readings/articles/informpost.html.

Postman, N. (1993). *Technopoly: the surrender of culture to technology*. New York, NY, USA: Vintage Books.
Reeves, T. (2006). How do you know they are learning?: *The importance of alignment in higher education. International Journal of Learning Technology*, 2(4), 294–309.
Rogers, E. (1995). *The diffusion of innovation*. New York, NY: The Free Press.
Smith, M. R., & Marx, L. (1994). Does technology drive history? *The dilemma of technological determinism Cambridge*, MA, USA: MIT Press.
Strobel, J. & H. Tillberg-Webb (2008). Applying a critical and humanizing framework of instructional technologies to educational practice. In Moller, L., Huett, J. and D. Harvey (Eds.), *Learning and instructional technologies for the 21st century*. NewYork: Springer.
Taylor, R. (2007). Learning after the end of knowledge. *Unpublished Masters Thesis*, Concordia University, Montreal, Canada.
Thorson, E., Chi, A., & Leavitt, C. (1992). Attention, memory, attitude, and conation: A test of the advertising hierarchy. *Advances in consumer research*, *19*, 366–379.
Tillberg-Webb, H. & Strobel, J. (2011). *Analysis of technological ideologies in education: a translation of lessons from technological dystopian literature into educational theory. Techne*, *15*(2), 170–181.
Warschauer, M. (2000). Technology and school reform: *A view from both sides of the track. education policy analysis archives*, *8*(4), 1–21.
Winner, L. (1980). Do artifacts have politics? *Journal of the American Academy of Arts and Sciences*, *109*(1).
Winner, L. (2004). Technology as forms of life. In D. M. Kaplan (Ed.), *Readings in the Philosophy of Technology* (pp. 103–113). Oxford: Rowman & Littlefield.
Yamagata-Lynch, L., Smaldino, S., & Click, A. (2008). Using activity systems analysis to trace changes in student reflections in Second Life courses. *Paper presented at the Association of Educational Communications and Technology International 2008 Convention*, Orlando, FL.
Yamagata-Lynch, L. C., & Smaldino, S. (2007). Using activity theory to evaluate and improve K-12 school and university partnerships. *Evaluation and Program Planning*, *30*(4), 364–380.

Heather Tillberg-Webb
Lesley University

Johannes Strobel
Institute for P-12 Engineering Research and Learning,
Purdue University

CHAPTER 17

JERRID W. KRUSE

IMPLICATIONS OF THE NATURE OF TECHNOLOGY FOR TEACHING AND TEACHER EDUCATION

INTRODUCTION

Beyond deeply understanding content and effective teaching strategies, teachers must understand the nature of their discipline as well as the nature of learning (Borko & Putnam, 1996). Genuinely understanding the nature of learning is the foundation for informed decision-making that takes into account students' struggles and the dynamic classroom context (Clough, Berg & Olson, 2009). Understanding the nature of their discipline helps teachers more cohesively structure their course and assist students in understanding the processes, characteristics, and philosophy of the discipline as well as the content.

Similarly, understanding the nature of technology (NOT) is of utmost importance for making wise decisions regarding teaching and learning with technology. That is, by understanding the NOT, teachers can more carefully structure technology use in their classrooms and decide when technology use might interfere with learning. Furthermore, NOT understanding is crucial so teachers can assist students in understanding the processes, characteristics, philosophy and content of technology. Unfortunately, when discussing what teachers should know about technology for instruction, the NOT is typically absent (e.g. Guerrero, 2005).

When first introduced to the NOT, some wrongly see it as a veiled way to resist technological advance or criticize technology use in schools. However, as Selber (2004) points out, taking a critical stance is important, but not enough because criticism does not prepare educators to work within the new reality they find themselves. A NOT framework is important for both critiquing and improving technology use in schools. Some important ideas emerging as necessary for considering the NOT in educational settings (Buckmiller & Kruse, 2011; DiGironimo, 2011; Kruse, 2009, 2012) include:

- Identifying technology
- The nature of technological progress
- The non-neutral, value-laden, or biased nature of technology
- The limitations of technology

Michael P. Clough, Joanne K. Olson and Dale S. Niederhauser (Eds.), The Nature of Technology: Implications for Learning and Teaching, 345–370.

- Technological trade-offs
- The interactions of technology and culture

The importance of technology is obvious, but less acknowledged is the emotional attachment many of us have to our technologies. These benefits and emotional attachments make critically assessing technology a difficult process and easily lead people to dismiss those raising such issues as Luddite or technophobe. However, Keen (2008) notes that familiarity with technology provides important insight for thinking more critically about technology. The ideas in this chapter draw from educational and technological literature as well as the author's extensive experience using technologies with K-12 students and lessons learned from teaching an educational technology course in the university setting.

No need exists to revisit the arguments for a including technology in the classroom. When educational technology is carefully contemplated and, where appropriate, effectively incorporated, it can improve teaching and learning. But the wisdom to effectively contemplate and incorporate technology demands an understanding of NOT issues, education goals, teaching and learning. This chapter will discuss aspects of the NOT useful for educators to make more prudent decisions about when and how to use various technologies, and when to eschew their use.

CORE NATURE OF TECHNOLOGY IDEAS

The nature of technology (NOT) is a multi-faceted construct. Further complicating the NOT construct is the manner in which NOT ideas interact with one another. The following discussion of key NOT ideas is necessarily brief and simplistic, but is sufficient for establishing a foundation to address the implications of these ideas for education.

Identifying Technology

When considering educational technology, too much emphasis is placed on electronic and digital technologies. When asking prospective administrators about their considerations for technology use in schools, Buckmiller and Kruse (2011) found that not a single participant mentioned any technology other than digital technologies and the Internet. This narrow focus ignores many other technological aspects of school and their effect on learning.

DiGironimo's (2011) conceptual framework includes technology as artifacts (e.g. computer) and technology as a creation process (e.g. engineering) when identifying technologies. The National Academy of Engineering (2009) reflects this view by noting that technology includes: practical knowledge, innovation, human activities, and systems of components. Expanding the commonly held, but narrow view of technology is important for discourse and decision-making regarding technology in schools.

The Nature of Technological Advance

New technology is developed in light of previous technologies (McArthur, 2007). That is, new technologies are often created from components and concepts of already existing technologies. As much as educators want to believe in the revolutionary power of technology, its advance more closely resembles evolution than revolution. That is, technological progress is "slow and cumulative" (DiGironimo, 2011, p. 1343). Consider, for example, the interactive whiteboard. This technology would not be possible without computer technology and touch screen technology first existing, and those technologies grew from previous technologies, and so on. New technologies are typically a unique application and recombination of already existing technologies. The crucial point here is to carefully consider how new technologies may simply be more sophisticated and expensive versions of already existing materials and approaches in education.

Technology development is but one aspect of technological advance. Adoption of new technologies is perhaps more important than development. What is developed is of little consequence if the technology is not adopted. Technologies that are too far removed from users' expected experience are not likely to be adopted. That is, new information, if not assimilated or accommodated into existing schemas, will be dismissed (Piaget, 1970; Posner, Strike, Hewson, & Gertzog, 1982). Returning to the interactive whiteboard, consider why such a technology makes so much sense to many educators. The chalkboard, the overhead projector, the whiteboard, PowerPoint and the interactive whiteboard are all intricately related. To some extent, the interactive whiteboard's development and use in schools reflects the prior ubiquitous use of chalkboards.

Technology is Value-laden

As with all human endeavors, technology is a value-laden enterprise (DiGironimo, 2011). As a result, technologies clearly value some goals and ideals over others. For example, text-based books value linear thinking over more branched, divergent thinking. Mobile smart phone technology is biased toward being constantly connected. These biases have benefits and detriments, but regardless, that technologies are value laden ought to be considered in education technology thinking. Unfortunately, the natures of these biases are often difficult to uncover.

The Limitations of Technology

Technology cannot fix all problems. Indeed, technology often causes new problems while helping solve others. Some of the world's most fundamental problems are not technological in nature. For example, we have sufficient agricultural technology to feed the world, but other factors having nothing to do with technology, keeps many people all over the world hungry. As Ely (1995, p. 12) notes:

> We have been brought up on the myth that almost any problem can be solved with a technological solution. In education, this assumption is dangerous and in terms of technology, it can be disastrous.

The most pressing education issues have little, if anything, to do with technology. Consider, for example, the role of standardized testing in education. While certain technological advances make gathering data about students, teachers, and districts possible, if and how such data are used will depend upon fundamental beliefs and assumptions about teaching and learning, not simply whether collecting such data is technologically feasible.

Technological Trade-Offs

All technologies come with trade-offs (AAAS, 2007; NAE, 2009). That is, along with technology's gains, come losses, or things given up. As Postman (1995, p 41) notes:

> [Technologies] are Faustian bargains, giving and taking away, sometimes in equal measure, sometimes more in one way than the other. It is strange—indeed, shocking—that with the twenty-first century so close on our heels, we can still talk of new technologies as if they were unmixed blessings, gifts, as it were, from the gods. Don't we all know what the combustion engine has done for us and against us? What television is doing for us and against us?

Consider, for example, the cell phone. The ability to make mobile phone calls has clear advantages for safety, convenience, and communication. However, such advantages come at the cost of privacy, rest, and resources.

Interactions of Technology and Culture

Technology impacts human cultures. Indeed, human historical eras are identified by their dominant technologies (NAE, 2009). Thus, not surprisingly, technology will impact the cultures of classrooms. Postman (1992, p. 19) summarizes the deep impact technology has on institutions and individuals:

> We need to know in what ways [the computer] is altering our conception of learning, and how, in conjunction with television, it undermines the old idea of school. Who cares how many boxes of cereal can be sold via television? We need to know if television changes our conception of reality, the relationship of the rich to the poor, the idea of happiness itself. A preacher who confines himself to considering how a medium can increase his audience will miss the significant question: In what sense do new media alter what is meant by religion, by church, even by God? And if the politician cannot think beyond the next election, then we must wonder about what new media do to the idea of political organization and to the conception of citizenship.

APPLICATION OF CORE NATURE OF TECHNOLOGY IDEAS TO TEACHING, LEARNING, AND SCHOOLING

Decisions regarding education technology often reflect simplistic assumptions and attitudes regarding technology. Understanding the influence technology has on schooling, teaching and learning and weighing both its pros and cons demands awareness of the NOT. This section explores ways in which the above NOT ideas can assist in making education technology decisions.

Technology's Influence

Asking what technologies educators and students should *use* and how they should *use* them ignores extensive examples in the NOT literature illustrating how technology has, without our awareness, ended up changing the way humans think and act. In this important sense, decisions regarding education technology must also consider how it may use us! That is, how might adopting a particular technology impact educators' and students' thinking and activities in ways that are undesirable? This question demands an understanding of the value-laden or biased nature of technology. Too often educators and education reformers speak of how technology ought to be used, but pay too little attention to how technology is actually used. As Selwyn (2010) notes:

> [T]he academic study of educational technology could be accused of having worked itself into an analytic corner – well-able to discuss how educational technologies could and should be used, but less competent and confident in discussing how and why educational technologies are actually being used.

Educators' intense focus on how technology "could" and "should" be used and not on the way the implicit cues of technologies influence thinking and action illustrates the lack of attention paid to the value-laden nature of technology. To make informed decisions regarding technology, educators must understand how belief systems impact technology implementation, how technology can change our beliefs in unexpected ways, and how technology affects teachers' actions in unexpected ways.

Beliefs impact technology implementation. Teachers' beliefs have always played an important role in classrooms (Fang, 1996; Haney, Czerniak, & Lumpe, 1996; Nespor, 1987). This is not surprising given what is known about how information is interpreted in light of prior schemas. Teachers' deeply held views of teaching and learning impact their choice and implementation of curricula, teaching models and strategies, and assessment strategies. Chen (2006, p. v) confirms this trend related to technology implementation:

> Teachers with more constructivist beliefs made efforts to allocate time for students to engage in problem- or project-based learning occasionally. Some of them used online discussion or presentation software to anchor and encourage

> discussion and interaction among teachers and students. Teachers who prioritized examination preparation mostly used technology to cover content, sometimes discarding technology when they considered technology not cost-effective or a distraction for student learning.

Unfortunately, new technologies are often developed in light of traditional views of teaching and learning. Therefore, although technologies do carry implicit messages about how they should be used, one should not expect technology to improve teacher beliefs or pedagogy. Far more likely, teachers will use technology in ways that fit their already existing pedagogy (Ely, 1995; Lazlo & Castro, 1995; Fraser & Deane, 1999; Selber, 2004). For example, when studying novice teachers' use of technology in science classrooms, Irving (2009) found that the technology was more often in the hands of teachers rather than students. The study noted that teachers most often used technology to provide visual images and models related to content. Rather than engage students in collaborative meaning making, the teachers used the technology in fairly mundane and teacher-centric ways. Although teachers are implementing technology, the actual classroom environment is not much different than traditional teaching.

Oftentimes technology is viewed as a means to create change in teachers' views of learning. Instead, resources might be better spent trying to change teachers' thinking about learning and teaching, and then introduce new technologies that support more mentally engaging learning environments. While techno-enthusiasts hope technology promotes educational change, the reality is that technology can and often does promote teaching and learning environments not very different from traditional learning spaces (Cobb, 1999; Fraser & Deane, 1999; Gance, 2002; Guzman-Rodriguez, 2007).

The impact of beliefs on technology implementation is unsurprising, yet acknowledging this impact is important for critically assessing the hyperbole regarding educational technology initiatives. Technology, infused into schools without consideration of how teachers conceptualize teaching and learning, will rarely result in significant change. For example, I observed a classroom where students were each on laptops and the teacher used an interactive whiteboard. However, students were simply interpreting a pre-made graph, answering fill-in-the-blank questions that required little thinking, and the teacher was using an interactive whiteboard to record lectures for future use in class. This teacher's view that teaching is telling and learning is repeating back information was evident in what occurred with the technology. When asked about recording his lectures, the teacher enthusiastically claimed, "I can record them once and then reuse them every year!" Clearly the teacher's use of technology did not encourage him to adopt more research-based teaching strategies.

Technology impacts beliefs. In the previous classroom example, the technology was also shaping the teacher's thinking and action regarding teaching, but not in the way techno-enthusiasts might hope. Because the technology makes recording information so easy, the technology encouraged the teacher to *not* reflect upon and improve his lessons from year to year! Thus, this teacher's *beliefs about teaching and ineffective teaching practices were further promoted by the technology*. Not only must educators consider the impact teacher beliefs have on technology implementation, they must also consider the unexpected, and possibly negative impact technology may have on teachers' beliefs.

Technologies come with inherent value systems. While these systems sometimes reflect the values of the user, they more often reflect the values of the designer and in some contexts may not reflect the value systems of either. For example, Postman (1992) notes how the printing press provided an avenue for the Bible to become available to a wide audience - effectively undermining some powers of the established church. Both Gutenberg (designer) and Luther (user) were surprised by the ability of the printed word to spread so quickly. Over time, the value system inherent in the printing press, that each person has access to text, became the dominant value system in society.

When considering education, Cuban (1998) notes how technology-based reforms too often shift values toward preparation for the technological workplace rather than preparation of critically thinking, engaged, and competent citizens. In Cuban's (1998, p. 7) words:

> The lure of higher productivity in teaching and learning via computer technologies, however, has seduced reformers into treating teaching like any other form of labor that has experienced productivity gains after automation.

Not only can technocentricity distract us from more meaningful goals in education, technology use raises new ethical concerns. For example, Ely (1995, p. 4) notes:

> [T]echnology sometimes raises new moral issues related to long-held goals that can now be achieved with unimagined effectiveness. For example, how fast do we want to move across the face of the earth? For how long shall we defer death?

Similarly, educators might ask, "To how much information do we want students exposed? How is information access different than learning? How short shall we make childhood?"

Beyond changes to value systems, technology is changing the demarcation between professional and amateur (Keen, 2008). This trend may have grave consequences for education. While professional educators should not dismiss amateur teaching web sites, they must proceed with caution. For example Khan Academy is an online resource for "educational" videos. This site is gaining great public and financial support. Yet, when watching the videos, professional educators easily recognize the emphasis on math as a skill to be practiced rather than concepts to be understood or superficial discussion of complex ideas divorced from all context. Khan's videos depict science as a simple rhetoric of conclusions and ignore the process and creativity of science, history, and mathematics. These messages inherent in the Khan Academy videos are dangerous and destructive.

The point here is not to demean Khan Academy, but to illustrate how technology can change long-held social structures, including schools, in ways that were not intended. These changes may be positive and/or negative, but too few think deeply about the social consequences that technology may have. Modern technologies have eroded the professional and amateur divide. While I might enjoy having access to medical information, I know that such access does not replace the value of seeing a professional physician. Likewise, access to information and access to a professional teacher should not be equated. Along with this erosion of the professional and amateur divide, modern technologies are taking important aspects of schooling. As access to information and even instruction becomes easier, in what ways are fundamental beliefs about schooling and even learning modified? In what ways are those modifications positive? In ways are they negative?

Okan (2003, p. 262) citing Salomon (1998, p.7) addresses such questions when noting how technology changes thinking:

> It is as if technology might take charge, demanding of constructivist philosophy and of the psychology of learning and instruction to follow suit and to adjust themselves to the technological affordances.

That is, much education technology demands that educators adopt a view of learning that fits with the values of the technology. If technology has its way, the idiosyncrasies and inefficiencies inherent in teaching for deep and meaningful learning will be replaced by an input/output view of teaching and learning, concomitant with a perspective of efficiency and economy of scale promoted by technology.

Technologies have shaped views of learning and knowledge for millennia. For example, many technologies, from Google to the written word, modify views on wisdom. While technology delivers instant information, it does not provide knowledge or engage in reflection (Ely, 1995). While school has long-valued trivia, current technologies have pushed us deeper into a rut where factoids are valued over deep conceptual understanding.

To illustrate how technology may shape views about learning, consider the student quoted below (Davidson et al., 2011, p. 44):

> My English professor got really angry about people texting and said, 'Don't you think it's rude while someone's talking?' But it's not. We've really become a generation where we have to do two things at once, and we can focus on each of them. It was a five-minute argument because he was losing.

To what extent has this student been set up to fail when engaging in complex learning tasks that requires singularly focused attention? To what extent will this student even choose such a task? Educators cannot ignore the impact technology is having on student views of learning and must consider the negative impact technology might have on their own view of learning.

Perhaps more insidious is how technology changes views of engagement in learning. While focused mental engagement is crucial for deep and meaningful learning, technologies change our definition of "engaged". Rather than mentally engage students, much educational technology simply entertains students. This notion sends dangerous implicit messages to students that learning is "a bitter medicine that needs the sugar-coating of entertainment to become palatable" (Resnick, 2004) or that only things that are fun are worth doing (Postman, 1985).

Rather than education, this "edutainment" causes an "inflated expectation in the learners that the process of learning should always be colourful and fun, and that they can acquire information without work and serious study" (Okan, 2003, p. 255). Indeed, Kazanci and Okan (2009) described a random sample of language software to be overly entertaining and "disneyfied". As Okan (2003, p. 259) notes, these messages being sent by technology and technology use are problematic because that…

> …meaningful learning may sometimes be difficult and requires cognitive and emotional effort should be kept in mind; this point is especially relevant in the light of the fact that post-secondary education is not usually a fun undertaking. On the other hand, recognising [sic] the serious nature of higher education does not necessarily mean that fun is an opposite of activities that are serious.

While educators might see students more occupied with the technologies used in classrooms, educators should wonder if students are truly mentally engaged or simply entertained.

Related to engagement and entertainment issues, educators must determine whether students are occupied with the technology or mentally engaged with the content to be learned. In many technology rich classrooms I have observed, teacher and student efforts are far more focused on how to operate the technology; not on the concepts the technology is supposedly designed to help students understand. The technology, far from being simply a tool, can easily distract students and teachers from desired learning.

Technology impact on actions. The previous section addressed how technologies can impact how and what we think, often in unanticipated and undesired ways. Views regarding who or what we think is qualified to teach, the nature of learning and teaching, and the nature of engagement are all impacted by technology. Technologies not only influence what we think about teaching and learning, they also impact educational practice. Selwyn (n.d., p. 6) cautions:

> Despite the immediate appeal of applications such as Facebook and Second Life it is necessary for educators to take time to reflect carefully upon the nature of these Web 2.0 applications as online learning environments and question the learning affordances they offer in practice.

While the critical stance toward educational affordances is important, asking educators to question the learning affordances may not be enough. Instead, we must consider the cues inherent in technology. For instance, the claw end of a hammer can be used as a flat-bladed screwdriver, but the very design of a hammer sends clear messages that it *should* be used to strike something. When techno-enthusiasts talk about the possibilities of technology they often ignore the subtle and sometimes not so subtle messages technologies provide about how they should be used. That is, technological affordances are discussed at length, but technological cues are nearly ignored. Because technology can have such great effect on beliefs and actions, educators ought to be wary of the subtle, but powerful hints, or cues, technology sends about how the technology *should* be implemented. While imaginations run wild with the possibilities, or affordances of technology, few (not even designers in some cases) consider the cues technology contains. For example, although textbooks can be used as a valuable tool in classrooms, the bolded words cue students (and teachers) to place emphasis on vocabulary acquisition over deep conceptual understanding. Also, consider how digital whiteboard technology cues the teacher to remain in one location, typically at the front of a room.

The notion that technology may impact the way humans act can be unsettling. Yet educators must recognize that technologies do constrain, and influence decisions. For example, if we expand our view of technology beyond modern electronics, the daily school schedule is an organizational technology to budget time. While the school schedule seems harmless, educators will likely recognize that it is the school schedule that decides how long they plan lessons or when instruction must cease for the day. School bell schedule technologies, in making fundamental decisions for us, may cause educators to make decisions not in the best interest of student learning (i.e.: cutting a lesson short when students are at the peak of curiosity or limiting the amount of time students explore a new phenomenon).

Technologies such as the bell schedule have likely become so engrained in our school cultures that we pay little attention to how these technologies shape our actions. Some educators may realize that the bell schedule is beyond their control anyway. So what of the technologies we choose to use in our classrooms; how might these technologies shape our teaching behaviors?

Waight and Abd-El-Khalick (2007), when studying a sixth grade teacher's classroom known for inquiry experiences and technological prowess, noted that when the teacher used computers, the inquiry in the classroom was diminished and students focused more on use of the computers than on the inquiry of the natural world. That is, students became more concerned with sharing computer time and accomplishing computer-related tasks rather than wrestling with the content at hand. Such results are likely linked to technological biases.

Computers, for instance, were created as personal devices and they send many cues to be used that way. For example, Kruse (2009) reports how his 8th grade students were unusually quiet one day while working on laptop computers. When he asked students why the class was so quiet, they blamed the computers and identified several physical cues (e.g. one mouse, one keyboard, and one screen) that *promoted* individual and silent work. Guzman-Rodriguez (2007) noted the same phenomena when first implementing a computer-based instructional model. Rather than working socially or collaboratively, the students worked individually. In both cases, the cues of the laptop encourage individual use, and "users" (actually, in an important sense, the computers "used" the students) did as the technology directed. However, the importance of understanding the NOT is illustrated by how the negative biases of technologies may be mitigated if those biases are understood, and purposeful instructional actions are taken. For instance, in subsequent activities Guzman-Rodriguez (2007) purposefully included discussion questions to encourage students to share their thinking with other students. Kruse (2009) chose to provide pairs of students with only one computer to encourage student-student interaction while using the computers.

Cuban (1998) further highlights how technology may change social relationships within classrooms. He writes how the balance of trust, caring and respect between teacher and student may be compromised in high-tech environments. For instance, a student who is struggling to understand a concept and near giving up is unlikely to be noted by an online teacher. Even if the student was noticed, an online teacher cannot look into that child's eyes and show empathy and give support while continuing to push the student to persevere through the difficulty. In education, as with many human endeavors, social relationships do matter, and technology often diminishes the potential meaningfulness of those relationships.

While educators might hope technology will create more collaborative and social learning environments, the examples above make clear that many of technology's biases are not conducive to effective learning environments. Technology emphasizes speed and efficiency – two terms with disastrous implications for deep, applicable, and meaningful learning. Mindful of how technology biases may impact learning, teachers can take steps to use technology effectively rather than permit it to use us. As Guzman-Rodriguez (2007, p. 345) notes, "It is up to the educator then, to promote in the classroom interaction among the learners." This statement and the examples above make clear that educators must understand the biases of technology, make decisions to overcome those biases, and when that is not possible, eschew that technology for the particular context.

Technology's Limited and Limiting Nature

Making wise educational technology decisions demands that educators understand that technology is developed and adopted in light of the past and new technologies or newly adopted technology rarely represent fundamental change. Because new technologies are developed based on past technologies, the ideologies on which new technologies are based are typically rooted in traditional views of teaching and learning. So while technology does influence views of education, the influence likely maintains the value systems educators are trying to overcome. For example, educators have long worked to overcome the notion that learning happens in isolation or is a matter of information transfer from teacher to students. Yet, the examples above in which students tended to work in greater isolation when working with computers make clear that technology pushes students and teachers to maintain the view that learning happens in isolation. Furthermore, many online education platforms (e.g. Khan Academy) reinforce the notion that teaching is simply a transmission of information from teacher to student.

Technology's reinforcement of traditional views of education is complicated by the additional impact of teacher beliefs on implementation of reform efforts. That is, teachers often enact reform efforts in a manner consistent with their deep-seated views of teaching and learning (Clough & Kruse, 2010). Even if technology were developed to promote more research-based views of learning, implementation of the technology will be mediated by teachers' beliefs about teaching and learning. Because technology may exacerbate traditional educational dilemmas and teachers' deeply held views mediate their use of technology, introducing new technology in educational settings rarely results in fundamental change (Cuban, 1986).

For the reasons above, educators must acknowledge that technology is limiting and limited. That is, technologies are limiting because of the values inherent in the technology. For example, as previously noted, Waight and Abd-El-Khalick (2007), Kruse (2009), and Guzman-Rodriguez (2007) found that technologies implemented in reforms-based classrooms hindered some of the goals of the teachers such as inquiry and student-student interaction. Technologies are limited because many educational issues are simply not technological in nature. That is, many educational issues are rooted in the fundamental disposition and beliefs of teachers. For example, providing a traditional teacher with modern technology simply means the technology will get used to reinforce traditional teaching (Ely, 1995; Lazlo & Castro, 1995; Fraser & Deane, 1999; Selber, 2004).

Guzman-Rodriguez (2007, p. 84–85) makes clear the limited nature of technology with respect to educational change:

> Gance (2002) argues that although computer technology could support constructivism, because of the nature of constructivism itself, the computers alone cannot provide a constructivist environment. He states that a constructivist environment requires engaged learners, real interaction with the materials, a real problem-solving context, and human interaction. Therefore it takes a very creative teacher not to fall in the constraints of technology and still create a constructivist classroom.

While techno-enthusiasts might hope for technological transformation, the problems in many classrooms lie beyond technology's reach. The technology has limits that educators and educational reformers must understand. Okan (2003, p. 255) notes when discussing effective learning environments that:

> [E]ducation is concerned with the development of cognitive structures and that educational technology is a medium, not a pedagogy that is useful in creating such learning environments.

Educators who constantly look to technology for enhancing learning environments may fall into the trap described by Ely (1995, p. 4).

> When technology makes it possible for people to do something, people do it, not always because it is necessary, but because it is possible.

Educators who fall into this trap likely see decisions regarding technology integration simply as cost issues (Buckmiller & Kruse, 2011) or as debate between do or do not rather than over core values (Cuban, 1998).

Once educators understand how technology is limited, consideration ought to be given to the ways in which technology is limiting our students. Given that technology is used to do something for us, educators ought to consider that the thing technology does for students may actually limit students' thinking about important aspects of a concept to be learned. That is, technology can effectively hide aspects of a phenomenon causing students to not mentally wrestle with important observations to develop skills or conceptual models (Olson & Clough, 2001; Potter & Kelly, 2006; Lunetta, Hofstein, & Clough, 2007).

For example, imagine students are learning about acid-base titrations using a computer simulation. This simulation will be useful in showing students the endpoint and they may even be able to add the titrant "drop by drop". Yet, the simulation will likely not show the need to carefully swirl the solution in between drops and will not provide an opportunity for a skilled teacher to ask, "Why does your solution stay pink longer and longer before going back to clear?" This question pushes students to consider the manner in which particles are interacting. Digital simulations hide this deep level of thinking about the particulate nature of matter. While the digital simulation could be modified *ad infinitum*, the simulation will never fully represent the actual task. Furthermore, students can dismiss a simulation as contrived or not representative of reality.

While student thinking might be hidden, much is gained from such digital simulations: students in schools without laboratory equipment can gain a sense of completing a titration, chemical spills and waste are minimized, and students can repeat the experience countless times. However, also consider what is lost: students do not see how the initial color change requires swirling to revert back to the original color, students do not need to consider the implications of going beyond the endpoint, and students do not need to consider what happens to chemical waste. Each of these "losses" represents a missed opportunity for deeper thought and learning.

As another example, consider the simple grammar and spelling checks accompanying most digital word processing programs. When students turn in a paper in which all grammar and spelling is correct, teachers may not realize the struggles students are having with grammar or spelling. Not only do students not have to learn about nuances of spelling and grammar, the teacher is unable to meaningfully assess students' ability to spell or use appropriate grammar as the technology effectively hides that portion of the students' thinking. That is, the technology is limiting the teacher's ability to accurately assess student thinking.

While these examples do not mean to imply educators should never use modern technology, they should give educators pause. Educators ought to consider what technologies do for their students that might actually limit the students' thinking or encourage superficial grasp of complex topics. Spencer (2011) notes that we must take the magic out of technology:

> As long as we focus tech integration on "how to use tools" we allow a mechanical process to remain an act of magic and the problem with pseudo-magic is that it inhibits kids from experiencing the majestic mysteries of what they can actually see, smell, taste, hear and feel. My children will some day use a computer, but not until they can strip away some of its magic.

Too often new technologies are adopted in schools thinking the pros, cons, and limitations have been carefully weighed. However, without understanding the NOT, this weighing is likely superficial. Once one recognizes what technology is, how technology progresses, the value-laden nature of technology and technological limitations, technological trade-offs run deeper than simple cost and time issues. Educators must carefully consider technological gains as well as losses for student thinking and learning.

Using NOT to Improve Educational Technology Use

The examples and discussion above are illustrative, but in no way exhaustive, of how understanding the NOT can inform decision-making concerning technology. They do, however, illuminate the kinds of questions that educators ought to ask about technology. Some of those questions are:

- What are the biases and cues of technologies, digital and otherwise, already in place, and how are they impacting teacher decision-making, teaching and student learning?
- What are the biases and cues of technologies under consideration for adoption, digital and otherwise, and how might they impact teacher decision-making, teaching and student learning?
- How are older classroom technologies related to new classroom technologies? In what ways might the use of older technologies impact the use of newer technologies?
- What does a particular technology do for students/teachers? How might this technology limit students'/teachers' thinking?
- What is gained and what is lost with the adoption of a particular technology?
- How might a given technology impact the culture of the classroom in both positive and negative ways?

For example, curriculum maps are a technology that impacts classroom teacher decision-making, teaching and student learning. Curriculum mapping may be thoughtfully accomplished, but too often they merely reflect the biases of another technology — textbooks. That is, educators often use textbooks to create curriculum maps, and this illustrates how an existing technology (textbook) may bias the development of subsequent technologies (curriculum map). Understanding the NOT and questions above, educators are in a position to recognize how a curriculum map may act to restrict teachers' willingness to explore students' thinking that deviates from the adopted map. When considering limitations of a curriculum map, educators should note how a curriculum map may inhibit teachers' creativity in planning and implementing instruction. The above questions, if seriously considered, require educators to acknowledge both gains and losses associated with the creation and implementation of curriculum maps. These gains and losses will undoubtedly be connected to the biases and limitations of curriculum maps. Lastly, the considerations of all these nature of technology ideas related to curriculum maps may help educators predict how curriculum maps may impact school cultures in both positive and negative ways. Perhaps the curriculum map encourages collaboration amongst teachers or results in "finger pointing" when students struggle with particular content.

The questions and thinking illustrated above also apply to electronic technologies. For example, consider the interactive whiteboard. This technology typically reinforces the traditional layout of a classroom, with the teacher at the front. Furthermore, the interactive whiteboard will likely be used as a novel way to present information – perpetuating the view that teaching is transmission of information from teacher to student. When teachers create interactive visuals on the interactive whiteboard, they may inadvertently be limiting students' creation of their own conceptualization of ideas. While many of these issues also apply to traditional whiteboards (illustrating the manner in which new technologies are so closely related to older technologies), teachers not considering the NOT are not likely to work against, or even recognize the ways in which either technology promotes ineffective educational practices.

PREPARING TEACHERS TO MORE THOUGHTFULLY CONSIDER EDUCATIONAL TECHNOLOGY USE

Understanding the NOT provides an important lens for assessing and making decisions regarding technology in education. Teacher preparation programs have an obligation to prepare educators who understand and value the NOT and apply that understanding to make effective technology education decisions. Most teacher preparatory programs devote a course or portion of a course to educational technology, but these experiences are typically directed at technical issues regarding technologies and how they might be implemented in education settings. Having students immersed in exploring technologies in education presents an ideal opportunity for addressing the NOT and its importance in technology education decision-making. In the context of learning about technologies, the NOT will not be seen as an "add on" and will provide additional perspective through which students can consider technology decisions in education.

NOT in an Educational Technology Course

Selber (2004) notes students should be users, questioners, and producers of technology. These objectives are useful for structuring preservice educational technology courses to help teachers be more effective users of educational technology. Understanding the NOT is necessary for effectively questioning technology and ensuring we use it rather than it using us. What follows is a description of a preservice educational technology course I teach that reflects Selber's (2004) framework and the core NOT ideas put forward earlier in this chapter.

Preservice teachers as questioners of technology. Because technology is ubiquitous and constantly changing, educators must not blindly adopt the latest innovations, but rather learn to critically examine technologies and how they may shape teaching and learning. Dreaming about how technology might improve student learning is not enough. Postman (1995) encourages educators to ask:

- For every advantage of technology, what is the corresponding disadvantage?
- How are the advantages and disadvantages of particular new technologies distributed unevenly?
- What is the underlying philosophy of particular technologies? For example, how do particular technologies change the way we think and act?
- What are the intellectual, emotional, sensory, social, and content biases of particular technologies?
- What goals are promoted and ignored by particular technologies?
- How does technology change the ways we view learning, teaching and schooling?
- How does the technology promote and inhibit thinking and learning?
- How does technology use us without our awareness?

Understanding and appropriately responding to these questions requires an understanding of the NOT. Many educators acknowledge that successful use of education technology requires effective pedagogical practices (Cuban & Kirkpatrick, 1998; Earle, 2002; Panel on Educational Technology, 1997; Rogers, 1999; Schwab & Foa, 2001), but as this chapter and others have made clear, technology is not simply a tool. The inherent aspects of any technology sends messages regarding how it should be used and these sometimes influence thinking and action in profound and unexpected ways. What this means for education is that technology can and often does impact pedagogical efforts and these influences may not be desirable (Olson & Clough, 2001). So while techno-enthusiasts want to focus on "how to use" technology, understanding and applying aspects of the NOT makes more likely we will use technology as opposed to technology using us.

I address the NOT consistently throughout my education technology course, purposely raising and drawing students' attention to NOT ideas during classroom discussions. My overarching approach borrows from Clough's (2006) framework for teaching the nature of science. In that work, he describes the importance of addressing the nature of science in settings running from decontextualized (i.e. activities, readings and discussions whose sole purpose is to illustrate important nature of science ideas), moderately contextualized (i.e. nature of science instruction is connected to classroom science content instruction and laboratory activities) to highly contextualized (nature of science instruction entangled in the actual work of scientists), and erecting extensive scaffolds between those contexts to help students make sense of difficult nature of science ideas and see them as accurately reflecting authentic science.

Similarly, I introduce the NOT to my preservice teacher education students through decontextualized NOT activities (e.g. an activity whose purpose is solely to illustrate a NOT idea) and ask students questions that draw out those targeted NOT ideas. For instance, I may have my students examine simple technologies such as a ruler and ask, "Of what use is a ruler?" Students reply, "measuring", and I follow with "How do you know a ruler is used for measuring?" or "What about the ruler indicates it is for measuring?" This discussion introduces students to important NOT ideas such as technology is more than electronics, the cues and biases inherent in technology, and how those cues and biases impact our thinking and actions. Most importantly, such a discussion about rulers rarely triggers an emotional response from students. When I have tried to start such discussion with more personal technologies (e.g. cell phones) students' emotional attachment to such technologies often interferes. However, when starting with less personal technologies, students are able to later apply the NOT ideas to more personal technologies.

As the semester progresses, the preservice teachers are asked to apply NOT ideas to more modern classroom technologies. These more contextualized examples are important to help preservice teachers apply the NOT ideas to their own teaching. In one example, I ask my students to take digital pictures of various plants around campus and upload the pictures to a shared website. Students are excited about how technology affords acquiring a tremendous amount of data that can be quickly collected for later analysis. After students have accomplished the task, I ask "What benefit might there be to having students draw the plant instead of snap a picture?" The resulting discussion highlights the lack of thought and careful observation when taking a picture compared to drawing a picture. While the preservice teachers easily recognize the gains from a particular technology, I must draw their attention to also think about trade-offs.

Later in the semester, preservice teachers are introduced to historical examples of technology (i.e. printing press, chalkboard, chalkboard, etc.) and how they have impacted individual thinking and action, values and culture. I also ask students how these technologies have changed education. For example, when discussing the chalkboard, I might ask, "How do you think the chalkboard has impacted education today?" This question requires attention to the nature of technological progress and the implications for current educational technology's impact on teaching and learning. These historical examples are crucial in assisting preservice teachers to take seriously how technology changes us, often without our awareness, and value the NOT as a lens for considering educational technology. Without compelling historical examples, teachers can easily dismiss claims regarding how current technologies may procure unintended consequences for education. However, the lessons from history are not so easily dismissed.

Importantly, these activities are peppered throughout the course, purposely sequenced, and extensively linked via questions I ask. For example, when addressing how chalkboards have impacted education, I ask students to link ideas they are discussing to previous discussions regarding the ruler, the plant photo activity, and other NOT experiences. Promoting these connections is an important scaffold to help the teachers understand complex NOT ideas and their implications for teaching and learning.

Preservice Teachers as Users and Producers of Technology

As Keen (2008) notes, critical analysis of technology is incomplete without familiarity with technology. Furthermore, because my course is designed to help teachers use technology more effectively in their classrooms, they must be introduced to educational technologies. Importantly, students are not simply using technologies, but becoming producers of educational technologies. That is, students learn to use technologies and also produce new technological artifacts (e.g. their own website). Introducing new technologies to students is an excellent opportunity to model how understanding and applying the NOT impacts practice. For example, when demonstrating the use of an online collaborative concept-mapping tool, I ask, "What might be some limitations of this technology?" My students typically note that an Internet connection is required or other relatively innocuous limitations. I press them further by asking, "In what way might this technology limit student thinking?" or "How might this technology limit students' ability to express their thinking?" These questions require going beyond superficial considerations to limitations imposed on learners' thinking; the heart of learning.

While examples and discussion are useful for introducing and developing NOT ideas, preservice teachers must learn to meaningfully apply NOT ideas on their own in their future classrooms. Toward this end, I have my preservice teachers complete five individual projects in which they identify and examine technologies they deem pertinent to their content area(s). Students are expected to consider how people learn and how the particular technology could aid and/or hinder learning. For instance, students must examine how their identified technologies might help and/or inhibit the representation of abstract concepts more concretely, how they might encourage and/or discourage student-student interaction and collaboration, and encourage and/or discourage reflective thought. Beyond precise ties to learning theory, my students are expected to carefully consider hurdles for implementation of the technology including systemic, infrastructural, and student struggles. What aspects of the school system might prevent effective implementation of a technology? What infrastructure must be in place for the technology to work? What struggles might students have when using the technology?

Requiring preservice teachers to consider the biases, trade-offs and limitations of technology, and explicitly link those to how people learn and their goals for students deeply engages the prospective teachers in the content they intend to teach with that technology. Some preservice teachers even claim that after considering the trade-offs, biases, and limitations of a particular technology, they changed their mind and decided the technology would not result in the desired learning they have for their students. For example, when exploring a game-like review program, a preservice teacher noted the manner in which the technology promoted trivia-based knowledge divorced from context and claimed they would not use the technology in the future.

These projects make apparent to me where and how my students' struggle to understand the NOT and its application to teaching and learning, Furthermore, because their analyses must be linked to their discipline specific content area and how people learn, I also am able to assess problems with their understanding of those areas. Not surprisingly, I find that preservice teachers often struggle most when applying NOT ideas to how new technologies may impact teaching and learning. Thus, students' work on these projects also provides me with information useful for assisting students where they struggle. Requiring students to apply the NOT in their instructional decisions makes clear to them the utility of the NOT and the utility of the NOT and helps them internalize those ideas for future decision-making.

Preparing Preservice Teachers to Teach the NOT

Teachers who understand and value the NOT are not only empowered to make more effective educational technology decisions, they are also in a position to help their students understand the NOT, and develop the attitude and cognitive tools necessary for weighing the pros and cons of technologies. Knowing how to use technology and how technology may shape thinking, values, institutions, culture and actions are all essential for effectively judging and knowledgably making use of technology. Although the benefits of understanding and applying the NOT are enormous, how that is to be accomplished is not clear to my preservice teachers.

Behind the scenes. To make preservice teachers more aware of how to teach the NOT, I draw their attention to what they were taught about the NOT and how I mentally engaged them in learning about it. This entails making my preservice teachers aware of the following:

- When planning lessons, consider where the NOT may be effectively addressed and plan for that inclusion;
- Wisely and effectively use technology in the classroom to reduce student ability to dismiss the teacher as technophobic;
- Use concrete activities that can be used to illustrate NOT ideas and *follow* these with questions to mentally engage students in thinking about those NOT ideas;

- NOT instruction should be initiated with impersonal technologies to reduce emotional hurdles;
- Apply NOT ideas to students' most cherished technologies after initial understanding of those NOT concepts has been developed;
- Use historical examples to move student understanding of NOT beyond speculation of how technology might affect culture to how technology *has* shaped culture; and
- Use effective questioning that requires students to think about the NOT rather than simply telling them about the NOT or rely on students to "pick up" NOT ideas on their own.

However, this bullet list is a summary of many complex ideas that are dependent on understanding how people learn, teacher decision-making and effective pedagogical practices. Helping teachers truly understand the significance and implementation of these NOT teaching ideas is itself complex. The following are merely examples of questions I ask at appropriate times to assist my preservice teachers in wrestling with each bullet point above.

- When planning lessons, consider where the NOT may be effectively addressed and plan for that inclusion
  - Why is planning for NOT instruction important?
  - What might happen if teachers do not plan for NOT instruction?
- Wisely and effectively use technology in the classroom to reduce student ability to dismiss the teacher as technophobic
  - How might students react to the NOT if technology is not used in class?
- Use concrete activities that can be used to illustrate NOT ideas and *follow* these with questions to mentally engage students in thinking about those NOT ideas
  - How can technology use in your classroom be used to address NOT ideas?
  - How might discussion of NOT issues while students are using technology help them make sense of the NOT ideas?
- NOT instruction should be initiated with impersonal technologies to reduce emotional hurdles
  - Why did I start NOT lessons with technologies such as a ruler?
  - What do you think might have happened had I started NOT discussions with cell phones?
  - How can you use this to introduce your students to the NOT?
- Apply NOT ideas to students' most cherished technologies after initial understanding of those NOT concepts has been developed
  - Why must students have a foundational understanding of NOT before applying the ideas to their personal technologies?
- Use historical examples to move student understanding of NOT beyond speculation of how technology might affect culture to how technology *has* shaped culture

- You read some historical accounts of how technology influences people and cultures. How were these accounts useful in developing your understanding of the NOT?
- Why do you think the questions embedded within the historical reading were so important?
- How could you use a similar strategy with your students?

• Use effective questioning that requires students to think about the NOT rather than simply telling them about the NOT or rely on students to "pick up" NOT ideas on their own.
  - Oftentimes teachers simply tell students information, but we know this is not effective for content instruction. How then should we approach NOT instruction?
  - What struggles might you encounter as you encourage students to apply NOT ideas to their own lives?

Many of these questions draw my preservice teachers' attention to how I taught the NOT, require a thoughtful rationale, and encourages teaching the NOT to others. As the preservice teachers come to understand the NOT more deeply throughout the semester, they engage more deeply with how they might teach their students about the NOT. Therefore, these questions are discussed throughout the semester and often revisited in appropriate contexts.

Practice teaching. To further promote my preservice teachers' teaching of the NOT, I have them incorporate NOT instruction in the context of teaching their discipline specific content. These opportunities take the form of a short lesson preservice teachers are expected to develop and deliver to their peers. Their lesson must use technology to engage their fellow preservice teachers with a content idea of their choice by representing the content more concretely, by encouraging reflective thought, and/or by encouraging meaningful social interaction. Furthermore, the lesson must include an explicit component in which the NOT is highlighted and the "students" are expected to engage with the NOT in ways reflective of effective NOT instruction outlined above. These practice sessions encourage the preservice teachers to make use of technology in teaching while seeking out opportunities to address the NOT with students. Furthermore, such practice sessions encourage the preservice teachers to give attention to effective NOT instruction, effective technology incorporation, and effective discipline specific content instruction — seamlessly joined together.

FINAL THOUGHTS

Olson, Clough, and Penning (2009) summarize essential NOT understandings by citing the *National Educational Technology Standards:*

> The *National Educational Technology Standards for Teachers* (ISTE, 2008) recommended understanding the social, ethical, and human issues inherent in

> technology, but many educators appear not to understand or take seriously that technology is not neutral or simply a matter of how people choose to use it. Any serious understanding of technology recognizes that it has inherent biases and promotes certain types of behaviors while suppressing others.

Unfortunately, educators too often believe technology *is* neutral — the impact merely a result of how we use it — and ignore the deeper and often insidious impact of technology on learners and learning environments. To understand these deep, and sometimes negative effects of technology, educators must have more robust understandings of the nature of technology.

Unfortunately, dominant discourses surrounding education technology inhibit critical analysis of technology in education. As Henry Becker, a University of California psychologist (cited in Ely, 1995, p.8) explains:

> [I]n education, our expectations for what can be done with computers are unduly inflated by our persistent tendency to publicize only our successes... Even worse is the widespread attention we give to partial anecdotal evidence that some children have achieved remarkable things using technology.

Becker's quote still rings true – perhaps more than ever. Techno-enthusiasts may be well intentioned, but they miss how educational technology carries fundamental values and assumptions about the nature of education. As Ely (1995, p. 4) notes:

> We do not seem to ask, "Why?" We have been swept up by the tide of technology without fully understanding what purpose it serves and the ultimate consequences of our adoption and use.

While the intent of this chapter is not to argue against educational technology, deeper deliberation regarding educational technology is hoped for. The nature of technology ideas outlined have clear implications for classroom use of technologies and provide an interconnected framework for helping educators be more critical of and make more informed decisions regarding technology use in schools.

Selber (2004, p. xii) powerfully points out the futility in taking sides regarding technology education:

> How-to guides teach useful information that can help students solve their most immediate and practical problems. Yet how-to guides succeed, in large part, by ignoring the terms and conditions under which computer technologies are imagined and created. And while theoretical critiques of computers point out their non-neutral aspects, these discussions typically look right through the complexities and uncertainties of actual situations of use. In addition, they frequently fail to provide realistic and constructive alternates to the circumstances being analyzed. So the end result of either emphasis is one and the same: students who are ill-prepared for the literacy challenges of the twenty-first century.

This chapter does not take sides regarding educational technology. Indeed, this chapter has been littered with real examples of technology use and misuse in educational settings. Furthermore, it has been made clear that teachers must come to be proficient users *and* critics of educational technology.

Only by acknowledging and wrestling with the nature of technology can educators move beyond passive consumers of educational technologies. Only by critically questioning technology can educators shake the foundational values on which new technology is developed and adopted. If educators recognize the way in which technology uses us, they may be able to choose more wisely what technology they allow into schools. By acknowledging the hidden ways in which technology influences decision-making, educators can more reasonably pursue worthwhile goals. Subverting the traditional school model will require subverting the technology that reinforces old structures. If education reformers understood the nature of technology, they may realize education reform will happen in the minds of educators, not in the implementation of new technologies.

REFERENCES

American Academy for the Advancement of Science (2007). The Nature of Technology, Chapter 3 in *Project 2061: Science for all Americans*, http://www.project2061.org/publications/bsl/online/.

Borko, H. & Putnam, R. T. (1996). Learning to teach. In D. C. Berliner & R. C. Calfee (Eds.), *Handbook of educational psychology* (pp. 673–708). New York: Macmillan.

Buckmiller, T. & Kruse, J. (2011). *Future school administrators' perceptions of gain and loss concerning technology adoption*. Paper presented at the Critical Questions in Education Conference, Kansas City, MO, October 17 & 18.

Chen, C. (2006). Investigating the Influences of Teacher Belief and Contextual Factors on the Technology Integration of Taiwanese High School Teachers. Unpublished Doctoral Dissertation, University of Texas at Austin.

Clough, M. P. (2006). Learners' Responses to the Demands of Conceptual Change: Considerations for Effective Nature of Science Instruction. *Science & Education, 15*(5), 463–494.

Clough, M. P., Berg, C. A. & Olson, J. K. (2009). Promoting Effective Science Teacher Education and Science Teaching: A Framework for Teacher Decision-Making. *International Journal of Science and Mathematics Education, 7*(4), 821–847.

Clough, M.P. & Kruse, J. (2010). Conceptual Change: It's not just for learning science. *Iowa Science Teachers Journal, 37*(1), 2–3.

Cobb, T. (1999). Applying constructivism: A test for the learner as scientist. *Educational Technology Research and Development, 47*(3), 15–31

Cuban, L. (1998). High-Tech Schools and Low-Tech Teaching. *Journal of Computing in Teacher Education. 14*(2), 6–7.

Cuban, L. (1986). *Teachers and machines: Classroom use of technology since 1920.* New York: Teachers College Press.

Cuban, L. & Kirkpatrick, H. (1998). Computers make kids smarter—right? *TECHNOS*, 7(2), 26- 31.

Davidson, P., Enzinna, A., Gannon, C., Hall, S., Hayward, C., Irondi, O., Magnifico, A., Perry, T. & Virag, M. (2011). What screenagers say about... *Educational Leadership, 68*(5), 44–47.

DiGironimo, N. (2011). What is Technology? Investigating student conceptions about the nature of technology. *International Journal of Science Education, 33*(10), 1337–1352.

Earle, R. S. (2002). The integration of instructional technology into public education: Promises and challenges. *Educational Technology, 42*(1), 5–13.

Ely, D.P. (1995). Technology is the answer! But what was the question? The James P. Curtis Distinguished Lecture, Capstone College of Education Society, College of Education, University of Alabama, April 14, 1995.

Fang, Z. (1996). A review of research on teacher beliefs and practices. *Educational Research, 38*(1), 47–65.

Fraser, S. P., & Deane, E. M. (1999). Educating tomorrow's scientists: IT as a tool, not an educator. *Teaching in Higher Education, 4*(1), 91–106.

Gance, S. (2002). Are constructivism and computer-based learning environments incompatible? *Journal of the Association for History and Computing*, 5(1), http://hdl.handle.net/2027/spo.3310410.0005.104.

Guerrero, S.M. (2005). Teacher Knowledge and A New Domain of Expertise: Pedagogical Technology Knowledge. *Journal of Educational Computing Research, 33*(3), 249–267.

Guzman-Rodriguez, S.M. (2007). Community College Students' Plant Biodiversity Learning Experience in an Introductory Biology Course: Exploring the value added by using a CD-ROM to develop inquiry lessons. Unpublished Doctoral Dissertation, Louisiana State University.

Haney, J.J., Czerniak, C.M. & Lumpe, A.T. (1996). Teacher beliefs and intentions regarding the implementation of science education reform strands. *Journal of Research in Science Teaching, 33*(9), 971–993.

Irving, K. (2009). Preservice Science Teachers' Use of Educational Technology in Student Teaching. *Journal of Computers in Mathematics and Science Teaching, 28*(1), 45–70.

Kazanci, Z. & Okan, Z. (2009). Evaluating English Language Teaching Software for Kids: Education or Entertainment or Both? *The Turkish Online Journal of Educational Technology, 8*(3), 30–38.

Keen, A. (2008). *The Cult of the Amateur*. Doubleday, New York.

Kruse, J.W. (2009). *Teaching and Assessing Philosophical Issues of Technology in an 8th Grade Science Classroom.* Paper presented at the Tenth International History, Philosophy & Science Teaching Conference, Notre Dame. June 24–28.

Kruse, J.W. (2012). Preservice Teachers' Views of the Nature of Technology. Paper presented at the Association for Science Teacher Education. Clearwater Beach, FL, Jan. 4–7.

Lazlo, A. & Castro, K. (1995). Technology and values: Interactive learning environments for future generations. *Educational Technology*, March/April p. 7–13.

Lunetta, V.N., Hofstein, A. & Clough, M.P. (2007). Learning the Teaching in the School Science Laboratory: An Analysis of Research, Theory, and Practice. Chapter 15, in S.K. Abell & N.G. Lederman (Eds.) *Handbook of Research on Science Education.* Lawrence Erlbaum Associates, New Jersey. 393–441.

National Academy of Engineering (2009). Technically Speaking. http://www.nae.edu/nae/techlithome.nsf/ (2/23/2009)

Nespor, J. (1987). The Role of Beliefs in the Practice of Teaching. *Journal of Curriculum Studies, 19*(4), 317–328.

Olson, J.K. & Clough, M.P. (2001). Technology's tendency to undermine serious study: A cautionary note. *The Clearing House, 75*(1), 8–13.

Olson, J., Clough, M., Penning, K. (2009). Prospective elementary teachers gone wild? An analysis of Facebook self-portrayals and expected dispositions of preservice elementary teachers. *Contemporary Issues in Technology and Teacher Education, 9*(4). Retrieved from http://www.citejournal.org/vol9/iss4/general/article1.cfm

Okan, Z. (2003). Edutainment: Is learning at risk? *British Journal of Educational Technology, 34*(3), 255–264.

Panel on Educational Technology. (1997, March). Report to the president on the use of technology to strengthen K-12 education in the United States. Retrieved October 13, 2000, from http://www.whitehouse.gov/WH/EOP/OSTP/N STC/PCAST/k-12ed.html

Piaget, J. (1970). Piaget's theory. In P. H. Mussen (Ed), *Carmichael's manual of child psychology* Vol 1 (3d ed) New York, Wiley.

Posner, G.J., Strike, K.A., Hewson, P.W., & Gertzog, W.A. (1982). Accommodation of a scientific conception: Toward a theory of conceptual change. *Science Education, 66*(2), 211–227.

Postman, N. (1985). *Amusing Ourselves to Death: Public Discourse in the Age of Show Business.* Penguin Books, New York.

Postman, N. (1992). *Technopoly: The surrender of culture to technology.* Vintage Books, New York.
Postman, N. (1995). *The End of Education: Redefining the Value of School.* Vintage Books, New York.
Potter, E. and Kelly, C.M. (2006). Circuits without Wires: Helping students understand the power of ions. *Iowa Science Teachers Journal, 33*(3), 22–27.
Resnick, M. (2004). Edutainment? No Thanks. I Prefer Playful Learning. Retrieved January, 2008, from http://parents-choice.org/article.cfm?art_id=172&the_page=editorials
Rogers, P. L. (1999). *Barriers to adopting emerging technologies in education.* Richmond, VA: Virginia Commonwealth University. (ERIC Document Reproduction Service No. ED429556)
Salomon, G. (1998) Novel constructivist learning environments and novel technologies: some issues to be concerned with. *Learning and Instruction, 8*(1), 3–12.
Schwab, R. & Foa, L. (2001). Integrating technologies throughout our schools. *Phi Delta Kappan, 82*(8), 620.
Selber, S.A. (2004). *Multiliteracies for a Digital Age.* Southern Illinois University Press, Carbondale.
Selwyn, N. (2010). Looking beyond learning: notes towards the critical study of educational technology. *Journal of Computer Assisted Learning, 26*, 65–73.
Selwyn, N. (n.d.). Web 2.0 applications as alternative environments for informal learning - a critical review. Paper for OECD-KERIS Expert Meeting - Session 6 - Alternative learning environments in practice: using ICT to change impact and outcomes.
Spencer, J.T. (2011). Skeptical of Computers. Spencer's Scratch Pad Weblog. Feb. 27. http://www.educationrethink.com/2011/02/afraid-of-computers.html
Waight, N., & Abd-El-Khalick, F. (2007). The impact of technology on the enactment of inquiry in a technology enthusiast's sixth grade science classroom. *Journal of Research in Science Teaching, 44*(1), 154–182.

Jerrid W. Kruse
School of Education,
Drake University

SECTION IV

TEACHING THE NATURE OF TECHNOLOGY

CHAPTER 18

MICHAEL P. CLOUGH

TEACHING ABOUT THE NATURE OF TECHNOLOGY

Issues and Pedagogical Practices

INTRODUCTION

That technology, without our awareness, changes the way we think and act is not at all obvious. When that idea is raised, people almost always consider it only in superficial ways, and focus exclusively on how they think technology has positively changed their lives – how they now spend more time on the internet, playing games, listening to music, talking on the phone, and texting friends. Rarely do their responses reflect awareness that technology also changes the way we think, that it can change our behavior in harmful ways, and that it can take us down paths we would not willingly have chosen. Perhaps the most significant and insidious bias of technology is how it promotes a forward looking mentality (full of wonderful possibilities) that suppresses a more balanced and accurate examination and reflection of its current and historical impact. The ubiquitous phrase "technological *progress*" without a parallel phrase conveying how technology may also set back individuals, culture and society reflects that bias.

MY PERSONAL JOURNEY COMING TO UNDERSTAND THE NATURE OF TECHNOLOGY

How our thinking and actions are unknowingly altered by technology is difficult to grasp. My own understanding of this idea came about slowly. While browsing at a bookstore in the late 1990s, I happened upon a book by Neil Postman (1995) with the provocative title *The End of Education*. The main thesis of this book, reflected in the title's play on words, is that without a "transcendent and honorable purpose" (p. xi) for schooling (i.e. the ends of education), that social institution is finished (i.e. the end of education). Postman argues why former compelling metaphysical purposes for schooling have lost their appeal, and he puts forward five possible transcendental narratives that might provide compelling purpose for schooling. Not until the last five pages of the book's final chapter, at the end of addressing his fifth narrative regarding how human beings shape themselves and the world with the

Michael P. Clough, Joanne K. Olson and Dale S. Niederhauser (Eds.), The Nature of Technology: Implications for Learning and Teaching, 373–390.

symbols we create, does Postman raise the nature of technology as an example. I found those five pages thought-provoking and meaningful, but not convincing. In time I moved on to Postman's *Amusing Ourselves to Death* (1985) and *The Disappearance of Childhood* (1982). My understanding of his ideas regarding the nature of technology grew, but I maintained that they were exaggerated.

That began to change with an experience I had one summer evening two years later when I was heading out on a bicycle ride and thought, "I don't have my cell phone." I was in a hurry to begin riding, resented the delay to find my phone, but did not want to leave without it. As I dismounted my bike, I thought, "Why do I need the phone?" For years prior to the availability of cell phones, I had ridden often and far from home with never a worry about being in phone contact with others. Why was I now bothered by not having my phone with me? I thought about that while riding that evening and the idea that technology changes the way we think and act. Incidentally, my reaction at that time to being out of communication is not unique. A biology colleague takes students to a wilderness area on the United States/ Canada border where no cell phone towers exist for miles. Students, he says, are very unsettled when they find their cell phones don't work and that they will be out of contact for the duration of the trip, despite being told this would be the case.

That fall, I had another experience that resulted in further pondering about how technology impacts thinking and action. A graduate student with whom I was having lunch noted that I was eating oatmeal. I said that at my last physical exam, my blood cholesterol had been above recommended levels and that after changing my diet, losing weight, and exercising more, I had dropped my number well into the normal range. I was surprised when my graduate student replied, "I just take my cholesterol lowering medicine and eat whatever I want." I thought about how the unintended consequence of such drugs (a technology) is to diminish in many individuals their personal responsibility for adopting healthier habits. That impact extends beyond individual responsibility to societal health care costs that, to a large extent, reflect the eschewing of prudent health decisions in favor of relying on current and possible future medical technology.

The following winter, while my young son and I shoveled deep snow from my driveway and sidewalk, I noticed that no other children were outside even though school was cancelled due to the snowstorm. This was in stark contrast to my childhood when neighborhood children always woke early to first shovel their own driveways and sidewalks and then head out to earn money shoveling snow for others. After that was a day full of playing outside. That is far different than what is generally now the case. For the past several years I have seen adults using snow blowers to clear snow, but rarely signs of children assisting in any way. Nor do I often see them emerge later in the day to play in the snow. Even on lovely spring, summer and fall days, neighborhood children largely stay inside. When I would tell my son he had to play outdoors, he protested, saying that he would have no one with whom to play. The varieties of entertainment technologies not only promote sedentary lifestyles that have led to a childhood obesity epidemic, they also create a culture that ostracizes those who do not or are not permitted to adopt that lifestyle.

Efforts such as Partnership for Play Every Day (http://www.playeveryday.org/) and NFL Rush Play 60 (http://www.nflrush.com/play60/?icampaign=rush_nav_play60) that now exist to encourage *children* to play sixty minutes each day reflect the way technology has changed childhood thinking and behavior. But here again, that impact goes beyond individuals to culture and society. As Richard Louv (2008) in *Last Child in the Woods* warns, the unforeseen impact of children shunning the outdoors is that, in not loving the natural world, they will not value and work to preserve it.

Many have noted how the internet, Twitter, texting and other electronic communication technologies have diminished many individuals' ability to follow extended logical arguments. Reflecting that, not too long ago, an article a colleague and I published in an electronic journal was piecemealed by the editors who said that on-line formatting requires shorter paragraphs. Echoing the tale of Emperor Joseph II's complaint to Mozart that one of his compositions contained too many notes, a graduate student in our program was recently told by a technology education faculty member that her written sentences were too long. These and many other experiences working with both undergraduate and graduate students have illustrated the unanticipated and unintended consequence of popular electronic communication technologies on writing and the ability of those who extensively utilize such communication technologies to follow lengthy arguments.

I now more easily see all around me the unacknowledged influence technology has on human thinking and behavior. For instance, school administrators and even science teachers increasingly see virtual labs and other technology replacing concrete science experiences with the natural world and with materials in a laboratory setting. Technology certainly can and should play a role in exploring and making more comprehensible phenomena too dangerous to directly explore (radioactivity for instance) or with theoretical entities like atoms and their behavior. However, in making possible this kind of engagement, the technology influences decisions regarding all science experiences. Full of good intentions, yet smitten with the technology, some make the claim that virtual labs are the new trend in teaching science. Ignoring what is well known about how children learn science and effective science teaching, hands-on science experiences are being marginalized in favor of virtual experience. Uncritically embracing technology, advocates of virtual labs see only positive future possibilities (including laboratory classroom and equipment cost savings) and neglect what will be lost.

As another example, many people are now tethered to their jobs even when not being paid. Had this been imposed upon us in an Orewellian sense (Orwell, 1949), workers would have risen up and demanded more fair working conditions and/or compensation. But in a Huxlean (Huxley, 1932) and Bradburean (Bradbury, 1951) sense, the nature of communication technologies and our unexamined adoption of them have resulted in our personal and work lives being inseparable. Communication technologies interfere with our personal lives in another way. Although they do make easier keeping in touch with those outside our immediate range, they are an assault on interaction with those in our immediate space. For example, while at an amusement park with my wife and son, I noticed a child begging for attention from his parents and grandparents who were all busy with their smartphones. That is why I found ironic the message in a recently aired television commercial showing individuals patiently glued to their evidently slow and outdated smartphones, ignoring those around them (http://www.youtube.com/watch?v=l9evyGr57hs). The answer to this dilemma was not to put down our phones and pay attention to those around us, but instead to purchase a faster smartphone.

The way technology development is directed at assisting humans with their interpersonal relationships, yet often has a chilling dehumanizing impact is illustrated in Turkle's (2011) book *Alone Together: Why We Expect More from Technology and Less from Each Other*. Even before reading her book, I had become irritated with electronic gadgets that say "Hello", "Welcome", "Good-bye" and other statements that ought to communicate sincere human feelings for those with whom we interact. A machine, like a person who says such things in a perfunctory way, has no such feelings. That we program our technology to appear to care debases the earnest remarks made by truly caring individuals. Turkle's book goes well beyond these concerns and reports on the advancements made in robotics and their advocates' hopes that future robots will serve as companions to the lonely and be of assistance to the infirm. Critically examining these developments, Turkle warns how such developments focus our thoughts squarely on the desired assistance for raising children, assisting the elderly and the infirm, and meeting our need for companionship. Whether we also consider our values and what *caring for someone* actually means, as opposed to a machine acting out its programmed actions, remains to be seen. How we answer such questions will impact the extent that futuristic assisting robots reduce our personal motives to reach out in loving ways to actually care for those we know as well as those we don't.

Of course, that most technologies have positive outcomes goes without saying, but this must be explicitly stated for two reasons. First, anyone who critically examines technology faces the simpleton response that he or she must be a Luddite. I have a son with Type 1 insulin-dependent diabetes and I am thankful for the medical technologies that his life and long-term health depend upon. So I am emphatically not anti-technology! The second and more important reason that the positive outcomes of technologies must be emphasized is because they are what influence us to not examine and thus miss the downside of those same technologies. Each of the technologies I have noted above were created or are being developed for a reason and that is what we employ them for and judge them on. That narrowing of our analysis creates a pervasive bias that causes us to ignore how those same technologies impact us in ways for which they were not developed. Thus, when not critically examined, technologies will have unanticipated and often undesirable consequences that are not recognized. For instance, as much as I value insulin pumps, without critical examination and restraint, they promote a mentality and behavior of eating high levels of carbohydrates which is an unhealthy practice for diabetics. The artificial pancreas project is directed at developing technologies that will more tightly control blood glucose levels, but if left unexamined, the downside of that positive future possibility is the downplaying or suppression of diabetics' responsibility to carefully monitor their diet. Future robotics development will assist us in many important ways, but if left unexamined, the downside is our own dehumanization. Thus, returning to the introduction of this chapter, the most significant and insidious bias of technology is how it promotes human thinking that sees in current and future technologies only positive possibilities, while silencing fair and more accurate assessments of its impact on thinking and action.

WHY ACCURATELY TEACH THE NATURE OF TECHNOLOGY?

The response to the question "Why teach the nature of technology?" is embedded in a larger issue regarding the purposes of schooling. Compelling reasons ought to exist for schooling children, for what is taught in schools, and for how we teach because, as Davson-Galle (2008) notes, compulsory schooling detains individuals, often against their will, for sustained periods of time. Moreover, schooling, when wisely considered and effectively accomplished, has an enormous positive influence on personal and societal well-being. I am purposely using the word *schooling* in place of *education* because while schooling *could* be directed at education, little of what goes on in schools actually resembles education. Instead, schooling as commonly enacted is primarily directed at *training* students to recall information and perform particular skills. If schooling was truly directed at educating children, then policies, curricula, teaching practices and assessments that promote and reflect the goals appearing in Table 1 would be far more pervasive.

Table 1. Commonly Suggested Education Goals for Students (adapted from Clough, Berg & Olson, 2009)

Students will:

1. Demonstrate deep and robust conceptual understanding of fundamentally important ideas.
2. Use critical thinking skills.
3. Convey an accurate understanding of the nature of disciplines being studied.
4. Effectively identify and solve problems.
5. Appropriately and effectively use communication and cooperative skills.
6. Actively participate in working towards solutions to local, national, and global problems.
7. Be creative and curious.
8. Set goals, make decisions, and accurately self-evaluate.
9. Convey a positive attitude about learning and wisdom.
10. Access, retrieve, and use existing knowledge in the process of inquiry.
11. Convey self-confidence and a positive self-image.
12. Demonstrate an awareness of the importance of what is being learned for personal and societal well-being.

Unfortunately, schooling is largely not about education. For example, despite extensive rhetoric about the rapid pace of technological change and the need to prepare individuals for a society and working world that will also swiftly change, policymakers, business leaders and technology enthusiasts foolishly promote technology training instead of technology education that would go much further and also prepare individuals to assess and appropriately respond to technology development. Technology instruction in schools is, with rare exceptions, directed solely at training students to operate technology and employ it for what it was designed to do while ignoring the goal of educating them about how to critically examine technology. Technology *education*, as opposed to a narrow technology training, would also assist learners in understanding what technology is, how and why technology is developed, the unanticipated impacts of technology, what is gained and lost by adopting any technology, and how society directs, reacts to, and is sometimes unwittingly changed by technology. Meaningfully addressing these and other important aspects of technology would promote habits of thought and action that ensure citizens are prepared to analyze technology so that they wisely use it rather than unknowingly permitting it to use them.

EDUCATING STUDENTS ABOUT THE NATURE OF TECHNOLOGY

We All Teach the Nature of Technology

Ironically, teachers do teach about the nature of technology whether or not that is their intent. Because the character of any subject matter is conveyed by the way it is taught, students develop ideas about a discipline even if the teacher does not purposely intend to convey those features. Consider for example the broad subject area of science taught in schools. School science instruction, generally speaking, consists of linear and factual-laden lectures, selected readings that report the end products of science research while ignoring how that knowledge was actually developed, cookbook laboratory activities and other activities where students primarily follow directions (Schmidt, McKnight, Cogan, Jakwerth & Houang, 1999; Weiss, 1993; Weiss, Pasley, Smith, Banilower & Heck; 2003). Those experiences along with the way teachers and instructional materials speak about science coalesce to convey mistaken ideas about the nature of science (Clough, 2006). Similarly, the very way technology is incorporated in lessons, the language teachers use when speaking about technology, the curricular materials employed, and what is left unexamined and unstated about technology coalesce to convey important messages about what technology is, how and why it is developed, and whether it is merely a tool we use for our desired ends or something more that ends up using us in the sense that it changes our thinking, actions and values. Thus, the issue is not whether teachers will teach about the nature of technology, only how accurate is the image that they portray.

What About the Nature of Technology Should A Robust Technology Education Address?

In deciding what to teach students about the nature of technology, we must keep in mind that the outcome of this effort will be a form of technology in the sense that a list is being developed to assist in accomplishing a task. This illustrates that technology is a much broader concept than is commonly envisioned (AAAS, 2007; NAE, 2009). The way technology can and does influence later action is illustrated by how learning objectives, when written as outcome statements, influence practice. For instance, learning objectives (what students should know or understand) influence teachers to focus primarily, if not solely, on the end product of instruction rather than the process of learning and teaching. Not surprisingly, a transmission and regurgitation process often ensues (Eisner, 2002).

Thus, the nature of a discipline should be seen as something to explore with students, not merely as a set of ideas students should learn. Writing about nature of science instruction, Eflin, Glennan, & Reisch (1999, p. 112) cautioned, "Just as science educators stress that science is more than a collection of facts, we emphasize that a philosophical position about the nature of science is more than a list of tenets." In the same way, students should come to deeply understand and apply nature of technology ideas, not simply know of them. Accurately analyzing technology and making appropriate decisions about it demands exploring questions like those proposed by Postman (1995) appearing in Table 2. These kinds of questions encourage both teachers and students to more deeply think about the nature of technology, its contextual nature, and promote thinking that takes into account context and complexity. If meaningful attention was given to these kinds of questions in schooling, then children would be well educated regarding the nature of technology and be far more likely to wisely use technology.

Table 2. Some nature of technology questions worth exploring (Postman, 1995)

- For every advantage of technology, what might be the corresponding disadvantage?
- How are the advantages and disadvantages of particular technologies distributed unevenly?
- How have particular technologies changed the way humans think and act?
- How might particular technologies now under development change the way humans think and act?
- What intellectual, emotional, sensory and social biases are inherent in particular technologies?
- What goals are promoted, ignored and dismissed by a particular technology?
- How does technology change the ways humans view learning, teaching and schooling?
- How does technology change for better and worse our interactions with one another?
- How does the technology promote and inhibit thinking and learning?
- How may technology change what humans value?

Teaching the Nature of Technology

While teachers do convey the nature of technology via the manner they teach when incorporating or referring to technology, once students have developed and codified ideas regarding the nature of technology, much concerted effort is required to alter those original ideas. This is because an important difference exists between initial conceptual development and later efforts to alter those concepts (Clough, 2006). Learners develop ideas to make sense of their everyday experiences. Even though children may not have been purposely and explicitly taught the nature of technology, they have developed many incorrect ideas to account for their many in and out-of-school experiences regarding technology. Those initial ideas regarding technology are initially tentative, but can and often do become well established and tightly held because they appear to accurately account for a wider array of experience. Over time, these ideas connect with other ideas and may form a tightly linked framework that is then used for seeing and interpreting their everyday world. For instance, from an early age children are bombarded by messages touting the purpose of various technologies and how they will solve some problem and make life better. While not overtly taught to see all future technological development as positive, that particular nature of technology misconception develops quite naturally. Tied to this is another misconception that technologies are mere tools, possessing no biases and certainly not influencing thinking and values. Students' use of technology is ubiquitous, and the ideas they have developed regarding the nature of technology are tacit, but become tightly held for both cognitive and emotional reasons.

What this means is that while students' initial conceptual frameworks are in part formed and reinforced by their implicit in and out-of-school experiences, once developed and strengthened, they are highly resistant to change (Posner, Strike, Hewson, & Gertzog, 1982). This is particularly the case regarding students' understanding of the nature of disciplines. The longer students have been immersed in a particular subject matter, the more developed and entangled are their notions regarding the nature of that discipline. Thus, students don't see their misconceptions as such, and employ them in making sense of new experiences. This is why accurately and effectively teaching the nature of technology demands that teachers overtly consider what nature of technology ideas should be explored with students, and how those ideas should be taught and assessed.

Deep and meaningful learning demands assiduous mental engagement. Learners must do more than simply attend to information; they must also overtly connect and compare that information to their prior knowledge. However, as previously noted, even when that kind of mental engagement occurs, learners often interpret and sometimes modify information so that it conforms to what they already think. Conceptual learning often demands not simply adding new information to what learners already think, but altering the way they think about their prior experiences and ideas (Driver, 1997). These and other reasons are why conceptual change and the teaching that promotes conceptual change are both far more complex and difficult than is commonly thought (Appleton, 1993 & 1997; Clough, 2006; Duschl & Hamilton, 1998; Limon & Mason, 2002; Pintrich, Marx & Boyle, 1993; Posner *et al.*, 1982; Strike & Posner, 1983 & 1992; Tyson, Venville, Harrison, & Treagust, 1997). Thus, moving students to a more honest understanding of the nature of technology is not merely a matter of presenting more accurate information or creating more accurate implicit experiences. Rather, teachers must overtly draw students' attention to nature of technology ideas, and in a way that has students think deeply about those ideas. Moreover, this must be done in a variety of contexts to convince students that their prior ideas regarding technology are mistaken.

Overtly drawing students' attention to the nature of technology does not mean teachers should simply lecture to students about it. Instead, teachers should ask questions like those appearing in Table 3. These kinds of questions overtly raise important nature of technology ideas, and they mentally engage students in thinking about those ideas. Teachers who understand the nature of technology and are proficient at asking questions like those found in Table 3 can teach about the nature of technology in most any lesson. Moreover, while teachers should at times purposely plan for instruction regarding the nature of technology, opportunities for raising nature of technology ideas often arise unexpectedly during classroom instruction. In both cases, teachers who can skillfully ask nature of technology questions are positioned to raise specific nature of technology ideas in a manner that engages students and scaffolds them to a deeper understanding of technology.

Nature of technology instruction can take place in a variety of instructional contexts. For instance, in chapter 19 of this book, Kruse (2013) notes the importance of introducing nature of technology ideas using technologies with which students have no close ties. Such technologies can be categorized as distal to students' emotional state. Lessons using these kinds of technologies isolate and emphasize nature of technology ideas in concrete ways, but do so using technologies that avert a thoughtless emotional rejection of the nature of technology idea being introduced. These kinds of lessons are important because they isolate and emphasize nature of technology ideas in concrete and plausible ways that students can begin to understand. However, such lessons may generate interest, but will unlikely impact students' deeply held nature of technology misconceptions. These lessons are, nonetheless, important because they make intelligible complex nature of technology ideas that previously have been invisible to students, and in doing so, a foundation is created for exploring these same issues with technologies that students obsessively employ and develop emotional attachments to, but have never judiciously examined. In addition to preparing students to benefit from further nature of technology instruction, such activities also raise students' interest in the nature of technology and communicate the importance that will be placed on it for the remainder of a course.

After introducing students to important nature of technology ideas in the manner described above, instruction should then make reference to technologies that students extensively and passionately employ (e.g. television, smart phones, video games, Facebook, the internet, tablets, modern medicine) and are thus more

Table 3. Example questions that draw students' attention to and encourage thinking about particular nature of technology ideas.

Nature of Technology Concepts	Example questions
Technology is a broad concept including both artifacts and the processes that created those artifacts. Examples of technology include, among other things, tools, machines, symbols, objects, and techniques.	• How is [insert tool, machine, symbol, object, technique, etc.] a form of technology? • How is democracy a social technology? • How is fire a technology?
Technology is developed for a particular purpose, but its impact may reach beyond its original purpose.	• For what purpose was [insert technology] developed? • For what other purposes is it being employed? • For what other purposes might it be applicable?
Biases are inherent in technology.	• How does the purpose and limitations of [insert technology] predispose you to employ it in particular ways, thus impacting decisions and other actions? • How does [insert technology] enhance creativity? • How does this same technology constrain creativity?
Technology is a Faustian bargain.	• What positive outcomes occur by employing this technology? (i.e. What is gained?) • What negative outcomes occur by employing this technology? (i.e. What is lost?)
Technology changes human behavior.	• How has [insert technology] changed human behavior in ways that were anticipated? • How has this same technology changed human behavior in ways that were not considered? (To ensure students understand how technology has changed their behavior, direct these same questions at students' behavior.)
Technology changes human thinking.	• How has [insert technology] changed the way humans think? • How has the development of certain medicines altered thinking regarding personal responsibility to make more prudent health care decisions? (To ensure students understand how technology has changed their thinking, direct these same questions at students' own thinking.)

(*Continued*)

Table 3. Continued

Nature of Technology Concepts	Example questions
Communication technologies impact privacy, personal space, and quiet time for reflection.	• How has [insert communication technology] changed where personal communication takes place? • How do communication and other technologies make difficult finding peaceful time for deep thinking and reflection? • What has been gained and what has been lost with communication technologies?
Technology promotes a positive forward looking mentality that suppresses a more balanced and accurate examination of its impact.	• How does the way we speak of technology bias us toward seeing it as primarily, perhaps only, in positive terms? • The phrase "technological progress" is commonly used. Why do we not have an equally common phrase for the downsides of technology?
The process by which technology is developed is linked, and thus constrained, by already existing technologies.	• How is this classroom interactive white board similar to chalk boards and white boards? Why do you think this is the case? • How is the development of new technologies linked to, and thus limited by, already existing technologies?
Technology influences human values.	• How have cell phones altered family values? • How has technology altered relationships? • How may the development of assistive robots erode human values of caring and compassion?

proximal to their emotional state. For example, assign students to analyze how much time passes on television before a new camera angle or scene appears. They will find that rarely does more than four seconds pass. Postman (1985) refers to this and other technology (foremost among these is the internet) that shortens our attention span as the "the Peek-a-boo world." Have students consider how this impacts attention span, the ability to focus and follow lengthy arguments, and meaningfully reflect on information. Draw students' attention to how texting and/or Twitter does the same while also assaulting formal writing that is necessary to convey complex thought in a logical, concise and clear manner. Countless other examples of how contemporary technology alters thinking and action can be purposely planned for as part of instruction — like those found in chapter 21 (Spencer, 2013) — or addressed when such opportunities arise in the course of everyday instruction. Asking questions similar to those appearing in Table 3 are again important for drawing students' attention to ways that contemporary technology impacts thinking and action. Students will likely struggle to understand or outright balk at any suggestion that their cherished technologies impact the way they think and act. To make these ideas more plausible, draw students' attention to the distal technologies examined earlier. Ask scaffolding questions that assist students in understanding how the same nature of technology issues that were raised with distal technologies apply as well to their proximal technologies. For example, ask questions like:

- "What cues or biases existed in [insert previously examined distal technology]?
- "How did that [insert previously examined distal technology] impact your thinking and action?"
- "What cues or biases exist with your [insert proximal technology]?
- "So how do those cues or biases influence your thinking and action?

Even as students begin to accept that technology does influence thinking and action, they will unlikely grasp and appreciate the full significance of the nature of technology and the importance of understanding it. Historical examples illustrating how technology has changed social institutions and values as well as individual thinking and behavior provides further evidence for nature of technology ideas. For example, Postman (1982) argues how the invention of the printing press resulted in the need for universal schooling so that individuals could learn to read and write. As the years of compulsory education grew, the passage to adult life was delayed thus creating a new social phenomenon — an extended childhood. Postman goes further to show how a new technology — the internet — has attacked basic notions of what childhood entails, promoting rapid movement into adolescence. At the same time, the technology of post-secondary schooling has delayed entry into adulthood resulting in an extended period of adolescence never before seen in history. These and other historical examples of how technology impacts individuals, society and culture (see this book's recommended readings) provide compelling evidence regarding the often invisible nature of technology.

Features of the three above contexts for nature of technology appear in Table 4. Figure 1 illustrates the scaffolding between these three contexts that assists students in developing a deep and robust nature of technology understanding that will more likely be applied to out-of-school technology experiences.

Assessing Students' NOS Understanding

While the above recommendations will ensure that the nature of technology is a consistent theme in a course, incorporating nature of technology questions as part of assessments throughout the school year is crucial. As Dall' Alba *et al.* (1993) and many others have stated, "assessment gives clear messages to students about what is important in the subject" (p. 633). To begin, teachers should determine

Table 4. Important contexts for nature of technology teaching and learning

	Emotively Distal Technology	**Emotively Proximal Technologies**	**Historical Examples**
Feature	Students have little emotional investment in the technology.	Students have high emotional investment in the technology.	Authentically documents and exemplifies how technology has impacted societal values and/or individual thinking and action.
Example	Ruler and marble used to illustrate nature of technology ideas (Kruse, 2013)	Mobile phones, Facebook, the internet, video games, etc. used to illustrate nature of technology ideas.	How the printing press brought forth the need for universal schooling thus influencing the length and concept of childhood (Postman, 1982).
Pros	Mitigates students' emotional response to and rejection of initially introduced nature of technology ideas.	Makes apparent how personal technologies alter students' everyday thinking and action.	Provides students with well documented evidence for nature of technology ideas making more difficult rejecting those ideas.
Cons	Nature of technology ideas will unlikely transfer to personal or societal technology contexts.	Nature of technology ideas will be emotionally rejected if technology examples are not carefully chosen or if scaffolds are not skillfully made to other contexts.	Students may not transfer such lessons to their own devices and personal thinking and action.

their students' prior ideas about the nature of technology early in the school year to enable more effective planning for such instruction. Once students' preconceptions regarding the nature of technology are identified, teachers should begin consistently incorporating instruction at appropriate times to create student dissatisfaction with their misconceptions and provide more accurate alternatives. The kinds of questions appearing in Table 3 can also serve as formative assessments that inform teachers of their students' developing NOS views. Thus, learning to ask these kinds of questions as a normal part of instruction is important for effectively

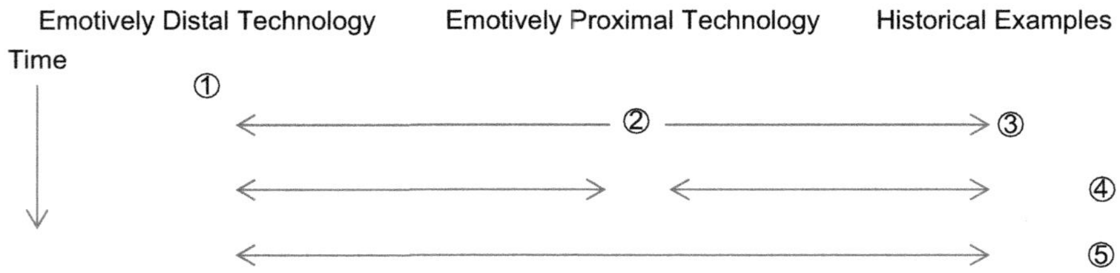

① Lesson addressing nature of technology using emotively distal technology. Teacher leads interactive presentation and discussion during lesson asking questions like those found in Table 3. Several of these kinds of lessons are advisable before moving on to nature of technology lessons addressing emotively proximal technology.

② Lesson addressing nature of technology ideas using emotively proximal technology. Teacher leads interactive presentation and discussion that includes questions that scaffold to ① and ③, thereby assisting in understanding and acceptance of ideas.

③ Lesson addressing historically accurate account of technology impacting societal values and/or individual thinking and action. Questions asked that have students compare nature of technology concepts in historical example to ①and ②.

④ Later lessons occurring in any of the three contexts should scaffold back and forth along all three contexts, making reference to previous nature of technology lessons.

⑤ Summative assessments may be embedded in any of the three contexts and seek links between contexts.

Figure 1. Teacher Scaffolding Across the Three Nature of Technology Contexts.

understanding students' nature of technology thinking, planning and incorporating instruction regarding the nature of technology, and assessing students' thinking as it develops.

Including nature of technology questions on summative assessments is important for making clear that understanding the nature of technology is an important part of students' education and must be taken seriously. However, because attention to teaching and learning about the nature of technology is relatively recent, formally developed assessments targeting the nature of technology are not readily available. Instruments assessing this important aspect of education are sorely needed for classroom implementation and research efforts. However, questions like those found in Table 3 can make for fine summative assessments. The downside to multiple-choice questions addressing the nature of technology is that context and important nuances are almost always lost. In whatever manner teachers summatively assess students' understanding of the nature of technology, students will realize that such understanding is important and will be assessed throughout the course.

EDUCATORS' RESPONSIBILITY TO ACCURATELY AND EFFECTIVELY TEACH ABOUT THE NATURE OF TECHNOLOGY

No doubt teachers already have extensive demands placed upon them. Adding yet another responsibility to their already overburdened workload may seem unfair. However, as noted earlier, all teachers teach the nature of technology by the very way they employ and talk about it in their classrooms. The issue is not whether teachers will teach about the nature of technology, only how accurately and effectively they will teach it.

Schools have largely welcomed the extensive infusion of technology for teaching children without examining the nature of technology and its Faustian bargain for teaching and learning. For instance, technology enthusiasts point to how technology enhances collaboration between students in class and with others well beyond the classroom walls. Enhanced communication that brings us closer together is, of course, what many technologies are designed to do. The unanticipated and often unexamined Faustian bargain has been that we increasingly form associations only with like-minded individuals and groups, more easily ignore and denigrate views we don't like, and substitute distant electronic communication for personal face-to-face interaction where we must acknowledge the whole person (Bauerlein, 2008). The extent of this can be seen everywhere if people would merely look up from their electronic gadgets long enough to see how alone they are while in the presence of others (Turkle, 2011).

Carefully examining technology is crucial for understanding what is gained and what is lost in blindly adopting particular technologies in and out of schools. This demands that educators first acknowledge and teach the most deceptive bias of technology — how it fosters an almost exclusive optimistic forward-looking mindset that suppresses a more fair and accurate examination of its historical, current and possible future impact. Any cursory review of the history of schooling makes clear that such optimism has always existed regarding how technology would improve both teaching and learning. But the results have been quite different. For instance, many current technologies have been seen as a way to motivate students, but the fallout from this entertainment approach to schooling has been a view that learning ought to be fun (or at least not demanding), and that important educational outcomes can be achieved on a wide scale without extensive teacher-student interaction, lengthy reading, or disciplined focus and reflection. Many educational technologies make easily accessible enormous amounts of information, and in doing so confuse information with learning and wisdom. The plethora of visual and auditory distractions, ease of point and click/touch, and the information overload that ensues promote short attention spans. The constant bombardment of stimuli from our nearly inescapable gadgets is an assault on the concerted time and attention that is required for reflection, deep learning and the development of wisdom.

What has been previously well-established regarding how people learn and the kind of teaching that promotes such deep thinking, reflection, and wisdom is not changed by the presence or absence of technology. Humans matter and humanizing teachers interact extensively with students to understand them and their thinking. This requires effective questioning, use of wait time, supportive non-verbal behaviors, active listening, and responding to students in ways that further thinking and reflection. These *human* and *humane* interactions are not simply about engaging students in meaning-making, but to convey that the children we teach are far more than a cog in a schooling factory and future economic machine. To what extent we acknowledge that technology is not neutral, analyze it for its biases, and then wisely employ it in and outside schooling will say much about our humanity or lack of it. Left to its own devices, education technology often implies that teachers are superfluous to student learning. Paraphrasing Arthur C. Clarke, any teacher whose interaction with children can be replaced by technology ought to be replaced.

Technology's influence on schooling, teaching and learning is not limited to electronic tools. *No Child Left Behind* and other outcomes-based technologies emphasize "scientifically based" education research, testing, and academic accountability. Such technologies bias thinking and action regarding the purposes of schooling and how children are taught. The biases of these tools direct us toward particular solutions for reaching targeted ends while ignoring others. Not surprisingly, the technologies promoted in schools follow quite naturally from reform documents' biases toward testing and academic accountability. The Faustian bargain is that we unwittingly agreed to be silent on the moral aspects that have throughout history been an inseparable part of education. That silence is deafening to those who see teaching as the sacred activity it can and should be. Research-based teacher decisions and practices are crucial for promoting many important ends of schooling, but alone they marginalize the sacred nature of teaching that is directed at helping children grow to be ethical, caring, altruistic, responsible, and mentally and physically healthy individuals. Our infatuation with technology further blinds us to the philosophical and moral aspects of schooling. Thus, research-based teaching can become mechanical and detached from children. Without attention to the sacred nature of teaching, teaching becomes mechanical and merely a job.

I am not seeking to place blame, only to bring to the forefront what makes a meaningful education worth having, and the sacred nature of teaching that brings about that kind of education. Children are far more than entities to be taught so that they can become cogs in an economic machine, and the sacred nature of teaching is far more than putting into place research-based strategies. Neil Postman (1995) emphatically argued that economic productivity alone does not provide a compelling justification for education. Nor does it provide a compelling rationale for the commitment that is required for effective teaching. The philosophical and moral reasons for education, and the sacred nature of teaching, are what compel teachers to put in the enormous time and effort helping children grow to be all we want for them. Without that sacred perspective, little reason exists to learn and implement effective teaching practices — to engage children in truly meaningful educational experiences rather than simply convey information and skills to them.

I once had a high school principal tell me that I take teaching too seriously. I don't think that is possible. Each of us, with great effort, can make a significant positive difference in the lives of students that will then spread well beyond our classroom, school and local community. That attitude is what compels teachers against great odds to educate (in its most noble sense) children about the nature of technology and so many other important ends ignored by policymakers. For only through deliberate and careful analysis of technology with equal deliberation regarding our values can we ensure that we use technology rather than permitting it to use us.

REFERENCES

American Academy for the Advancement of Science (2007). Project 2061: *Science For All Americans – Chapter 3 The Nature of Technology*. http://www.project2061.org/publications/bsl/online/.

Appleton, K. (1993). Using Theory to Guide Practice: *Teaching Science from a Constructivist Perspective, School Science and Mathematics*, *93*(5), 269–274.

Appleton, K. (1997). Analysis and Description of Students' Learning During Science Classes Using a Constructivist-Based Model, *Journal of Research in Science Teaching*, *34*(3), 303–318.

Bauerlein, M. (2008). *The Dumbest Generation: How the Digital Age Stupefies Young Americans and Jeopardizes our Future*. Penguin Group, New York.

Bradbury, R. (1953). Fahrenheit 451. *Simon & Schuster*, New York.

Clough, M. P. (2006). Learners' Responses to the Demands of Conceptual Change: Considerations for Effective Nature of Science Instruction. *Science & Education*, *15*(5), 463–494.

Clough, M. P., Berg, C. A. & Olson, J. K. (2009). Promoting Effective Science Teacher Education and Science Teaching: A Framework for Teacher Decision-Making. *International Journal of Science and Mathematics Education*, *7*(4), 821–847.

Dall' Alba, G., Walsh, E., Bowden, J., Martin, E., Masters, G., Ramsden, P. & Stephanou, A. (1993). Textbook Treatments and Students' Understanding of Acceleration. *Journal of Research in Science Teaching*, *30*(7), 621–645.

Davson-Galle, P. (2008). Why Compulsory Science Education Should Not Include Philosophy of Science, *Science Education*, *17*, 677–716.

Driver, R. (1997). In Annenberg/CPB Minds of Our Own Videotape Program One: Can We Believe Our Eyes, *Math and Science Collection*, PO. Box 2345, South Burlington, VT 05407–2345.

Duschl, R. & Hamilton, R. (1998). Conceptual Change in Science and in the Learning of Science. In B. Fraser & K. Tobin (Eds) *International Handbook of Science Education*, pp 1047–1065. Kluwer, Dordrecht, The Netherlands.

Eflin, J.T., Glennan, S. & Reisch, G. (1999). The Nature of Science: A Perspective from the Philosophy of Science, *Journal of Research in Science Teaching*, *36*(1),107–117.

Eisner, E. W. (2002). *The Educational Imagination: On the Design and Evaluation of School Programs, 3rd Ed.* Pearson Education, Upper Saddle River, NJ.

Huxley, A. (1932). *Brave New World.* HarperCollins, New York.

Kruse, J. W. (2013). Promoting Middle School Students' Understanding of the Nature of Technology. Chapter 19 in M. P. Clough, J. K. Olson & D. S. Niederhauser (Eds.) *The Nature of Technology: Implications for Learning and Teaching, Sense Publishers*, Rotterdam, The Netherlands.

Limon, M. & Mason, L. (2002). *Reconsidering Conceptual Change: Issues in Theory and Practice*, Kluwer, Norwell MA.

Louv, R. (2008). *Last Child in the Woods: Saving Our Children from Nature-Deficit Disorder*. Algonquin Books, Chapel Hill, NC.

National Academy of Engineering (2009). *Technically Speaking*. http://www.nae.edu/nae/techlithome.nsf/ (Retrieved 1/14/2013)

Orwell, G. (1949). *Nineteen Eighty-Four*, Penguin Books, New York.

Pintrich, P.R., Marx, R.W. & Boyle, R.A. (1993). Beyond Cold Conceptual Change: The Role of Motivational Beliefs and Classroom Contextual Factors in the Process of Conceptual Change, *Review of Educational Research, 63*(2), 167–199.
Posner, G.J., Strike, K.A., Hewson, P.W. & Gertzog, W.A. (1982). Accommodation of a Scientific Conception: Toward a Theory of Conceptual Change, *Science Education, 66*(2), 211–227.
Postman, N. (1982). *The Disappearance of Childhood.* Vintage, New York.
Postman, N. (1985). *Amusing Ourselves to Death: Public Discourse in the Age of Show Business.* Penguin, New York.
Postman, N. (1992). *Technopoly: The Surrender of Culture to Technology.* Vintage, New York.
Postman, N. (1995). *The End of Education: Redefining the Value of School.* Vintage, New York.
Schmidt, W., McKnight, C., Cogan, L., Jakwerth, P. & Houang, R. (1999). *Facing the Consequences: Using TIMSS for A Closer Look at US Mathematics and Science Education.* Kluwer: Dordrecht, The Netherlands.
Spencer, J. T. (2013). Technology Criticism in the Classroom. Chapter 21 in M. P. Clough, J. K. Olson & D. S. Niederhauser (Eds.) *The Nature of Technology: Implications for Learning and Teaching, Sense Publishers*, Rotterdam, The Netherlands.
Strike, K.A. & Posner, G.J. (1983). On Rationality and Learning: A Reply to West and Pines. *Science Education, 67(*1), 41–43.
Strike, K.A. & Posner, G.J. (1992). A Revisionist Theory of Conceptual Change. In R. Duschl & R. Hamilton (Eds.), *Philosophy of Science, Cognitive Psychology, and Educational Theory and Practice*, pp. 147–176, SUNY, New York.
Turkle, S. (2011). *Alone Together: Why We Expect More from Technology and Less from Each Other.* Basic Books, New York.
Tyson, L.M., Venville, G.J., Harrison, A.G. & Treagust, D.F. (1997). A Multidimensional Framework for Interpreting Conceptual Change Events in the Classroom, *Science Education, 81*(4), 387–404.
Weiss, I.R. (1993). Science teachers rely on the textbook. In R. E. Yager (Ed.) *What research says to the science teacher, volume seven: The science, technology, society movement.* Washington, D.C.: National Science Teachers Association.
Weiss, I.R., Pasley, J.D., Smith, P.S., Banilower, E.R. & Heck, D.J. (2003). *Looking inside the classroom: A study of K-12 mathematics and science education in the United States.* Chapel Hill, NC: Horizon Research, Inc.

Michael P. Clough
School of Education,
Iowa State University

CHAPTER 19

JERRID W. KRUSE

PROMOTING MIDDLE SCHOOL STUDENTS' UNDERSTANDING OF THE NATURE OF TECHNOLOGY

INTRODUCTION

Integration of "21st century skills" into all curricular areas has received much attention. The need for these 21st century skills is often linked to the changing nature of our society and economy due to new technologies. However, the 21st century skills construct is vague and unoriginal to the 21st century. That is, educators have typically sought critical thinking, creativity, collaboration, information literacy, and other common goals for students. Yet, as technology plays an ever-increasing role in our lives, the ability to think critically about technology may be the most critical 21st century skill. While students will need to be users and producers of technology, teachers must also help their students become critical questioners of technology (Selber, 2004).

By highlighting philosophical underpinnings and social consequences of technology, the nature of technology (NOT) provides a useful framework for helping students more effectively question and more meaningfully reflect upon technology use. Questions that illuminate some NOT ideas include:

- What is technology?
- How does technology both enhance and limit human activity and thinking?
- In what way does technology create bias and in what way does technology enhance bias?
- What are the gains and losses of technology use? What are technological trade-offs?
- How does society affect technology and how does technology affect society?
- What factors affect the development and adoption of new technologies?

While understanding these philosophical ideas and social consequences of technology use has been called for (NRC, 1996; National Academy of Engineering, 2009; AAAS, 2007; ISTE, 2007), few teachers or programs explicitly address the

Michael P. Clough, Joanne K. Olson and Dale S. Niederhauser (Eds.), The Nature of Technology: Implications for Learning and Teaching, 391–410.

NOT. Instead, technology education is too often limited to helping students become proficient users of technology.

Paradoxically, as educators have become increasingly occupied with teaching students to use technology, technology designers have worked to make technologies more and more intuitive. Perhaps, as technology becomes easier and more seamless, educators should increase emphasis on bringing to light the dangers of unexamined technology used and the ways technology changes the way humans think and act rather than continue to focus on technical proficiency. This chapter describes how, in an 8th grade science course, I helped students develop deeper technology literacy through engagement with the NOT.

ENGAGING STUDENTS IN NATURE OF TECHNOLOGY INSTRUCTION

Discussing NOT ideas concerning modern technologies is important, but engaging students in such discussions can be challenging. For example, when I asked students to consider the downsides of cell phones, they typically struggle with or dismiss the idea that trade-offs exist by citing only positive impacts cell phones have had on their lives. The emotional reaction of students to such questions indicates the deep relationship students have with their cherished devices, and these emotions can interfere with NOT learning. This is not surprising given the manner in which emotions moderate conceptual learning (Pintrich, Marx, & Boyle, 1993).

Given students' emotional attachment to their personal technological devices, introducing NOT ideas using technology examples for which students have little affection is important. This and the abstract nature of NOT issues also demand extensive scaffolding. Thus, single units or lessons on the NOT will not suffice for achieving deep understanding of the NOT. However, NOT instruction should not and need not detract from discipline-specific content instruction. Instead, as with effective instruction addressing the nature of a subject, NOT ideas ought to be introduced and revisited throughout the year in the context of content instruction. For example, when introducing students to a new technology (e.g. a microscope in a science class), exploring what the microscope does for us and what the microscope does not do for us will help students further reflect on the limited nature of this technology. Additionally, asking students to consider how the microscope might impact scientists' thinking about nature helps students wrestle with important science content related to microscope use, the nature of science, and NOT ideas.

Introducing NOT Ideas

Before beginning a unit on the structure of the Earth, I first engage students in an activity meant to simulate how scientists investigate the interior of the Earth using seismic waves (Kruse & Wilcox, in press). Students are given a "tray" with unknown structures hidden beneath a cardboard piece in the center (see Figure 1). The tray is a paper-box lid with a flat piece of cardboard raised approximately 1cm inch above the base of the tray so that students cannot see the unknown structure under the cardboard. Students are directed to determine as much as they can about what is under the cardboard piece. The only rules for their investigation are that the tray must remain intact and the only object allowed under the cardboard piece is a marble. Students quickly realize that they can learn much about the unknown structure by bouncing the marble around the inside of the tray – much like scientists learn about the inside of the Earth by observing how seismic waves behave.

Each group is given only one marble to conduct tests and told that before they will be given another marble, they must first provide a compelling rationale for why another marble is needed. Almost every group asks for additional marbles illustrating that they believe more of the device will make their task easier. However, I never provide additional marbles, regardless of students' rationale (the rationale for this decision is explained to students later in the activity). As groups roll their marbles around their trays, I walked through the room listening in on their conversations. I posed questions to help students improve their technique, but I avoided telling them what to do. For example, I have asked, "I notice you are rolling your marble around randomly. How might you set your tray so that you can collect more consistent data?" This prompt has typically caused students to prop one end of the tray up on some books and roll the marble down the slope in different locations.

After students explored the trays for some time, I called for their attention and asked, "What have you learned about the structure under the cardboard piece?" As students shared their ideas, I acknowledge what they say without confirming or rejecting their thinking. After students ideas were exhausted, I asked, "How will you document your data for other groups to see?" This question led to many different ideas and I made clear that each group would have to decide how to best represent their data. Before sending them back to their task, I introduced several NOT ideas discussed below.

An important aspect of technological literacy is having dynamic definitions of technology (AAAS, 2007; NAE, 2009). To develop more robust characterizations of technology I asked, "In what sense is the marble you are using a technology?" Initially, students were puzzled by this question, so I asked, "What do technologies do for us?" Once students noted that technologies make certain things easier, more efficient, or even possible, I then asked, "What is the marble making easier/possible?" Now students recognized what the marble was doing for them so I asked, "Why did you not at first think of the marble as a piece of technology?" Students claimed that they usually think of technology as gadgets and electronics and I asked, "Given your previous definition of technology, what would be considered technology other than electronics?" After this guidance and reflection, students named a variety of objects from the wheel to the pencil as technology.

I also used this early opportunity to introduce students to two other aspect of the NOT — the limited (AAAS, 2007; NAE, 2009) and non-neutral nature of technology (Carr, 2009; Keen, 2008; Postman, 1992). While students have little difficulty recognizing that current technology is limited, they often think that, given enough time, technology will solve most all problems. Furthermore, the notion that technology is not neutral and influences our decisions in profound ways can be unsettling to students. Therefore, introducing these ideas using a dispassionate example of technology, such as a marble, is helpful.

To start I asked, "Why haven't any of you wondered about the color of the structure under the piece of cardboard?" Students were a bit caught off guard by this question so I was sure to provide a bit more wait time and looked at them expectantly while smiling to encourage them to share their thinking. Students typically said, "We just didn't think of it?" So I asked, "How might the marble have caused you to ignore the color of the structure?" Students saw where this dialogue was going and dismissed the logic by saying, "The marble didn't affect us. We just can't see the structure so didn't think about the color?" So I asked, "When you think about the structure, how do you think about it?" Students indicated that they picture it in their mind, and I then asked, "Why do you not picture it in color?" If students continued to struggle with this line of thinking I might ask, "What kinds of information is the marble giving us about the structure?" What kinds of information is the marble not able to give us about the structure?" and "How did using the marble (a technology) impact your thinking about only the structure of the unknown object and not its color?" To further the point, we would make lists for the first two questions and discus how any technology does certain things well and is severely limited in other ways. The purpose and limitations of any technology are a form of bias inherent in the technology and may impact how we think.

One other approach I have used is to ask students why more than one marble would make this task nearly impossible. They quickly note that following the trajectory of one marble alone is difficult. I then asked, "While you recognize that problem now, what might your initial desire for more marbles indicate about your view towards technology?" Students typically recognize that they think more is better, and I use this to illustrate that more technology can cause new problems. Through these and other conversations, students are introduced to the notion that technology cannot solve all problems and often creates new ones (AAAS, 2007).

After these discussions, students used the remainder of the first day to continue their investigation. On the second day of the students' investigation I told them they would receive a new piece of technology to aid in their investigations and that they may use the technology however they wished as long as only the marble goes under the piece of cardboard. When handed a ruler with a channel/ridge down the middle (Figure 2), most students used the ruler to mark the points where they start the marble rolling or made other length measurements they thought might be important.

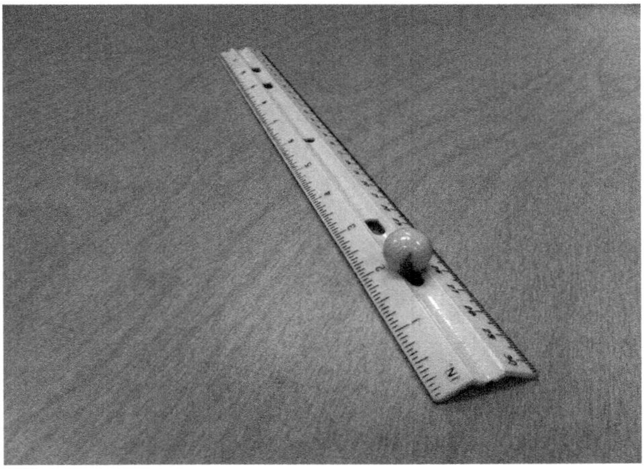

That students use the ruler in such predictable ways afforded another opportunity to engage in NOT discussions concerning the non-neutral or biased nature of technology. After students used their rulers for a few minutes, I called for their attention and asked, "How are you using your rulers to aid your investigation?" After students said they are measuring aspects of the tray to create accurate drawings or to ensure symmetrical data collection, I asked, "Why are you using the ruler to measure things?" Students answered that the ruler is "made for" measuring. I replied, "How do you know it is made for measuring?" Students typically gave one of two responses: "That is how we were taught to use the ruler" or "The lines on the ruler are for measuring stuff".

In response to students' rationale that they were taught to use the ruler to measure, I encouraged them to revisit the idea of bias regarding technology. I reminded students that the previous day we had discussed how the marble is biased toward certain kinds of information, and then asked "In what way do our biases as human beings also affect the way we use technology?"

In response to the idea that the lines on the ruler are for measuring, I introduced the notion of cues and affordances. I asked, "In what way do the lines serve as a cue (i.e. bias our interpretation) for how the ruler should be used?" Students quickly noted the lines are evenly spaced and are numbered to make measuring easier. Then I asked, "In what way do such cues result in technologies telling us what to do?" This question raised some red flags for students so I asked, "If technology does not tell us what to do and we could do anything with the technology, why did you all use the ruler to measure instead of using it like I will now demonstrate?" I then used the channel in the middle of the ruler as a ramp to guide the marble's motion in a straight line. Students were still a bit uneasy, but started to grasp that technologies do have significant influence over us. I then introduced the term affordances (what technology allows) and compared it to the subtle cues technology gives has that indicate how the technology wants to be used.

Students typically continue the tray investigation for another class period before I make explicit connections between the trays and the way scientists investigate the structure of the Earth. As a reminder, the tray activity is primarily directed at teaching science content, but asking the questions noted above also raises important NOT ideas without detracting from the science content. The activity and ensuing discussion also avoid raising the intense emotions that students exhibit when discussing their cherished technological devices. Finally, the activity also provides a common concrete experience that I often refer to when assigning NOT readings and discussing NOT issues during the course.

Linking NOT Ideas to Modern Technologies

The above experience and others like it, while drawing students' attention to important NOT issues, can be easily dismissed by students because of its in-school and contrived nature, an argument that Clough (2006) notes is also a problem in instruction targeting the nature of science. For students to deeply understand and apply the NOT, these same NOT issues must be brought to their attention using more modern technologies ubiquitous in their lives. For this reason, modern technology use within the classroom serves an important purpose. Teachers who are technology savvy, yet who also make apparent their awareness of the NOT and their critical analysis and use of technology are crucial for at least two reasons: (a) students will unlikely dismiss NOT instruction thinking it reflects a teacher's inability to appreciate and use technology; and (b) such teachers provide a role-model for the kinds of thinking and action NOT instruction is intended to promote. Furthermore, using technology in the classroom provides fertile experiences on which teachers can encourage students to reflect regarding the NOT.

For instance, I had my students collaborate online with classmates from other class periods. Unsurprisingly, they chose to work with those whom were already friends. Later in the year when discussing the NOT, I asked, "Many people say the internet will encourage us to work with different people who have different perspectives than us. Yet, when you had the opportunity to work with whomever you wanted via the Internet, you all chose people with whom you were already close friends. What does this demonstrate about people's belief that the Internet supports getting to know people who are different from us?" Students mirrored the logic of Keen (2008) when noting that the increased access to everyone allows for us to more easily ignore those next to us who might have different points of view or experiences than us. Instead, students recognized the access afforded by the Internet can easily cause increased isolation.

To further illuminate the manner in which the access afforded by the Internet can reinforce bias, students were encouraged to research online about a topic seen as controversial in the public's view (i.e.: stem cell research, global warming) and write a summary of their ideas before and after their research. Most students only paid attention to sites advocating the position they already held, and their views regarding the topic became more extreme. After having students compare their two summaries I asked, "The Internet is supposed to give us access to information from many different perspectives, yet most of you only sought research that supported your initial view. Why do you think that is the case? What does this indicate about the Internet?" From this activity and discussion, students come to understand how information technologies, while providing greater access to information, can encourage biased thinking through increased ease of locating like-minded views.

Later in the course, after students had spent several weeks studying the moon, I used a widely available presentation software program to quickly summarize all they had learned about the moon. While presentation software may be used in an educationally effective manner, the purpose of my presentation was to have students reflect on the pros and cons of the presentation. Students easily recognized how the efficiency of the presentation lacked detail and only made sense to them because of their past study. In an effort to draw students' attention to the NOT, I asked, "What about the technology encourages the kind of presentation you witnessed?" Students noted how the program's templates and even the manner in which the presentation is projected to the front of the room encourages a more teacher-centered approach and the presentation software encourages reduction of concepts to easily digestible chunks.

Technology's non-neutrality was highlighted often during class discussion because students struggled to grasp that every technology has an inherent bias, or is value-laden. Other NOT ideas were raised and discussed when opportunities arose. For instance, to draw students' attention to the limits of any technology, I first noted that presentation software programs can do many things, and then asked "What things is the technology not able to do regarding the creation of your presentation?" As another example, after using their cell phone cameras to collect images of rocks found on school grounds, I prompted consideration of technological trade-offs by asking students, "What are the pros and cons of using your cell phones to collect images of the rocks you observed?"

Again, using technology as part of instruction while simultaneously raising NOT issues makes clear to students that a teacher is not simply "anti-technology" and that the teacher carefully considers the NOT when making decisions regarding technology. While these instances provide opportunities to highlight the NOT related to more modern technology, students might not transfer the ideas to more personal technologies they treasure outside of school. Only after students have begun understanding NOT ideas do I push them to consider their more treasured personal technologies. For example, when I judged that students were grasping the notion that technologies come with trade-offs, I asked "In what ways have cell phones both been a good thing and a bad thing in your relationships with your friends?" Students typically note how cell phones encourage more frequent communication or communication with those who live far away as positive aspects, but acknowledged the lack of body language and facial expression when talking on the phone. Students also admitted how text messages are easily taken out of context or how they sometimes spend more time texting friends in another location than talking to the friends sitting at a table with them. That is, the students gain access to their friends at all times, but lose the impetus to engage with people sitting in the same room. While students could articulate how their cell phones come with gains and losses, they remarked how they had not considered the negative side of their cell phones before. To push them further I asked, "How can we use this knowledge to make better decisions about how we use our cell phones?" Similar conversations were had about the trade-offs, limitations, and biases of cell phones, Facebook, Skype, and YouTube.

As with any sophisticated idea, students often struggled to understand some NOT ideas and/or apply them to other technologies – particularly ones they used without reflection. At these times, I referred students back to a previous activity and NOT discussion (e.g. the tray activity, the presentation software discussion) to help scaffold students' in using their previous NOT understanding and apply that to new situations.

Historical and Contemporary Examples

Historical and contemporary examples serve to further elucidate complex NOT ideas. These examples provide needed context and broader scope to illustrate both the importance and the applicability of NOT ideas. Perhaps most importantly, historical examples can provide empirical evidence for the deep and long-term impact technology has on human beings.

Most of the historical and contemporary examples I used with the 8th graders came from my personal reading and research. These examples were often shared as anecdotes or short stories told in the context of instruction. That is, as students wrestled with particular NOT ideas, the examples were used to help students make meaning of the ideas being discussed. I sometimes verbally shared anecdotes and examples, at times had students read excerpts of readings, and at other times read contemporary news articles that illustrated the targeted NOT ideas. Below I present several examples I used with students.

a. To illustrate how the side effects and trade-offs of technology are often long-range and difficult to foresee, I addressed the manner in which technology impacts public discourse. For example, Keen (2008) notes that the web 2.0 revolution has increased the immediacy of news and no longer are we limited to major networks' points-of-view. Yet, bloggers fail to acknowledge that the very entity (major news outlets) they seek to abolish is the source for their livelihood. Even more far reaching is Keen's prediction that as news continues to move online and becomes a matter of choice, consumers of information will ignore news topics important for an effective democracy. Eventually, if those news stories no longer receive attention, they may cease to be published. Postman (1985 &1992) had similar concerns when he noted that the journalistic value of television news was sacrificed in favor of entertainment value.
b. When discussing the non-neutral NOT, I highlighted how the advantages/disadvantages and side effects of technology are rarely dispersed equitably. For instance, Carr (2009) notes that initially only wealthy individuals have access to new technologies. Eventually, a critical mass of consumers results in decreased cost, but the first wealthy users have greater market share, user capital, and often control how the services are distributed. Another anecdote noted how the decision to stop using DDT in pesticides in the United States had grave consequences for the control of malaria in other parts of the world.
c. To help students understand how technologies have both explicit and implicit values, and how the desired objective of technology may be undermined by its implicit values, I shared with them Postman's (1992, pp. 3–4) story of how Socrates' Thamus opposed the invention of writing because, in contrast to its stated purpose, writing would be an assault on wisdom. During discussion, I stopped and asked students how Google, or the Internet more generally, mirrors the predictions of Thamus. No longer do we value deep knowledge and wisdom, but instead place undo value on having isolated trivia at our fingertips, divorced from context. Using an historical example, I then asked students how the printing press favored some values over others. While the tight logical linear thinking demanded in printed text has clear value, more divergent thinking serves important functions as well. During this discussion, I drew students' attention to how we were using our science textbook in the course compared to how the book "called out' to be used. The students noted that textbooks promote a chapter-by-chapter approach to learning. Not surprisingly for those who understand the non-neutrality of technology, much research exists regarding how science textbooks determine the scope, sequence and pedagogy in science courses (e.g. Weiss, 1993 & Weiss, Pasley, Smith, Banilower & Heck, 2003). In my class, I purposely eschewed that

inherent bias of textbooks, instead addressing ideas from multiple perspectives in order to develop deep conceptual understanding of ideas. However, this approach begins with understanding how the nature of textbooks (a technology) values and promotes a particular approach to teaching and then making a conscious decision to not permit the textbook to determine how I teach a course.

d. During the initial Internet revolution, the easy access to information was expected to bring enlightenment and open minds. The ability to interact with diverse cultures and diverse perspectives beyond our physical limitation was to bring harmony among people. Both Keen (2008) and Carr (2009) note the removal of physical boundaries via the Internet has not had these predicted effects. Instead, what the Internet seems to have done is provide a new and easier avenue to *avoid those* who are different from us, and to downplay the need for face-to-face contact. The internet promotes ways of thinking and acting that result in less interaction with neighbors and more narrow perspectives on issues. We now jump online and interact with those who hold similar views to our own, or rudely attach opposing views. Rather than opening minds, people have seemingly become more entrenched in their one-sided views. United States political blogs have not opened bipartisan communication, but become increasingly more polarized (Keen, 2008). What this illustrates is that beyond supporting some goals over others, without explicit attention to the bias of technology (just one crucial goal of effective NOT instruction), it may even make some decisions for us. Of course, few of us would admit to this notion, yet cannot deny the powerful role technology plays in our lives. When making decisions (e.g. should I e-mail someone nearby or make the effort to meet with them face-to-face, we often do not acknowledge the role technology plays in these decisions.

e. To help students understand how technology bias impacts our society, I revisited insights regarding public discourse. Postman and Powers (1992) warn that rich context cannot be developed within a 22-minute news program and that "watchability" has replaced journalistic credibility. Keen (2008) argues this same concern when noting that legitimate news sources have begun catering to Internet culture and lower attention spans. Unfortunately, social media continues to push political discourse toward sound bites rather than well-articulated and logically coherent arguments. To make connections to the science content, I encourage students to consider how the media's discourse concerning global warming might contribute to popular opinion regarding global warming.

f. Many 8th grade students may not be interested in technology's impact on political discourse, so I also addressed how other cultural indicators such as business, religion, and education have been affected by technology. For instance, new technologies often result in companies altering the way they do business and employ people. The assembly line model increased production efficiency, but was an assault on craftsmanship. New technologies often create new jobs, but destroy former forms of employment (often those that required less education). Keen (2008) points out that the web 2.0 shift has led to decrease in some jobs, but has not produced new jobs. Alternatively, rather than jobs disappearing, the web revolution has led to increased outsourcing to other countries (Friedman, 2006).

g. Technology's impact on religion and value systems has been far reaching. The printing press was responsible for Martin Luther's ideas being spread widely and quickly during the Reformation. No longer did the public have to rely on priests for religious council. After the printing press, each person became his or her own theologian. The effect of which has been both positive and negative (Postman, 1992). Similarly, community value systems can no longer be enforced. Neither the public nor the individual conscience has as much power as it once did. Illegally downloading music or viewing obscenities online are easier to do when there is no shopkeeper to look in the eye or to check identification (Keen, 2008).
h. As part of the presentation software activity discussed in the previous section, I also had my students consider how this and other technologies have had far reaching effects on education. Clearly face-to-face small group discussion is a powerful form of instruction, but technological advances have promoted a business/economic efficiency approach to education that downplays its importance. Extensive pressures now exist to place classes online so that we might serve larger populations of students more conveniently at less expense. Had these new technologies not been available, the questions we ask about education and the answers we entertain would be different.
i. In addition to cultural indicators, technology affects individuals including how they live, work, think and act. Automobiles encouraged movement further away from ones place of work. Communication technology promotes 24/7 availability, and that expectation now exists among those we know and many employers. Nor does space limit the number of people who can live within an area. The sky is literally the limit. The issue is not whether or not technology has advantages and disadvantages — it has both. But I emphasize to my students that understanding the NOT is crucial for thinking deeply about these advantages and disadvantages, many which are not easily discerned, and then making appropriate decisions.

Importantly, the above anecdotes, examples, and stories are addressed within the context of everyday science content instruction. That is, when appropriate, I raised NOT issues in the context of the science ideas and activities occurring. At these times, when students were wrestling with a particular NOT idea, the examples were given to encourage students to understand the greater implications of the NOT beyond the classroom. After sharing the stories, the students were expected to make connections to their own experiences with technology or the activities they had done in class. To encourage understanding and application of NOT ideas, I asked questions like those I have included above, and those below:

- How does this example compare to what you experienced [insert activity in class]?
- In what way does this story reflect your own use of technology?
- How might this example be used to make better decisions about technology?

- How does this story/example illustrate that technologies often have unforeseen negative consequences?
- Based on the example from history, why do you think understanding how technology is value-laden (biased) is so important?

Summary of NOT Instruction

While the three previous sections discuss three separate approaches to NOT instruction, the approaches were highly integrated. Linking students' experiences using technology to their experience with previous activities (like the tray activity) encourages a willingness to critique technology. Few students have trouble critiquing the trays/ marbles, and when parallels are drawn to technologies they enjoy using, they are more cognitively and emotionally ready to take part in such discussions. Having students hear or read historical examples encourages students to see how the issues highlighted concerning the tray and their own technology use applies more broadly. Using historical examples helps students understand the long-term impact of technology and contemporary examples demonstrate the need to raise these issues in today's society.

Again, my NOT instructional efforts typically occurred in the context of science content instruction. For example, when discussing Galileo's observations of the moon with a telescope, I noted that some people dismissed his ideas claiming the telescope caused his observations of mountains and that the moon was perfectly smooth. This instance allowed for discussions about how technology shapes our thinking. Another example is when students want to use beakers for precise measurement of liquid. The graduations of the beakers cue students as to the appropriate use of the beaker to measure volume despite the inaccuracies of the graduations.

The examples above are representative of, but do not exhaust, the ways in which NOT instruction occurred in my 8th grade science course. The examples serve to illustrates how NOT instruction can be implemented as a normal part of instruction. Importantly, raising NOT issues may be more important than resolving them. The complexities of the NOT, like many important topics, are difficult to grasp, and may differ depending on the technology in question. Yet, understanding the importance of such questions, and raising them are the first steps to making more informed decisions regarding technology.

IMPACT ON STUDENTS' NOT UNDERSTANDING

Having put significant effort into NOT instruction, I was interested in determining what impact, if any, my efforts had on my 8th grade students' thinking. Toward that end, I investigated the NOT views of twenty students in the same class period. Students in this class attend a medium sized Midwestern middle school situated in an urban area. The student demographics matched those of the school in which the study took place, 60% Hispanic, 40% white non-hispanic.

Because understanding the NOT is such a new area of concern for education, little prior research has been conducted in this area. Unable to find any pre-existing instrument to assess students' NOT thinking, I created four open-ended NOT questions designed so that prior to and after NOT instruction, students would find them intelligible and be in a position to provide answers that accurately reflected their thinking. To ensure students accurately interpreted the questions, they were pilot tested using a group of students from a different class period. The pilot group was interviewed to seek idiosyncratic interpretations of the questions. Problematic questions were modified based on pilot group information. Based on this process, the four questions below and were used in this study:

1. What is technology? Provide examples if you can.
2. How has technology been negative (detrimental) for people (both individuals and society)? Provide examples to support your points if you can.
3. What does it mean to be technologically literate? (What should people understand about technology? Why?)
4. How do you think technology has or will affect your education both positively and negatively? Provide examples if you can.

These questions were chosen because they seemed to highlight students' misconceptions and struggles with the nature of technology. For example, question two does not ask about positive impacts of technology because students were already very aware of technology's positive impact. Therefore, asking the question on the pre/post assessments would likely yield few, if any, interesting observations.

The students were given these questions as a pretest approximately one month before nature of technology instruction took place and followed with semi-structured interviews regarding their pre-test responses. During the interviews, students were asked to explain their responses on the pre-test, and follow-up questions were asked to clarify their written and verbal responses. After approximately two months of NOT instruction described in the previous section, the students were given the same post-test and again interviewed to ensure accurate interpretation of their post-treatment responses. For each question, the students' written responses and interviews were analyzed and their views were categorized as described below.

Categorizing Nature of Technology Views

Student views for each question were categorized as either naïve, partially informed, or informed. All student responses and interview transcripts were read and descriptive notes were made to identify indicators of student levels of understanding. These notes were then used to form hierarchical categories. Student responses/transcripts were then re-read and the students' views were identified as naïve, partially informed, or informed. Each category of student views is described below for each of the four NOT questions used in this study with supporting excerpts of student words.

What is technology? Student views categorized as naïve usually limited technology to only electronic devices or claimed that technology is "everything".

> Technology is like cellphones or video games. Technology is the only things that a human can make. Technology is the things that make a difference in some peoples' lives. (SV, posttest)

Student views categorized as partially informed expressed a mixed view on technology. They might state generally that technology can be anything to help people, but then limit their examples to electronics. Some responses in this category are accurate, but lack sufficient detail or nuance to be in the informed category.

> Technology is anything we use to help us. Technology is electricity that we made to help us. Pencil, paper, books, silverware and such. (CW, posttest)

Student views considered informed clearly understood that technology is more than gadgets or electronics and noted that technology's purpose is to assist humans in some way. They also noted more complex understandings such as the role of systems in technology.

> Technology is anything designed by people to help or "ease a burden" for other people. Technology can be tools or systems or electronics. Anything that a person had to make. Ex: railroad system as well as railroads in general. Ex2: Electricity as well as the machinery that you plug in. (MS, posttest)

Negative aspects of technology. Naïve conceptions of the negative aspects of technology included the notion that technology is only bad if it is broken or used inappropriately. Also, views that focused solely on people becoming lazy or addicted to the technology were categorized as naïve.

> The things like tv's and computers are things that some people can't live without. They sit around and play videogames or watch tv all day. And to society some people would go crazy if they didn't have it. (JW, pretest)

Partially informed views were characterized by having some naïve views and some informed views. On the post-test, responses in this category might note examples from class discussions, but did not generalize these views beyond those examples.

> I think technology has been negative because we don't do things on our own. Like on a computer if we spell something wrong or have a grammar mistake, the computer automatically fixes it, so you never learn to do it yourself. (CR, pretest)

Student views categorized as informed identified multiple examples regarding negative aspects of technology. Also, these students included statements that generalized the examples to all technologies. Informed responses also made clear the nature of technological trade-offs. While some aspects of technology are negative, other aspects are positive.

> Technology has caused a lot of new issues and questions to come up. Things like abortion and who owns what on the internet. We may think a technology is going to help us but it really doesn't. Computers were designed to connect people but in a way they did not. They kept us from being with people DIFFERENT from ourselves. We go on the internet and talk to people just like us. Ethical issues like abortion are causing people to debate all around the world. If the technology had not been invented we would not have the problems. For every new technology we lose something. It can be big or small. One example is cars which causes pollution, which contributes to global warming. But at the same time we get to go places quicker and easier allowing us to do more things during the day. (JW, posttest)

Defining technology literacy. The views considered naïve focused on people being able to use technology and understanding how the technology works. These students made no reference to NOT issues discussed above and any reference to "problems" was referring to fixing the technology.

> Being able to understand technology. Like when you get a cellphone and it's different and complicated, someone who is technologically literate would be able to figure it out easily. (MS, pretest)

Partially informed views included students who hinted that technology has a negative side or that technology literacy is more than just knowing how to use technology. Yet, these students did not provide enough detail or were unable to explain their thinking.

> You should understand technology because it helps you and hurts you (TP, pretest)

> Technologically literate means to have a great understanding about technology like if you know a lot about how people make the technology and what can be used for other than its base point (JM, pretest)

Informed views included the perspective that people should know how to use technology, but also raised specific concerns regarding negative aspects of technology and how it might affect society. They explicitly noted that technology must be questioned or critically examined.

> I think that to be technologically literate means to know what the technology is, what it's meant to be used for, its pros/cons, its difficulty of access, negative and positive effects, side effects and so on. Because if you going to be tech. literate you might as well know how to use tech. And of course people should know the negatives and positives of the thing because they need to make decisions on if the positives of using the device are worth having the negatives. Because it's like an exchange. (CR, posttest)

Pros/cons of technology in education. Naïve views focused exclusively on positive effects of technology in education. Student views in the naïve category also might express that only the lack of technology in education is a problem.

> I think if we don't have no technology like things we need for education such as computers and things we need in school. And if we don't have it, then that would make it harder to do things on our own without no technology. (MR, pretest)

> I think it has only affected it positively by allowing students to get more information. Technology can also be positive in education because if you have a lot of it I think that the students would be more likely to learn more than what they need to. (JM, pretest)

Partially informed views noted that technology has been both positive and negative in their education. These students often contradicted themselves demonstrating incomplete conceptual change. They might claim technology helps them learn faster, but they don't really learn the material as well. Additionally, students in this group may note accurate ideas, but be overly vague.

> Technology will affect my education positively because it helps me learn better and faster. Technology will affect my education negatively because it might be too fast like the PowerPoint, it is meant for quickness. (EH, posttest)

Students with informed views could articulate how technology could be both beneficial and negative without contradicting themselves. Many of these students brought in several NOT ideas to demonstrate a robust understanding of NOT ideas applied to education.

> Technology can help us learn more, but at the same time less. We can do more things, but it is too easy to get the answers. Computers have sites like google and yahoo that allow us to get the answers we want quick. And not all schools have technology. One day I might have a teacher that has barely used technology before. It could bring me back in my education or it might help me more. The teacher might be used to explaining things instead of us memorizing. That would make me think about what I'm learning. (JW, posttest)

RESULTS

The pre and post treatment categorizations of students' responses on the four-item NOT instrument appear in Table 1. The results clearly illustrate that students' NOT understanding generally improved after the treatment, but that many students' post-test responses fell short of being informed.

Most students started out with naïve views regarding all aspects of the NOT. While this is not surprising given the nature of public discourse surrounding technology, this initial finding makes clear the need to address NOT issues with students. Students initially viewed technology as gadgets and electronics and

Table 1 Numbers of students in views on nature of technology (N=20)

Nature of Technology Item	Pre-Instruction			Post Instruction		
	Naïve	Partially Informed	Informed	Naïve	Partially Informed	Informed
Defining technology	10	9	1	1	14	5
Negative aspects of technology	13	7	0	3	10	7
Defining technology literacy	14	6	0	5	8	7
Pros/cons of technology in education	15	5	0	7	8	

struggled to articulate nuanced negative aspects associated with technology. Indeed, prior to NOT instruction, students expressed that the most significant problem with technology was not having enough of it. Students' naïve thinking was further demonstrated by their assertion that the essence of technological literacy was simply the proficient use of technology.

After instruction, only one student expressed a naïve view regarding examples of technology. However, more nuanced views of what technology is were less frequent despite discussions related to knowledge and systems as technologies (NAE, 2009). Students' understanding that technologies have negative consequences improved after NOT instruction. Post-test responses articulated more negative aspects beyond technological malfunction, but some students still struggled to fully understand and express the gains and losses regarding technological affordances or could not generalize to more far-reaching negative consequences of technology such as changes to public discourse and society.

More students struggled to synthesize what they had learned about the NOT in expressing what being technologically literate means and applying what they had learned in expressing the pros and cons of technology in education. While several students expressed coherent ideas that went beyond what was discussed in class, many students limited their responses to more obvious or previously discussed examples.

DISCUSSION

Given the pretest results and students' reaction to class NOT activities and discussions, they clearly had not previously been encouraged to think about NOT ideas in any depth. Yet, after only two months of instruction, most of these 8th grade students achieved partially informed views, and many expressed, for the targeted age group, informed views. However, due to the commonly conveyed, yet insufficient, views about technology in our society, the complex and abstract character of NOT ideas, and the resilience of misconceptions, two months is likely inadequate for deep conceptual change regarding the NOT. Not only is much abstract thinking demanded, students must also confront a lifetime of societal cues (e.g. television, the internet, policymakers, teachers, etc.) that express only the glory of technology.

Although the NOT is fraught with abstraction, the data previously presented supports the contention that 8th grade students can engage in and improve their understanding of NOT issues. The major roadblock to engaging individuals with these difficult and abstract concepts is not that they are unable. Rather, in the Huxleyan world in which we live, most people wrongly believe they have critically examined the technology they use. Few are aware of the more subtle issues at work. Further complicating students' difficulties in critically examine technology are the emotional ties they have to their technology. Most people see technology as a staunch friend, but rarely consider the costs of that friendship (Postman, 1992).

Given the emotional ties that many people have with particular technologies, NOT ideas are best introduced using everyday items such as rulers, scales, and pencils, and then scaffolding students' emerging NOT understandings to Facebook, cell phones, the internet, and other treasured technologies. Using concrete experiences, like that illustrated in several examples earlier in this chapter, assists students in grappling with and understanding abstract NOT ideas. Lastly, providing more far-reaching context for students through historical examples encourages students to recognize the instructive power of the NOT beyond their own experiences. Through all this, the teacher's role is crucial, because the questions asked of students are the scaffolds that promote mental engagement and understanding. Not surprisingly, all this reflects what has been written about effective nature of science instruction (Clough, 2006). These strategies, taken together and used throughout students' education, will better prepare students to be conscientious of technology's deep impact.

The strategies above help to explicitly draw students' attention to the NOT, but must be reinforced via teachers modeling the careful consideration of technology use. This modeling requires that teachers not shy away from using technology in the classroom, but that teachers make explicit their thinking about technology use. Using technology with students makes clear the teacher is not simply anti-technology and provides fertile ground in which NOT discussions should be rooted. For example, when deciding how to record data for a class investigation the teacher might ask students, "What might be gained from using a spread sheet for our data and what might we lose?"

While teachers should integrate the NOT into instruction, they must do so knowing that political and economic factors are causing a narrowing of the curriculum and assaulting the very meaning of what being educated means. For good reasons, teachers may feel that little room exists to address the NOT. Schools increasingly mirror Postman and Powers' (1992) concern regarding information glut stemming from television. That is, the public knows "of many things, but about very little" (Postman and Powers, 1992, p. 156). As professionals, teachers ought to actively push back against shortsighted views of education and work to create learning experiences that help students become more thoughtful individuals as well as learn traditional content. Including the nature of technology within content courses provides a lens through which students can become more thoughtful. Yet, when considering the impact of technology in various disciplines, whether language arts, mathematics, science, art, music, or history, students will be required to think more deeply about the content, resulting in improved content understanding as well.

In today's high stakes testing environment, the trend to make learning more about fact acquisition than understanding must be actively resisted. Education ought to provide students with the reasoning capacity to make informed decisions. While the NOT could be reduced to a list of tenets to be memorized, that would not promote teaching and learning directed at rational decision-making. Rather than focusing on how well students articulate a prescriptive set of NOT ideas, educators should care most about how students use the ideas to make more reasoned decisions about technology. If a student says that technology has bias, but cannot meaningfully discuss possible biases of a particular technology, the knowledge has little value. Understanding the NOT requires context, thus the NOT might be better discussed as questions to be raised as Clough (2007) noted concerning the nature of science.

FINAL THOUGHTS

When new technologies debut, much fanfare follows regarding how they will revolutionize our work or play, and efforts focus on adopting and using the technology. Rarely do we hear serious discussion regarding the issues presented in this chapter. Yet, acknowledging that all technologies come with trade-offs and side-effects is important to thinking critically about technology and making truly informed decisions regarding its use (AAAS, 2007; NAE, 2009). Schools and teachers have a responsibility to *educate* students so they become critical consumers of technology. Aspects of the NOT are essential tools for developing such a critical stance and are crucial components of technological literacy.

Clearly from the prevalence of student initial naïve ideas, the students had not previously been encouraged to consider the negative aspects of technology. Yet, these NOT ideas are pervasive in reform documents. The American Academy for the Advancement of Science's *Atlas for Scientific Literacy* even dedicates an entire chapter to the subject. The lack of empirical literature further indicates that the nature of technology has largely been ignored in science education and education in general. As STEM (Science, Technology, Engineering, & Mathematics) initiatives continue to pervade educational discourse, educators must not lose sight of the natures of these disciplines. Understanding the natures of these disciplines is requisite for the deep literacy necessary for our students to meaningfully participate in informed dialogue and decision-making.

If we are to be educating students for the technological world in which we live, we must encourage them to develop the thinking strategies that will help them more meaningfully evaluate the role of technology in their lives. We cannot predict where technology will lead, but we can work to ensure a technologically literate public raises critical questions. If students are only taught how to use the newest technologies, the implicit message is clear that using technology is more important than reflecting on best use, possible side effects, and the manner in which technology uses us. If technology is not critically examined, the sci-fi worlds described in *1984* (Orwell, 1949), *Brave New World* (Huxley, 1932) and others may not be as far-fetched as many people think.

REFERENCES

American Academy for the Advancement of Science (2007). Project 2061: Science For All Americans – Chapter 3 The Nature of Technology. http://www.project2061.org/publications/bsl/online/.

Carr, N. (2009). *The big switch: Rewiring the world, from Edison to Google*. W. W. Norton & Company, New York.

Clough, M. P. (2006). Learners' Responses to the Demands of Conceptual Change: Considerations for Effective Nature of Science Instruction. *Science & Education, 15*(5), 463–494.

Clough, M. P. (2007). Teaching the Nature of Science to Secondary and Post-Secondary Students: Questions Rather Than Tenets, The Pantaneto Forum, Issue 25, http://www.pantaneto.co.uk/issue25/front25.htm, January. Republished (2008) in the *California Journal of Science Education, 8*(2), 31–40.

Friedman, T. L. (2006). *The World is Flat: A brief history of the twenty-first century*. Farrar, Straus and Giroux, New York.

Huxley, A. (1932). *Brave New World*, RosettaBooks LLC, New York.

International Society for Technology in Education (2007). *National Educational Technology Standards.*

Keen, A. (2008). *The Cult of the Amateur*. Doubleday, New York.

Kruse, J. W. (2008). Integrating the Nature of Science Throughout the Entire School Year. *Iowa Science Teachers Journal. 35*(2). Spring.

Kruse, J.W. & Wilcox, J.L. (In Press). Engaging Students with the Nature of Science and the Nature of Technology by Modeling the Work of Scientists. *The Clearinghouse.*

National Academy of Engineering (2009). *Technically Speaking.* http://www.nae.edu/nae/techlithome.nsf/ (Retrieved 2/23/2009)

National Research Council (1996). *National Science Education Standards.* National Academy Press, Washington, D.C.

Orwell, G. (1949). *Nineteen Eighty-Four*. Harcourt Inc, Orlando, FL.

Pintrich, P.R., Marx, R.W. & Boyle, R.A. (1993). Beyond Cold Conceptual Change: The Role of Motivational Beliefs and Classroom Contextual Factors in the Process of Conceptual Change. *Review of Educational Research. 63*(2), 167–199.

Postman, N. (1985). *Amusing Ourselves to Death: Public Discourse in the Age of Show Business.* Penguin Books, New York.

Postman, N. (1992). *Technopoly: The Surrender of Culture to Technology.* Vintage Books, New York.

Postman, N. & Powers, S. (1992). *How to Watch TV News.* Penguin Books, New York.

Selber, S.A. (2004). *Multiliteracies for a digital age*. Southern Illinois University Press, Carbondale.

Weiss, I.R. (1993). Science teachers rely on the textbook. In Yager, R.E. (Ed.) *What research says to the science teacher, volume seven: The science, technology, society movement.* Washington, D.C.: National Science Teachers Association.

Weiss, I.R., Pasley, J.D., Smith, P.S., Banilower, E.R. & Heck, D.J. (2003). *Looking inside the classroom: A study of K-12 mathematics and science education in the United States.* Chapel Hill, NC: Horizon Research, Inc.

Jerrid W. Kruse
School of Education,
Drake University

CHAPTER 20

JAMES JADRICH & CRYSTAL BRUXVOORT

CONFUSION IN THE CLASSROOM ABOUT THE NATURES OF SCIENCE AND TECHNOLOGY

Implications for Scientific & Technological Literacy

INTRODUCTION

While philosophers, historians, sociologists, cognitive scientists, and practitioners of science and technology have laboured to explicate differences between the natures of science and technology, those distinctions are not typically understood or recognized by students (Constantinou, Hadjilouca, & Papadouris, 2010; Gardner, 1994; Roy, 1990). The resultant confusion poses unique challenges for science and technology teachers who seek to have students understand and utilize science and technology in ways that are consistent with their particular natures and purposes. In this chapter we describe evidence to support the claim that students do indeed confuse the natures of technology and science and, in particular, do not act as if they clearly understand the distinct *purposes* of these separate but related disciplines. Further, we outline some of the probable causes of this confusion, explore the undesirable effects of this misunderstanding, and end with pedagogical implications for science and technology education.

As others have done before us, we begin by defining technology and science in terms of the goals and purposes of the two disciplines. While academicians of science and technology are not settled on how to precisely characterize all aspects of these two different fields, we maintain that there is uniform agreement that science (pure and applied) and technology pursue different end goals and exist for fundamentally different purposes, although at times these differences may be quite subtle and difficult to parse in practice. Specifically, the primary goal of science is to examine evidence in the natural world in order to generate scientific models (and theories) (Gilbert, 1991; Harkema, Jadrich, & Bruxvoort, 2009; Jadrich & Bruxvoort, 2011; Seok Oh & Jin Oh, 2011). Scientific models can take different forms, including explanatory models, such as Einstein's theory of relativity; descriptive models, such as Jane Goodall's description of the behaviour of chimpanzees; mathematical models, such as the mathematical formula for calculating the period of a pendulum;

Michael P. Clough, Joanne K. Olson and Dale S. Niederhauser (Eds.), The Nature of Technology: Implications for Learning and Teaching, 411–426.

and similarly, predictive models, such as a climate or weather model. Scientific models and theories are related in that models are considered to be subsets of scientific theories—theories being more comprehensive systems of explanation. Correspondingly, the types of questions pursued in science include such ones as: "Do reaction kinetics vary as a function of pressure?" or "Is free fall acceleration independent of mass?" Pursuit of answers to such questions with the end goal being generation of scientific models is the work of persons doing basic or pure science (Clough, 2004).

Relatedly, scientists doing what is called applied science are also working to generate models. However, applied science is constrained differently than what has been described above in that it is pursued because of the likelihood that such work in the end will provide a particular societal benefit or benefit the scientific community. In other words, the end goal of applied scientists, like scientists doing basic science, is to develop scientific models. Nevertheless, in the case of applied science, the motivation for pursuing particular studies is the development of new technologies or for solving engineering or societal problems. For example, many areas of study in condensed-matter physics (the study of crystalline solids, liquids, supercooled liquids like glass, amorphous materials like ceramics and polymer compounds) are driven by the pursuit of novel materials with technological applications. Faster and smaller transistors, photovoltaic cells, high temperature superconductors, and fiber-optic communication devices are among the various products developed by engineers as a result of applied research in condensed-matter physics.

In contrast to basic and applied science, the end goal of technology is to produce products or solutions aimed at generating a desired outcome. The outcomes or products developed are typically targeted towards societal problems, needs and desires or for scientific purposes. Questions or problems that we might recognize as technological in nature include: "Can a vaccine be designed to fight against a pandemic illness;" "can a wind turbine be built to generate cost-effective electrical energy for an urban center," or "can software be developed to inhibit hackers from breaking into a governmental agency's records" (AAAS, 1990; 1993; Clough, 2004; Constantinou et al., 2010; Gilbert, 1991; Hammer, Russ, Mikeska, & Scherr, 2008; ITEA, 2000; McComas, 1996; NRC, 1996; Schauble, Klopfer, & Raghavan, 1991).

Notably, as we have sought here to distinguish the purposes of science and technology, we have portrayed the essences of these disciplines with respect to the (normative) purposes or end goals of the practitioners and not in reference to the practitioners' particular activities. We chose this approach with due consideration of the literature-supported position that any attempt to distinguish science from other disciplines by describing the actions of the participants (e.g., collecting observations, taking measurements, doing experiments, etc.) or by the cognitive processes utilized (e.g., hypothetical, inductive, deductive, and analogical reasoning) must inevitably fail, because *all* human activities include similar features of thought and practice (Koslowski, 1996; Dunbar, 2002). This is especially evident when the actions and processes of engineering and technology development are considered along side the actions of scientists. In other words, we contend that a comprehensive listing of all the types of actions and thinking required of scientists doing science (basic or applied) is indistinguishable from that of engineers and technologists developing new technologies. At the same time, and more importantly in view of our ultimate aim in writing this chapter, we also assert that careful observations of the ways students select and perform activities consistent with a science or engineering task is demonstrative of how often students confuse the purposes of science and technology.

CONFUSING THE NATURES OF SCIENCE AND TECHNOLOGY

An examination of two sources—published literature and our own research on scientific reasoning with children—enlighten this discussion as to how and why students display confusion between the purposes of science and technology. The main premise that we wish to establish in this section is the following: Students demonstrate a misunderstanding of the purpose of science (confusing science with technology) when they wrongly employ strategies consistent with engineering and developing technologies rather than science; even when they have been specifically tasked to reason with a scientific end goal in mind.

Literature Support

Multiple studies exist to support the claim that students do not easily distinguish between the purposes of science and technology (Kuhn & Phelps, 1982; Schauble, 1990; Schauble, Glaser, Raghavan, & Reiner, 1991; Schauble, Klopfer, & Raghavan, 1991; Tshirgi, 1980). Specifically, when students should be reasoning with a scientific end goal in mind (i.e. developing scientific models to account for natural events), they often reason instead toward a technological end (i.e. they attempt to maximize an outcome or achieve a desirable result). Students working toward a technological goal when they should be doing otherwise are said to be *engineering* (Schauble, Glaser, Raghavan, & Reiner, 1991; Schauble, Klopfer, & Raghavan, 1991); a term that is meant to reflect the normative activities of engineering practitioners.

Evidence of students' predisposition toward reasoning with engineering goals instead of scientific ones is illustrated in the following two accounts. When Schauble and Glaser (1991) tasked students to investigate the effects of a boat's design on speed, students became wrongly preoccupied with constructing fast boats (i.e., engineering). In other words, although the intention of the activity was to have students investigate the effects of boat design (e.g., weight, height, narrowness) on speed, students interpreted the purpose of the activity as the optimization of a desired outcome—that is, designing fast boats. Similar results were obtained in a different context when students were tasked to investigate the effects of car design on speed. Instead of designing controlled experiments to explore important and unimportant variables, students interpreted the purpose of the task to be the construction of a car that would go as fast as possible (Schauble et al, 1991). Just as before, students worked to achieve a desired outcome (technology), as opposed to developing a model (science) of car speed as a function of car design. In both of the cases cited here, students' actions were more coherent with a technological purpose rather than a scientific one, despite having been explicitly tasked to pursue the latter.

Evidence from Our Own Research

Our own research examining students' employment of scientific reasoning skills and strategies lends support to the assertion that students commonly confuse the purposes of science with those of technology. In the following section, we describe two different tasks—a bouncy balls question and a germination question—where students display engineering tendencies rather than scientific ones, consistent with the literature-based phenomena described above.

Bouncy balls question. In one particular study with 120 students (grades 4–8), participants were shown a set of 5 bouncy balls, all of differing sizes and weights. Students were told that each ball had been dropped (not thrown) from the same height, and that measurements were made to determine how high the balls bounced off the ground. After it was clear that the students understood the experimental procedure as described, they were shown a data table that listed the size and weight of each ball and the height to which it had bounced. Students were then asked to respond to the following two questions: (1) Was this a fair way to test if size affects how high a ball bounces? (2) Was this a fair way to test if weight affects how high a ball bounces? Since neither size nor weight had been controlled in the experiment, there was no way to disentangle the affects of size or weight on how high the balls bounced. Consequently, the correct response to both questions should have been that this was an unfair way to test.

Consider one student's response to the questions of whether or not this was a fair way to test if size or weight affects how high a ball bounces. As a matter of background, this student (pseudonym, Erin) has already indicated to the interviewer that she believes that size should make a difference–larger balls should bounce higher. Consequently, she acts confused by the results that do not confirm her expectations.

Erin: I think it's the weight.

Interviewer: You think it's the weight. Do you think they did this in a fair way to test the weight?

Erin: Yeah.

Interviewer: But not the size?

Erin: Not the size.

Interviewer: All right, now tell me again, so, why do you think the weight was a good way to do it, but not the size?

> Erin: Well, because, um, the size didn't work because, like, that ball was smaller [pointing to a particular test] and, um, that ball was bigger [pointing to a different test], and the ball that was smaller bounced higher. So, that wasn't fair.
>
> (Source: EH, TRIAGE)

Erin appears to interpret the purpose of this experiment (for ball size) as an attempt to demonstrate that larger balls bounce higher than smaller balls. When she examines the test results and finds that a small ball bounced higher than a large ball, she declares that the test for size was not fair. In her words, "size didn't work …", and therefore she deems the test to be unfair.

Like Erin, many other students in this study (nearly 45% in all) adjudicated fairness based on the results obtained rather than a study of experimental design. They would declare a test "unfair" if they expected a positive correlation between a variable and outcome but such a result was not achieved; or, in other instances, they declared a test to be "fair" if they expected a positive correlation between say size (or weight) and bounce and felt that the data showed such a correlation (although there actually were no correlations in the data between the variables and the results). Erin likely had strong views about what should be observed (as scientists do at times), and she was attempting to save the appearances of what she expected. (Indeed, when scientists operate in this way, colleagues frequently accuse them of "engineering their results.") All together, the tendency to assess the integrity and merit of a procedure based on the result or success in achieving a particular outcome—likely the outcome that is expected or desired—is the essence of "engineering" as it has been described in the literature.

Still other students determined fairness simply based on whether or not the results of the experiment appeared to co-vary positively (or negatively) with either of the experimental variables. In those cases, the students did not attempt to impose a specific relationship between a variable and the bounce height; they simply looked for clear patterns in the results (height measurements). Put in another way, if a variable (input) did not yield a discernible outcome (output), then the test was judged to be unfair.

One could argue that the students' responses are not indicative of engineering tendencies, but rather an inability or an unfamiliarity with what is meant by fair testing in scientific investigations. However, other aspects of our research effort disconfirm this possibility. When asked similar, but non-identical questions on fair testing, these same students previously demonstrated an understanding of and need for fair testing in certain scientific investigations. In other words, when these same students were asked similar questions such as this one involving bouncy balls, they successfully sorted through the given descriptions and identified valid and invalid experimental designs, including successfully identifying those procedures that controlled variables properly. Furthermore, even children younger than students in our study understand the need for fairness when testing, so long as they are familiar with the context of the test and with the materials involved in the test (Irwin & Moore, 1971; Piaget, 1932). There seems then little question that these students understood the meanings of "fair testing" and "fairness." However, an important difference to note when comparing the questions asked of students on fair testing is that in the latter case involving the bouncy balls question described above students were told not only about the experimental procedures for each investigation, but they were also given explicit information about the results that had been obtained. Arguably, the presence of results was essential for revealing the extent to which students focus on engineering over scientific pursuits. In turn, many students in this study were unable to decide tests were fair or not unless the results of the tests were given, thus adding credence to our assertion that students often evaluate scientific tests from an engineering perspective.

Considering again Erin's reaction (as presented above) to the bouncy balls question on fair testing, arguably such a response cannot be considered cogent when contextualized in a framework involving a scientific purpose. As stated before, in science experimental "fairness" depends on whether or not the experimental design allows one to relatively unambiguously determine if one variable necessarily co-varies with another. Fairness is not judged based on whether or not a preferred, an expected, or even an obvious outcome has been produced; for the goal of science is to express the relationships naturally existing amongst the variables under study, and not to produce a particular result. However, when Erin's judgment on fairness is contextualized in a framework involving a technological purpose, her decision to focus on outcomes and link "fairness" to whether or not an outcome was achieved is arguably sensible (but inappropriate), since she had been tasked to reason with a scientific goal in mind (as she had correctly done for previous questions). All together, an engineering framework as described in literature brings cogency to a response such as Erin's and sheds light on why students (from a scientific perspective) may appear to be inconsistent in their ability to judge the fairness of a scientific test.

Germination question. Notice a similar result when we asked these same students a different question. In this case students were asked to evaluate results related to whether or not common factors (e.g., light, darkness, amount of water) affect germination rates of radish seeds (e.g., Does temperature affect germination rates? Does amount of water affect germination rates?). In doing so, many of the students turned the conversation away from the question, "What can you conclude about light (and amount of water, etc.)?" toward the question, "What does it take to get all the seeds to germinate?" Consider the response posed by Sandy (a pseudonym):

> Sandy: Um, I know the problem with, this one [points to test 1 where seeds were placed in sunlight with one cup of water and germinated 14 of 15 seeds]. I thought this [test 1] was the best one to grow with.
>
> Interviewer (I): Okay. What's drawing you to that?

Sandy: Except with all of them, I think, this one [points to test 1] probably could have gotten 15 and this one probably could have gotten 14 [test 5 where seeds were placed in darkness with 1 cup of water and 13 of 15 seeds reportedly germinated] and maybe even 15. Maybe if they opened the bags so the plants could get some air.

I: Okay.

Sandy: And, um

I: They have a little bit of air in the bags, just so you know.

Sandy: I know, I know that, but, or maybe they should have put a little more air in the bags.

I: Ah, okay, if they had done that, what do you think this number [test 1] would have been?

Sandy: 15.

I: Okay, got it.

Sandy: And for that one [points to test 5], 14.

(Source: SG, TRIAGE, Spring 2007)

Responses like Sandy's display a focus on getting as many seeds to germinate as possible rather than determining which factors (e.g., amount of water, salty water, sunlight, warmth) affect germination rates. Again, these responses are coherent when viewed through an engineering framework, which focuses on maximizing an outcome (e.g., getting all of the seeds to germinate). However, students weren't told that they had to get as many seeds to germinate as possible; they were told that these tests were done to see what affects germination rates. Students again were predisposed to employ an engineering strategy even when tasked to do otherwise, which lends support to the claim that students are not sure how to identify the opportune time for employing a scientific versus an engineering pursuit and indeed may confuse the pursuits of engineering and science.

We pause here to speak to potential concerns that may have arisen in the reader's mind with regard to our perspective on the importance of both scientific and engineering pursuits. To be clear, our point here is not to imply any diminution of engineering or the associated purposes therein. It is not our assertion that engineering is of lesser importance or somehow a handmaiden to a scientific purpose. Most assuredly, an effective engineer employs both scientific and engineering perspectives at times, and both are important to that field. Similarly, scientists also utilize engineering approaches at times as well as scientific ones in their work. Indeed, a great deal of overlap exists between these two domains. Our hope is that the literature and research described here are enlightening with regard to the complexities involved in better educating students on the primary purposes of both domains which, if not heeded, could threaten students' potential success as a budding scientist or engineer (or citizen).

WHAT ACCOUNTS FOR THE CONFUSION CONCERNING THE NATURES OF TECHNOLOGY AND SCIENCE?

Abstractions versus Concreteness

The output of scientific work takes the form of scientific models that are representations of systems existing in the natural world (Gilbert, 1991). In general, these models are abstractions requiring mental leaps of logic; mental manipulation of spatial relationships; conceptions of large spans of time, distance, speed, or their relationships to one another, etc. (Inhelder & Piaget, 1958). As such, the output of scientific work is, by and large, abstract and often intangible. In contrast, the public can actually see or directly experience the tangible output of technological pursuits such as new vaccines or wind turbines. Technological artifacts are, in general, less abstract than scientific ones.

Researchers in developmental and cognitive psychology (e.g., Inhelder & Piaget, 1958; Willingham, 2009) would argue that people generally come to understand abstract concepts only after becoming familiar with more tangible objects and experiences that may be used to represent the more abstract understanding. In other words, people more easily conceptualize ideas that are more concrete and familiar rather than abstract. Given that technology by its nature and pursuit is more concrete than are scientific models, it is understandable that students think more easily and readily in ways that are technological in nature, even when tasked to do otherwise.

How are Scientists Remembered? Consider a ranking of the most influential scientists past and present as reported in a book titled, *The Scientific 100: A Ranking of the Most Influential Scientists, Past and Present*, by John Galbraith Simmons (2000). In this list, Albert Einstein (described in the book as associated with atomic theory and 20th century science), Louis Pasteur (described as associated with germ theory of disease), and Galileo Galilei (described as associated with motion of celestial bodies) are ranked as the second, fifth, and seventh most influential scientists, respectively. When examining the work of such scientists, connections are evident between their science and subsequent technological innovations stemming from their research efforts. For example, Albert Einstein's theory of special relativity was applied to the development of the atomic bomb; Louis Pasteur's work on germ theory and disease transmission led to pasteurization. Although the work of these famous scientists was mainly scientific in purpose, the general public is far less likely to associate these persons with their science and much more likely to remember them in view of the innovations stemming from their scientific pursuits, even if the scientists themselves did not work at all on the associated technologies. Here again we see how science and technology are intertwined together, but not necessarily portrayed accurately. For instance, Einstein's revolutionary contribution did not and would not have conceivably come about from an effort to build a bomb or develop a new energy source. Rather, his sole concern was to better understand the natural world and his thinking was motivated solely by that goal. In other words, the knowledge that he developed also raised technological possibilities that could not have previously been imagined. And yet, the general public is unlikely to recognize this interplay and far more likely to remember the associated technologies than the basic science.

Science portrayals in the media. Media presentations arguably influence the public's perceptions of what is done by scientists and engineers. These portrayals often (albeit perhaps unwittingly) conflate the purposes of both science and engineering and perpetuate confusion. Such confusion is portrayed in various popular media sources, such as movies (e.g., Frankenstein, Back to the Future), newspapers, and television shows (e.g., shows focused on criminal investigations). For example, imagine a news story on a new invention, such as a vaccine or a telescope. When the media presents such stories, they often credit the invention to the work of a scientist—not an engineer or not a person who acted as both a scientist and an engineer. Likewise, when physicists at our institution recently published on a previously unreported pattern of asteroid activity, media sources reported on the story with the scientists pictured beside the tool that they used in their work—a telescope—perhaps unintentionally suggesting that the concrete telescope was fundamental to understanding the essence of the scientists' abstract asteroid model.

The comic below (Figure 1) recognizes this same problem as the "mad scientist" isn't working toward a scientific model or theory, but rather is working as an engineer to create some sort of technology—a death ray (Kulkarni, 2007).

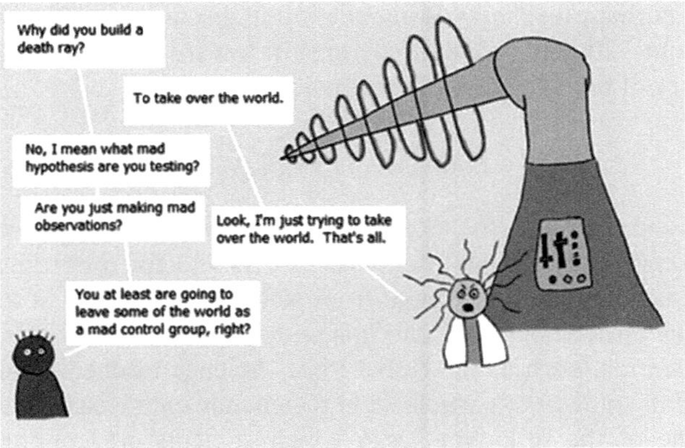

This is not to say that scientists in practice don't ever apply their findings or that engineers don't ever think about how the natural world works. Indeed, scientists often seek to apply their findings. For example, as scientists compete for grant funding to support their work doing pure or applied science, they often work to generate a compelling argument by calling out what they believe to be the possibility of technological gain (e.g., medical interests, environmental interests) made possible through their work. Similarly, engineers who are engineering effectively must first work out or select appropriate scientific models for a particular situation and then apply such model(s) for the purpose of generating a technological product or outcome. However, the point of emphasis here in this section is that invariably a scientist is portrayed in the media as working toward a technological end and not as one who is attempting to develop a scientific model or theory. While a more correct and nuanced portrayal would be cumbersome, it would be less likely to reinforce the false notion that science and engineering are interchangeable words and pursuits when this is not the case.

Science Teaching Methods Often Unintentionally Reinforce Misconceptions

Science classes, in general, are structured in ways that often inaccurately conflate the purposes of science and engineering. The goal of too many school science laboratory experiences is to achieve a desired result (e.g., build a bridge that holds the most weight) or to get the correct answer to within a certain percentage (e.g., determine the experimental value for the acceleration of gravity or determine the density of various metal objects). These sorts of activities are more in keeping with the goals of engineering rather than science (Harkema et al., 2009; Jadrich & Bruxvoort, 2011). While such classroom activities may have some merit, too often they are sources of misconceptions regarding the nature of science and technology, especially when they are referred to as "science activities" or, even worse, "science experiments" when that is clearly not the purpose perceived by students (and perhaps teachers). Without explicit discussion on these purposefully different pursuits, students are unlikely to shift toward meaningful understanding of the purposes of both science and technology.

Influence of Competition

Competitions and contests associated with science exist for the purpose of generating interest in science and technology-related careers and for rewarding stellar work in the field. Such competitions range from smaller events held at the local level, such as science fair competitions and invention conventions, to events that garner worldwide attention, such as the Nobel Prize. Science teachers commonly create contests for their students to participate in to generate enthusiasm in the classroom. Consider some of the following tasks which students are commonly asked to complete as "projects" or "challenges" or "assignments" in science classrooms: (a) design a launching device that keeps an egg in the air for the longest time without breaking the egg, (b) create a fan out of a single piece of paper (8 1/2 x 11) that completes the most revolutions possible, (c) design a catapult that launches a tennis ball a certain distance with certain accuracy, (d) design a series of simple machines that are connected and continue to set off one another, (e) design a machine powered by a mousetrap that goes the longest distance. Examination of these assignments illustrates a greater emphasis on a technological pursuit rather than science even though the competition is advertised as a "science competition" or "science contest." Such events may generate interest in science and technology while at the same time breed confusion in students' minds as to the distinctions between such pursuits.

WHY IS THIS CONFUSION A PROBLEM?

A number of reasons exist to substantiate why confusion with regard to the purposes of science and technology is problematic.

Perpetuating Misconceptions Related to Nature of Science

Confusion in students' minds with regard to the purposes of science and technology perpetuates misunderstanding of certain aspects of the nature of science. One misunderstanding is an undervaluing of basic science. This undervaluing manifests in a general public much more willing to support the potential for development of technologies and devices rather than basic science (Ryan & Aikenhead, 1992; Clough, 2004). For example, in 2007, the U.S. government spent nearly 375 billion dollars on research and development (R&D), and during this year, the fraction of the U.S. budget for R&D allotted to projects deemed as basic science was 18 percent, whereas 22 percent was dedicated to applied science and 60 percent to the development of projects (Boroush, 2007). Lost on the general public is the usefulness of basic science itself and, furthermore, its relationship to the development of technologies and devices (Clough, 2004).

Doing Good Science and Good Engineering

Doing good science involves understanding the purpose of science as the process of constructing models that account for events in the natural world (Gilbert, 1991). If students are confused, as we have been discussing, then judging if "good science" has occurred is predicated on whether or not positive "outcomes" are attained rather than fair analysis in search of a mechanism. In other words, evaluation of success would be based on the extent to which a maximum or expected output is attained rather than the importance of developing and refining a model or theory. In addition, completion of an engineering project should be achieved via a careful consideration of the important factors affecting an outcome (a scientific purpose) and then applying those models to derive a desired outcome (a technological purpose). However, if a person is outcome driven, then completion of such a project can be sought through employing an unsystematic, trial-and-error approach until the desired outcome is achieved, thereby bypassing use of a proper engineering approach. (Indeed, that is the unfortunate outcome for many students when tasked with "scientific" challenges such as building a mousetrap-powered car or maximizing the range of a catapult.) Engineers disparage unsystematic, trial-and-error approaches, labelling them pejoratively as "tinkering" or "inventing."

Accurately Understanding the Conditional Nature of Scientific Claims

Misunderstanding the purpose of science likely yields confusion with regard to the conditional (but durable) nature of scientific claims. Such confusion manifests in the taking of an extreme position on this issue. One such extreme is the expectation that evidence *must* exist to prove scientific models and theories are true, and if such evidence doesn't exist, then the science under study is suspect and should be thrown out. For example, a person taking this position demands absolute scientific proof that evolution has occurred or that the earth is warming." The expectation is that absolute proof can and should exist, and if it does not, then the science under study can be dismissed. Alternatively, a second extreme position is the expectation that because scientific support exists, then the associated phenomenon will *absolutely* occur. A person holding this belief might argue, "Since oats reduce cholesterol, if I eat oats, my cholesterol level will definitely go down;" or "Tests have shown that acetaminophen can damage livers; therefore, if I take acetaminophen, I will damage my liver." Either extreme described here reflects a lack of understanding of the nature of scientific models and theories as human constructions that are inherently conditional (but durable) in nature (Gilbert, 1991). If one understood the purpose of science as generation of scientific knowledge with an inherently conditional nature, then perhaps such understanding could undo the tendency to act on either of the previously mentioned extreme positions.

Scientific Veracity is Deemed Negotiable as a Relative Commodity

Imagine one wants to send a message across a long distance. Many different options exist to accomplish this task. One could conceivably mail a letter, send an email message, send a fax, or make a phone call. All of these options would be valid applications of technology, because they all could be used to accomplish the intended outcome—relaying a message from one person to another. Your selection of one of these methods over the others might be based on many factors such as cost; timing; the personal affectations you wish to communicate; the availability of the technology desired; and your concern about the sustainability of natural resources, but you would *not* make your selection based on how well a particular technology seems to represent naturally occurring processes or how well it might be used to predict naturally occurring phenomena. Those things simply do not matter. If a technology allows you to achieve what you set out to accomplish within the constraints that are personal to you, then it is a valid option.

On the other hand, the products of science (models and theories) *are* constrained by how well they describe or predict natural phenomena. Although many different scientific models may exist for a single phenomenon, what all of those models must have in common is general agreement within the scientific community that they represent (or can be used to predict) that phenomenon "reasonably well" and in a "reproducible way." (By "reasonably" we mean to within some level of uncertainty that is generally acceptable to the scientific community invested in the study of that phenomenon. By "reproducible way" we mean that others within the community can independently achieve the same level of agreement between the model and the phenomenon.) If this level of agreement between the model (or theory) and the natural phenomenon does not exist, then the model or theory is discredited. Importantly, students operating with an engineering model of science will not come to this same conclusion.

For example, consider a classroom scenario recorded by Hammer et al. (2008) in which students were instructed to drop a book and a piece of paper to determine which falls more quickly. When the students presented their findings to each other, all but one student reported that the book fell faster. One child, Ebony, reported the opposite. He reported that the paper fell faster. (Unbeknownst to the teacher and the other students, a videotape of the activity revealed that Ebony had placed his paper under the book that he dropped, thus ensuring that the paper hit the ground first. All the other students had dropped their books and papers independently of each other; as the teacher had assumed that they would.) The students' immediate reaction to Ebony's report was disbelief, because they had all obtained a different result. However, neither the students nor the teacher asked Ebony to repeat his test, and in the end the class concluded that his result must be as valid as their own.

Now, if one considers this classroom scenario from a scientific perspective, then the accord between the class and Ebony is unjustified. Scientifically speaking, everyone should observe the same result if they all do the test in the same way, because it is assumed that scientific results are reproducible. Therefore, the class should have asked to see how Ebony had done his test in order to get to the bottom of the discrepancy. They did not, and therefore they ceased to do science.

On the other hand, if one considers this scenario from a technological perspective, then the students' acceptance of Ebony's result is quite reasonable. Since Ebony *did* get the paper to hit the ground first (and assuming that he is not lying), then his result is valid. Technological success is judged based on the attainment of an outcome, such as successfully sending a message, and not on the extent to which you have developed a model or theory of a natural phenomenon. Reproducibility is unimportant, because everyone can have their own way of doing things, and scientific models, then, become relative to the individual. The pedagogical danger here is obvious. The conclusion students may reach when they confuse the technological perspective for the scientific one is that ideas in science are personal and relative – not open to external critique or consensus. Personal belief and unsubstantiated opinion carry as much weight as scientific models and theories that have been rigorously scrutinized and tested. In the end, scientific models and theories are judged based more on the individual's likes and preferences than on how well they represent and predict natural phenomena.

ADDRESSING HOW TO MINIMIZE THE CONFUSION

A number of changes to typical science instruction could serve to minimize confusion in students' minds with regard to the purposes of science and technology. These changes include explicitly drawing students' attention in the science classroom to the development and assessment of scientific models, altering the objectives of many laboratory science activities, and explicitly teaching the purposes, interactions, and value of both science and technology.

Progress towards disentangling the purposes of science and technology in students' minds may be made if science teachers (and perhaps technologists) explicitly teach on the development of scientific models and theories as the primary purpose of science itself. Important discussions on the characteristics of scientific models and theories, their limitations, the need for consensus of acceptance, the existence of multiple models and theories, the inherent incompleteness of scientific models and theories, the importance of model revision through continued testing (i.e., Is my model still useful and working given the evidence I have?), and acceptance of a certain amount of unavoidable error would serve to dispel false notions about science and its purpose. Further, students ought to spend time where appropriate developing and using models themselves in the context of learning fundamental science content (e.g., magnetic behaviour, solubility, particle motion). Such model development would involve explicitly challenging current models by seeking disconfirming as well as confirming evidence that would present science content development in a way that is consistent with how scientists themselves conduct investigations (Harkema et al., 2009; Jadrich & Bruxvoort, 2011). In addition, when designing laboratory experiences, careful attention to the language used when posing research questions and satisfactory criteria for assessing ideas developed to answer those questions would serve to emphasize the distinctions between science and engineering. For example, rather than telling students that today's laboratory will show how increasing temperature increases reaction rates (outcome-focused), students could be told that they will be investigating the questions, "Are reaction rates a function of temperature, and how will our results impact our model for molecular dynamics?" The latter question emphasizes development of a model (science) rather than achievement of a particular, expected outcome. When students are told what should happen in a "scientific investigation" and then asked to verify this expectation, they are not working in a manner that is consistent with how scientists themselves work in practice. Careful balance in science classrooms needs to exist between outcome-focused activities (e.g., design a mousetrap car) and investigations which pursue models (e.g., Does germination of radish seeds depend on light, and what does this tell us about carbohydrate storage in seeds?). All the while, when activities of any kind are occurring in science classrooms, teachers need to make more explicit mention of the goals of technology and science, the interrelationships between these two pursuits, and consistencies between these said purposes and the specific activities going on.

We are at a crossroads in science and engineering education where an emphasis on improving students' understanding of science and technology is of utmost importance on the national and international scene. (See, for example, the new national science framework in the U.S. that explicitly incorporates ideas of engineering into the science curriculum.) Students' adeptness at understanding what is considered valid "scientific practice," done with a scientific purpose in mind, as well as what is considered valid "engineering practice," done with a technological end in mind, will serve them well as they try to ascertain the advantages and disadvantages of either pursuit. The hope in the final analysis is that students would move away from thinking of science and engineering as interchangeable and move toward understanding the goals of both, how they interact, and explicitly deconstructing the way both shape civilization. Only then can prudent decisions be made to shape society as we wish, rather than blindly following where our science and technology lead us.

ACKNOWLEDGEMENTS

Dr. Rachel Reed, the National Science Foundation (NSF AYS Grant: 0639694), Van Andel Education Institute (Grand Rapids, MI), Samuel Sportel, Jennifer Koning, Allyson Green, Nathan DeHaan, Jeff Bucchianeri, and Katie VanderHeide.

REFERENCES

American Association for the Advancement of Science (AAAS). (1990). *Science for all Americans.* New York: Oxford University Press.

American Association for the Advancement of Science (AAAS). (1993). *Benchmarks for science literacy: A project 2061 report.* New York: Oxford University Press.

Boroush, Mark. (2007). National patterns of R&D resources: 2008 data update. Data Brief, National Science Foundation (NSF 10–314). Retrieved 1 July 2011 from http://www.nsf.gov/statistics/nsf10314/content.cfm?pub_id=4000&id=2

Clough, M. P. (2004). The nature of science: Understanding how the game is played. In J. Weld (Ed.), *The game of science education* (pp. 200–227). Boston, MA: Pearson Education Inc.

Constantinou, C., Hadjilouca, R., & Papadouris, N. (2010). Students' epistemological awareness concerning the distinction between science and technology, *International Journal of Science Education, 32*(2), 143–172.

Dunbar, K.N. (2002). Understanding the role of cognition in science. In P. Carruthers, S. Stich, & Siegal, M. (Eds.), *The cognitive basis of science* (pp. 154–170). Cambridge: Cambridge University Press.

Gardner, P. (1994). Representations of the relationship between science and technology in the curriculum. *Studies in Science Education, 24*(1), 1–28.

Gilbert, S.W. (1991). Model building and a definition of science. *Journal of Research in Science Teaching, 28*(1), 73–79.

Hammer, D., Russ, R., Mikeska, J., & Scherr, R. (2008). Identifying inquiry and conceptualizing students' abilities. In R. A. Duschl & R. E. Grandy (Eds.), *Teaching scientific inquiry: Recommendations for*

research and implementation (pp. 138–156). Rotterdam: Sense Publishers.
Harkema, J., Jadrich, J., & Bruxvoort, C. (2009). Science and engineering: Two models of laboratory investigation. *The Science Teacher, 76*(9), 27–31.
Inhelder, B., & Piaget, J. (1958). *The growth of logical thinking from childhood to adolescence.* New York: Basic Books, Inc.
Irwin, D. M., & Moore, S. (1971). The young child's understanding of social justice. *Developmental Psychology, 5*, 406–410.
ITEA (International Technology Education Association). (2000). *Standards for technological literacy: Content for the study of technology.* Reston, VA: International Technology Education Association.
Jadrich, J. & Bruxvoort, C. (2011). Learning & Teaching Scientific Inquiry: *Research and Applications. Arlington, VA: NSTA* Press.
Koslowski, B. (1996). *Theory and evidence: The development of scientific reasoning.* Cambridge, MA: Massachusetts Institute of Technology Press.
Kuhn, D., & Phelps, E. (1982). The development of problem-solving strategies. In H. Reese (Ed.), *Advances in child development and behavior* (pp. 1–44). New York: Academic.
Kulkarni, S. (2007, December). 46. Death ray. Retrieved July 23, 2009, from http://cowbirdsinlove.com/46
McComas, W.F. (1996). 15 myths of science. *School Science and Mathematics, 96*(1), 10–16.
NRC (National Research Council). (1996). *National science education standards.* Washington, DC: National Academy Press.
Piaget, J. (1932). *The moral judgment of the child.* New York: Harcourt Brace.
Roy, R. (1990). The relationship of technology to science and the teaching of technology. *Journal of Technology Education, 1*(2), 5–18.
Ryan, A.G., & Aikenhead, G.S. (1992). Students' preconceptions about the epistemology of science. *Science Education, 76*(6), 559–580.
Schauble, L. (1990). Belief revision in children: The role of prior knowledge and strategies for generating evidence. *Journal of Experimental Child Psychology, 49*, 31–57.
Schauble, L., Glaser, R., Raghavan, K., & Reiner, M. (1991). Causal models and experimentation strategies in scientific reasoning. *Journal of the Learning Sciences, 1*(2), 201–238.
Schauble, L., Klopfer, L. E., & Raghavan, K. (1991). Students' transition from an engineering model to a science model of experimentation. *Journal of Research in Science Teaching, 28*(9), 859–882.
Seok Oh, P., & Jin Oh, S. (2011). What teachers of science need to know about models: An overview. *International Journal of Science Education, 13*(8), 1109–1130.
Simmons, J.G. (2000). *The Scientific 100: A Ranking of the Most Influential Scientists, Past and Present,* New York, NY:Kensington Publishing Corp.
Willingham, D. T. (2009). *Why don't students like school?* San Francisco, Jossey Bass.

James Jadrich
Department of Physics and Astronomy,
Calvin College

Crystal N. Bruxvoort
Department of Chemistry and Biochemistry,
Calvin College

CHAPTER 21

JOHN T. SPENCER

TECHNOLOGY CRITICISM IN THE CLASSROOM

I first heard about a tragedy in Tucson, not from major television news networks, but from a direct message sent by a politically-active friend who was attending the political gathering where a mass shooting took place, including the shooting of an Arizona congresswoman, Gabrielle Giffords. While the television news sputtered around trying to offer details (initially wrongly claiming that she was dead, likely from pressure to be the first to report big news), I found myself reading Google News, piecing together Facebook posts, e-mailing friends and reading Twitter updates.

I turned the television off when the same information was repeated—the news helicopter circling the same grocery store and reporters using the same peppy intonation they would use to announce the final score of a basketball game. While I couldn't see the "expert witnesses" using my methods, I could keep up on National Public Radio's website and then ask medical questions to a friend of mine who is a trauma surgeon.

Meanwhile, I engaged in philosophical conversations with friends and family members about speech and freedom and safety. I watched the tone turn ugly at times and I found myself caught up in it as well. Yet, when we learned of the death of a nine year old girl in that shooting rampage, the tone of the conversations changed. My response was to post on my blog. I probably put things online too soon, but the online environment is where I went. Students sent me messages asking if that was the same Gabrielle Giffords we had interviewed in class. They sent links to YouTube videos and asked hard questions about insanity, justice and the universe itself.

Then I turned it off. All of it. I took a break from Twitter and from Facebook and from YouTube and Google News and Blogger and I walked outside and played baseball with my sons. My son pulled me aside and asked, "Was that real or was it just TV?"

So, how would I handle that in the classroom? I would grieve with the students and ask questions and we would blend social media and face-to-face conversation. We would share our emotions, our thoughts, the information we find, the bias we see and together we would try to piece together the story and how those events relate to our lives. Dealing with traumatic events is not something I can organize in advance.

At one point, I asked my students the question that my son had asked me. "Was that real or was that just the media?" We discussed the nature of reality, of truth, of

Michael P. Clough, Joanne K. Olson and Dale S. Niederhauser (Eds.), The Nature of Technology: Implications for Learning and Teaching, 427–440.

tragedy when we feel it locally but also at a distance. We argued about whether the media had enhanced or inhibited human connection.

Many ways exist to be social, and children will choose different methods to maneuver social media. My middle school students are learning, not only how to maneuver social media, but how to think critically about information they encounter. Far too often, children's formal education in technology is about how to use the technology, and if issues are raised that transcend skills, those issues are usually limited to being safe online (or a being a good online citizen) and evaluating website credibility. Images conveyed through modern media blur the distinction between real and "just TV," and we do our students a disservice if we do not address issues in the philosophy of technology and the ways that the media often unknowingly shape our thinking. This chapter illustrates how I raise such issues with my students. In order to make clear that I do not compartmentalize technology criticism into a single unit, I purposely convey my approach to teaching technology criticism in a non-systematic manner. What follows is intended to make clear how I integrate such criticism in the context of what is being studied at any given time.

MEDIA NON-NEUTRALITY

Students walk into my classroom with a sense of excitement. Many of them approach the computers, netbooks and iPods through a lens of entertainment. A few recognize that these will be learning devices. Fewer still understand the negative effects of technology. Therefore, I ask students to complete a technology literacy survey, covering attitudes, beliefs, self-efficacy, experience and skills related to various media. Within the last two years, less than three percent of my students have questioned when it is wise or unwise to use technology. The annual data reflects a larger socio-cultural paradigm of an uncritical embrace of technology. My students' responses and the larger societal context have been the impetus for an intentional, curriculum-embedded approach to technology criticism.

Initially, students struggle to recognize that the power, convenience and efficiency offered by multimedia devices do not necessarily mean an increase in the quality of work, the authenticity of an interaction or the effectiveness of an endeavor. Soon students are able to assess the pros and cons of technology through debate, discussion, reflection and brainstorming. One approach we have used is to engage in a Twitter chat about technology, followed by an in-person debate. Afterward, small groups discuss the pros and cons of both approaches. Other times, we use art work, videos, podcasts and blog posts about the nature of power, technology and "capturing" life through a medium. After awhile, technology criticism becomes another filter that students use when analyzing literature, history, math contexts and science experiments. Suddenly the visuals from Hiroshima and Nagasaki, the tale of John Henry, biographical sketch of Henry Ford and the student-created solar ovens take on a new meaning.

However, it is not enough to simply evaluate the trade-offs of a medium. For a deeper level of technology criticism, students need to internalize and articulate the paradoxical nature of technology and its impact on society.

When students first bring their iPods and the class takes out their Chromebooks, I ask them how they have used the technology. After four years of taking an informal survey, just two out of four hundred students have used devices for photo editing, film-making or blogging. None of them have collaborated on a project outside of their local context. We talk about the business models of Google, Apple and Amazon and the notion of a "consumer" device. From there, we discuss the notion of "hacking" and ask what it would mean to create rather than consume. Students engage in a "gallery walk" where they brainstorm ways they could use the devices creatively.

As students begin to rethink their devices, I start to model questions about the limitations of the devices. In Twitter chats, blog posts, videos and live debates, students talk about how technology is changing the social context in negative and positive ways. I add a second layer of reflection as students discuss why they chose their specific technology tool and how it would vary in negative and positive ways if they had used a more traditional tool. When we turn the devices off, students confess that they didn't pay as close attention to body language and they felt anxious with the constant moving of text and images. I'm never entirely sure where the discussion will go, but students are often poetic about the loss of humanity that they experience. One girl said, "My mom looks at her screen more than my little brother. She holds it up to her face when she should be holding him. It bothers me." Students lament the loss of context and physical geography. They lose the audible contact with surrounding events. This is a middle-ground position that avoids an unrealistic fear of technology at one extreme, and unquestioning technophilia on the other. These devices are not neutral; we always lose something, and we are often unaware of what we lose since we are usually focused on what we gain. This is a central concept my students need to understand if they are to make informed decisions about how to spend their time and what technologies to use.

TECHNOLOGY DECENTRALIZES AND CENTRALIZES POWER

We begin with the question, "In a globalized society, who holds more power: corporations or nations?" Students post their answers first to their blogs and then move to a class Twitter chat. A few students choose to read one another's blogs and leave comments. One group debates this issue in-person, while attempting to find facts online to support their positions.

> "Who owns the school?" I ask the students.
>
> "The community. The taxpayers, right? It's public, so it belongs to us. That's why I say a country still has more power. Technology can pull us away, but the government can coerce us to be here the entire time?"
>
> "But were you really here?" another student asks. "You were on Blogger and Twitter."

"Maybe it's an issue of who had more power if you were two places at once."

"I think Mr. Spencer had the power. He was the one who gave us the question and told us that we had to answer it."

"Yeah, but who organized the information?"

"True, Mr. Spencer didn't limit us to one-hundred and forty characters."

"Yeah, but we organized it ourselves," another student suggests.

Thus, before viewing power through military or political force, students are able to analyze the role of a medium in shaping, organizing and censoring one's thoughts. Students analyze the organizational structure of Facebook and Twitter through a sociological and anthropological framework. The class reads a journal article about the potential dangers of auto-fill in search.

In the process, students analyze the business and ideological models of various technology platforms. If Apple is essentially a hardware, software and multimedia company, how do they benefit from closed information systems? How do they influence a classroom space? If Google is essentially a transnational advertising agency, how do they benefit from customized search and auto-fill? How does "free" become a deceptive phrase? Students analyze the oft-contentious political battles that exist within democratic movements of Linux, open source and Creative Commons. It is important for students to recognize the social and political forces that shape the development of any media platform.

Students analyze the Roman concept of *panem et circenses* (bread and circuses) as they compare and contrast methods of population control in *Brave New World* and *1984*. In the process, students develop a metaphor for smart phones. The most common metaphors include drugs and weapons, suggesting that students see technology as inherently dangerous, but also necessary. One student points out, "Do we really need a better metaphor for an iPhone than an apple. That forbidden fruit that lets us have the whole world in our hands. We have knowledge of good and evil, but we use it to hide."

Students use a historical framework by analyzing the role of the printing press in the rise of the nation-state, asking whether an Enlightenment worldview, the rise of a modernism and the development of a nation-state were tied directly to the instant access and dissemination of language-specific texts. Students analyze the symbiotic relationship of an Enlightenment philosophy and the printing press. They begin to recognize that those who own the technology often own the collective voice and the public memory.

Students analyze the ways in which democratic movements of the Arab Spring utilized social media to spread democratic ideas and the ways that official state-run media continue to stifle many of the reforms they once hoped for. We get into a debate about whether the medium created the democratic impulse or if it was simply a product of it.

ANTICIPATE THE UNPREDICTABLE

When students shift from uncritical acceptance of technology toward a more critical approach, our class grapples with a litmus test for implementing a new medium. If we are not careful, students will fall into the fallacy that humanity can use technology wisely by predicting its costs and benefits in advance. However, a brief glimpse into the last century suggests a failure in the human imagination to predict the costs of technology. Scientists had an accurate assessment on the destructive capacities of splitting an atom. However, many in society failed to grasp the larger social fallout from a world changed by the existence of nuclear warfare. Students watch a haunting video of Openheimer saying, "We have become death," and suddenly the raw power of technology becomes something they are forced to wrestle with.

For this reason I require my students to study ancient mythology in connection to the role of technology in society; this is an ideal context to address the social consequences of technology. Students debate the themes of the biblical story of the Tower of Babel, in recognizing that we are speaking a binary language to collaborate on towers that move beyond the clouds, with satellites offering instant access to trans-geographic communication. Some students see the story as a sacred text speaking boldly to our culture. Others see it as a paranoid, tribalistic story that we can now move past as we embrace technology.

Students answer the question, "To what extent have smart phones and social media improved the way we communicate?" This is followed by questions of what we have lost in constantly filtering communication through a technological medium. Students also discuss the unpredictability of technology in the stories of Pandora and Prometheus (both used as names in technology companies). From there, the class studies instances when communities have anticipated the pros and cons of a medium in an effort to use a medium wisely.

Students then analyze current social issues related to social media, including: the loss of privacy within a culture of self-surveillance, social castes based upon digital inequity, inability to empathize, deterioration of wisdom in an age of instant information, speech without context, the push toward being audacious in order to gain a fringe market (also the push toward audacity to gain attention among peers) and the loss of identity with personal branding and informational overload. Afterward, students analyze articles from the early stages of social media with the driving question, "How well did the collective imagination at the time anticipate the issues we are now dealing with?"

Finally, students make predictions regarding nano-technology, the Singularity movement, genetic engineering and artificial intelligence. It is difficult for them to see that their failure in imagination will most likely be on the side of overly conservative estimations of change.

As we create the digital boundaries for our own class, I ask students to keep in mind that the true costs and benefits of using various media are often unpredictable.

However, it is better to anticipate potential changes and monitor for unanticipated changes than to blindly accept technology. Students examine case studies of both neo-Luddite and technocratic communities as we attempt to create a class-wide litmus test for technology usage. As the year progresses, students examine how multimedia shapes our class sense of ethics, privacy, voice, identity and community. In this sense, the class becomes a microcosm of the larger global debate regarding the effects of technology on our sense of humanity.

TECHNOLOGY AS A HUMANIZING AND DEHUMANIZING FORCE

It's a myth of modernity that humanity can simply wield technology in such a way that we access all the benefits while avoiding all the drawbacks. Often, teachers use the term "tool" as a metaphor for technology integration, assuming that the user is the one responsible for the success or failure of the tool, and that the tool is simply neutral. However, technology is perhaps better conceptualized as a double-edged sword that shapes the user along the way.

It is easy for students to identify the man versus machine motif in both classic and contemporary stories. The most accessible starting point has been the question of whether the human spirit truly triumphed in the legend of John Henry. From there, students analyze a persuasive piece suggesting that a deeply human endeavor is the creation of tools and use of them to transform our sense of reality. As a result, students begin to question the paradox of technology as a humanizing and dehumanizing force (often at the same time).

I also have students analyze a video clip of Oppenheimer stating, "We have become death, the destroyer of worlds" (http://www.youtube.com/watch?v=26YLehuMydo). The question is raised whether the nuclear bomb Oppenheimer speaks of was an inhuman invention, and whether his words would have been different had the interview not been filmed for posterity.

In order to recognize the larger historical context, students read about the loss of oral language in the movement toward print (through a Socratic dialogue recorded, perhaps hypocritically by Plato). Students also analyze the role of technology in saving lives alongside the costs of such life-saving efforts with the Green Revolution, hydroelectric power and modern medicine. While choosing their own multimedia format, students then communicate the extent to which technology is humanizing and also dehumanizes.

The question, "Does this change our sense of humanity?" is an essential question students ask throughout the school year. As students write blog posts, they address how their writing changes without the use of hand-written drafts and questions about what is lost on a personal level when we no longer have an identifiable handwriting style. When they use concept maps, we ask whether this is an authentic reflection of a mental process. When students record their voices, they ask about the potentially dehumanizing effects of compression, editing and surveillance in the auditory media.

Technology criticism varies in its level of explicitness. After a robotics unit, I ask students to analyze how they unintentionally humanized the robot. They ask about the gender assigned to the robots and the social implications of treating machines as humans. They examine the algorithms we use to define our choice processes in search engines, internet media players, and "relevance" selectors among social media. Students also watch selected interviews from proponents of the Singularity Movement.

When learning about personification, I ask students to write letters to personified technology. While many students initially scoff at this idea, they soon realize how often they have personified technology throughout their lives. The letter-writing assignment helps students come to terms with the often-intimate relationship they have with their technological devices. As one student writes, "Oh Android, you are becoming more human than you realize. Who receives more attention than you? Who is allowed closer to my lips? I keep an arms distance from others, but we are constantly holding hands."

TECHNOLOGY REFLECTS AND CONSTRUCTS TRUTH

I ask students to view a picture of Abraham Lincoln and read a primary source description of the former president. Afterward, students use Twitter to discuss which method is more accurate.

"The picture can't be altered, but you can always change your words," a student writes.

"I think a picture is more accurate, because it shows you rather than telling you."

"Yeah, you can see Lincoln. It's more accurate."

"I don't know. He's posing for the camera. Maybe it's actually more fake?" a student asks with a careful question mark.

We move from Twitter to a short class discussion. The students are shocked to find that the picture is an altered photograph, often displayed in textbooks without any explanation of the disingenuous image.

"So, if we can manufacture truth, how do we know something is real? How do we know what is true?"

After the short discussion, I ask students to respond to a discussion question on our class blog: In what ways does the medium itself fail to capture reality compared to oral or written language? It takes a few minutes of think time before they point out the danger in the camera of framing reality and the lack of visual language to capture abstract concepts. Some suggest that symbols work just as well, while others see symbolism as being more dangerous than abstract writing. The discussion is part of a unit on primary and secondary sources, public memory and the way a medium transforms the stories we tell.

When students create a documentary, they compare and contrast the way the medium (video versus verbal) changes the way people answer questions. Students also discuss the role of the medium in making the documentary more entertaining and the question of whether in-depth information can be conveyed in a multimedia format. Moreover, the act of editing audio and video becomes a chance for students to see the dangers that the medium imposes in providing accurate information.

However, such criticism is not limited to social studies. In science, students can analyze the myth that pictures or videos "capture" the truth better, when in fact they often lead us to pay less attention to the natural phenomenon and oversimplify abstract concepts. For example, the spatial limitations of a diagram lead artists to de-emphasize the relative distance between the parts of an atom or the distance between planets. Thus the medium itself creates an inaccurate conceptual model that children internalize.

In math, students can examine the non-neutrality of data and the manipulation of numbers through graphical representations. Students can analyze graphs from the White House, Fox News, CNN and MSNBC for bias and propaganda. However, students can also analyze cultural attitudes toward statistics, the propensity to value quantitative over qualitative metrics and the worldview that results in an image-based, graphical mindset.

SOCIAL MEDIA AS A TRANS-GEOGRAPHIC AND GEOGRAPHIC FORCE

Social media exists as both a medium and a location. Society borrows from both geographical and procedural language when referring to social networks. One way to analyze the sense of space-less space is for students to create a semantic environment inventory. Students annotate a series of articles related to social media using one color to represent the media-related language and another to represent place-based language. I approach this as a lesson in vocabulary and expository text rather than technology criticism. Next, the class discusses the question, "Where am I when I'm on Facebook?"

From there, students identify the values, norms, tokens and customs of a specific online space and then compare and contrast it to their own location. Afterward, they compare and contrast social media interaction to print-based and oral interaction. Finally, students find a non-verbal way to communicate whether the media element enhances the social interaction and whether the social element improves the ability to express language.

Students analyze the sense of space-less space when they engage in a service learning, problem-based learning activity with other students across the globe. This leads to a discussion about the way the medium creates a difference sense of space while still allowing them to feel like they are in the current location. Each year, students analyze the culture conflict inherent in a collaboration method that allows users to slip into a sense of neutrality based upon the assumption that online space is somehow a non-geographic "other space."

SOCIAL MEDIA MAKES US TRANSPARENTLY OPAQUE

I ask my students, "How do you change when you interact online?"
"I can't be myself," a student says. I watch as the class nods in unison.
"Grown-ups are so concerned with cyber bullying. They think we're being awful online. And sometimes we are. Sometimes we post bad stuff to Facebook. But a lot of the time I feel like I hold back. I can't be myself."

"It's not just Facebook. It's life," another student responds.

"What do you mean?" I ask.

> "You never know when someone is videotaping you. It's like we're all celebrities in a reality TV show. But we're the viewers and we're the show."
>
> "So, you're saying it's a culture of surveillance. Is that such a bad thing? I mean, maybe people are being held accountable," I push back.
>
> "Not if they can't be themselves," a student responds.

When I ask students to craft a metaphor for their online identity, the most common answers are masks, brands, labels and gags. However, one student offers a more dramatic, *Harry Potter*-inspired metaphor, "Social media is my Horcrux. I get to be immortal. I get to be in two places at once. But there's a cost. I can't be myself. I've lost my soul."

Often digitial citizenship has been presented as a form of personal public relations, where students manage liability and promote a personalized brand. Rather than using media to develop their voice authentically, schools implore students to behave nicely and avoid leaving an offensive cyber footprint. As a result, students learn to hide online. Some students engage in passive-aggressive anonymous flaming and cyber bullying. Others present a squeaky clean self-image with the hopes of impressing future employers and college entrance screeners.

While a certain level of self-censorship occurs in all social contexts, the permanence and transparency of social media lead students to over-correct. What is gained in being pleasant is lost in the inability to be real. In a culture where anything can be recorded, mixed and displayed publically, we now have the potential to be entirely transparent in a way that forces people to be vigilantly opaque.

CRITICIZE WHAT YOU EMBRACE AND EMBRACE WHAT YOU CRITICIZE

In order to think critically about technology, students need to avoid the polar extremes of blind acceptance or absolute rejection of a medium. However, instead of looking for a Hegelian synthesis or even a "happy medium," I ask students to approach technology dualistically as both technophiles and neo-Luddites. I present these concepts with two prototypes: the geek and the guru.

The geek is someone who not only uncritically embraces technology, but also uses multimedia tools to create something innovative. Geeks can solve social and political problems through the use of new technology. They are creative, passionate and interested in moving society forward through the constant integration of new tools. In contrast, the guru is someone who thinks critically about technology, who values tradition who embraces vintage ideas, who wants to know how technology changes society and who weighs what is gained and lost in using technology. Geeks run the risk of missing the social impact of technology and constantly "progressing" without defining a sustainable meaning of progress. Gurus run the risk of being irrelevant, cynical and unable to recognize the benefits of new technology to humanize society.

By embracing both concepts, students must wrestle with a certain level of cognitive dissonance. However, seeing the two sides as paradoxical and complementary, students are able to make sense out of the aforementioned paradoxes of technology. I do not approach technology criticism systematically or relegate it to a singular unit of study. Instead, it is an integrated approach; like reading and writing across the curriculum. I choose this method, because I want students to engage in technology criticism within a context and thus avoid compartmentalization.

The end result is often messy, confusing and disorienting to students. However, it is also authentic, holistic and organic in nature. By embracing the nuance of each paradox, students recognize both the limitations and the opportunities of each medium. In the process, they not only learn to be digital citizens, but they move toward being critical thinking, democratic citizens, better equipped to make sense of a rapidly changing globalized world.

APPENDIX C

ADDITIONAL TOPICS AND GUIDING QUESTIONS WHEN TEACHING DIGITAL CITIZENSHIP

Digital citizenship is more than simply playing it safe, just as a democratic citizenship is more than wearing an "I Voted Today" sticker or chanting slogans. The following are examples of the kinds of questions I ask students in order to raise nature of technology issues inherent in everyday technologies. At first, I sound like Red from *One Flew Over the Cuckoo's Nest*, but eventually students catch on to the kind of thinking I am promoting.

Social Networks
How do sites like Myspace and Facebook shape how we interact with each other? How do those sites make money? Are we becoming desensitized by advertising? Have we made social interaction into a commodity?

Digital Identity
In what ways do you create a digital identity for yourself? What are some of the dangers in being transparent? What are some of the dangers in being anonymous? Are we becoming more image-conscious? Does this make us more arrogant? Are we losing what it means to be human? How does the constant obsession with "new" cause us to mistake novelty for importance?

Search Engines
How does auto-fill change the way we think? What are the dangers in allowing a computer or an algorithm organize our thoughts?

Images
How do images shape your view of concepts? Are pictures more accurate than words? What are the dangers in photo-editing software and our ability to believe what we see? Is a "made-up" picture less real than what you actually saw (especially if your mind is able to misrepresent it as well)? What are the dangers in "capturing" life on camera? Are there people, places or ideas that should not be "captured" on camera? Does the use of digital photography make people less careful about the pictures they choose to take? Does the quantity change the quality?

Video

How do people change when they are on video? What are the dangers of having to be entertaining? In what ways do we live in an entertainment culture? What are the costs of editing a person's words and chopping it up? How does the narrative change? In what ways does the act of video force people to be more amusing? Do Americans trust pretty people more than ugly people as a result of the video-culture demanding good-looking people for things like news and talk shows?

Music

Is the album dead? Is that a good or a bad thing? Are songs going to get shorter or longer as a result of digitization? Do you think the instant availability of recording technology will increase or decrease the overall quality of music? Do you ever feel like you know a lot of songs, but don't know any songs really deeply? Does music have more or less power when it is portable? People listen to music in isolation. They used to listen to it in groups. What is the purpose of music? Why do you think that previous generations have been said to be defined by their music? What did we lose in the process of digitizing music? We have no shared canon of music. What does that mean for our ability to have collective storytelling as a culture?

Intellectual Property

Does creative commons actually destroy innovation? If property should be shared, why not resources? What makes an idea "yours" in a world where so many ideas are synthesized and customized so quickly? What are ways you can be careful about respecting intellectual property?

Wikis

How do wikis fail to safeguard against errors? What are the dangers in wiki anonymity? What are the benefits of a wiki? How does your voice change when you write a wiki?

Blogs

How do people change their tone of voice or their style of writing when it becomes public? How does the structure of the blog change the length that a person writes? If we can easily edit blogs, does that make us more careless in choosing words than if it were on paper? What is the downside of a society where everyone can be a blogger? Is there a danger in a world where anyone can be "right" and no one has to be an expert? What are the dangers of libel? Do most bloggers consider the credibility of their sources? Is a blog a publishing tool or a communication tool? If people can comment at any time and the conversation isn't bound to time or space, what do we sacrifice in terms of space and presence? How does that shape our communication?

Communication Tools

If anyone can access you at any time, are you ever really present when you are with someone? How do communication tools make us more human or less human? Are people lonelier when they are more connected? Or does the instant connection allow people to feel a deeper sense of connection to people? How have communication tools changed our syntax? our grammar? our vocabulary? What is more real to you: an instant message or a face-to-face conversation? Why does it seem like we're not talking as much anymore?

Information

Does the instant availability of information change how we view truth? In an age where it's so easy to manufacture and publish lies, is there any way to know what is true? How does a website's structure affect your ability to decide if it is true? Is it possible to have too much information? What happens to the value we place on knowledge if it is so readily available? Are we getting smarter or dumber, or do we simply think differently than before?

Cyber Footprint

How does your online identity and interaction live on even after you have deleted it? Will that change how you interact online? Is it worth the lack of privacy in order to access the convenience of "living in the cloud?" Have you made mistakes that are now recorded online? How does that make you feel?

Operating Systems

How do operating systems manipulate you? How have you changed the way you think based upon the desktop environment you use? In what ways does your computer itself change your attention span? Is it true (or simply a myth) that operating systems are designed to make people multi-taskers? Have computers changed our work ethic?

Social Decorum

What is the tone of comment posts that follow news items, videos, and blogs? Why does the tenor of these conversations so quickly become toxic? How does the nature of the medium enable this to occur? Why wouldn't such comments be tolerated in a face-to-face setting?

John T. Spencer
Middle School Teacher,
Phoenix, AZ

AUTHOR BIOGRAPHIES

Warren J. Blumenfeld, Ed.D., is an associate professor in the School of Education at Iowa State University in Ames, Iowa specializing in Multicultural and International Curriculum Studies; & Lesbian, Gay, Bisexual, Transgender, and Queer Studies. He is author of *Warren's Words: Smart Commentary on Social Justice*; co-editor of *Investigating Christian Privilege and Religious Oppression in the United States;* co-editor of *Readings for Diversity and Social Justice; Editor of Homophobia: How We All Pay the Price*; co-author of *Looking at Gay and Lesbian Life*; Co-editor of *Butler Matters: Judith Butlers Impact on Feminist and Queer Studies*; Author of *AIDS and Your Religious Community*; & co-researcher and co-author: *2010 State of Higher Education for Lesbian, Gay, Bisexual, and Transgender People*. He also serves as an editorial blogger for The Huffington Post.

Crystal Bruxvoort is an associate professor of science education and chemistry at Calvin College where she teaches undergraduate science content courses for elementary education, introductory chemistry, secondary science methods (I and II), secondary student teacher supervision, and graduate education courses on the nature of science, scientific reasoning, and conceptual change. Her scholarship focuses on the nature and development of scientific reasoning and its implications for science learning, teaching, and teacher education. She is the co-author of a book published by NSTA Press titled, *Learning & Teaching Scientific Inquiry: Research and Applications.*

Michael Bugeja, director of the Greenlee School of Journalism and Communication at Iowa State University, is author of *Interpersonal Divide: The Search for Community in a Technological Age* and *Living Ethics Across Media Platforms*, both by Oxford University Press and both winners of the Clifford G. Christians Award for Research in Media Ethics. He is a frequent contributor about technology and education to the Chronicle of Higher Education and Inside Higher Ed. His scholarship focuses on technology's impact on society.

Michael P. Clough is an associate professor of science education at Iowa State University where he teaches The Nature of Science and Science Education, Secondary Science Methods I, Secondary Science Methods II, and Restructuring Science Activities. He is the recipient of several awards for his teaching (at both the university and secondary school level), scholarship and service. His scholarship is directed at the nature of science and its implications for science learning, teaching, and teacher education; and the synthesis, criticism, and clarification of extant knowledge and research in science education. He currently serves as president of the International History, Philosophy and Science Teaching (IHPST) organization.

Michael Todd Edwards is an associate professor of mathematics education at Miami University. He is the co-editor of the *Ohio Journal of School Mathematics* and the *GeoGebra North American Journal* and co-director of the GeoGebra Institute of Ohio. Dr. Edwards' research interests focus on the teaching and learning of school mathematics with technology (specifically dynamic mathematics software), ethical issues surrounding the use of free software and the free software movement, and writing as a vehicle to learn mathematics at all levels of instruction.

Suzanne R. Harper is an associate professor of mathematics education at Miami University where she teaches various content courses for prospective mathematics teachers including *Mathematical Problem Solving with Technology*, *Secondary Mathematics from an Advanced Perspective*, and *Geometry for Middle Childhood Teachers*. A common theme that permeates her teaching and research activities is the development of effective uses of technology in the teaching and learning of mathematics. Dr. Harper has received several awards for her teaching, scholarship and service, including the National Technology Leadership Initiative Fellowship.

Benjamin C. Herman is an assistant professor at the University of South Florida where he teaches Communication Skills in the Science Classroom, Reading and Communication in Science Education, Teaching Secondary Biology, The Philosophy and Nature of Science, and Theories and Practice of Science Teaching and Learning. Ben's research is primarily directed at promoting accurate and effective nature of science instruction and factors impacting teachers' NOS instructional implementation. His scholarship also addresses socio-scientific issues instruction and environmental education.

Gordon Hull is assistant professor of philosophy at UNC Charlotte, where he teaches courses at the intersection of philosophy, law and technology, as well as in political thought more generally. He is the author of *Hobbes and the Making of Modern Political Thought* (Continuum, 2009), and has published articles both in the history of philosophy and on contemporary issues. His current research focuses on issues in intellectual property and privacy raised by new information technologies, and he is currently working on a book tentatively titled *The Biopolitics of Intellectual Property*.

James Jadrich is a professor of science education and physics and the chair of the Science Education Group at Calvin College, where he teaches undergraduate science content and methods courses for elementary education students and graduate education courses on scientific reasoning, conceptual change, and the nature of science. He is a recipient of Calvin College's Presidential Award for Exemplary Teaching. His scholarship is directed towards the nature and the development of scientific reasoning. He is the co-author of the NSTA Press book, *Learning & Teaching Scientific Inquiry: Research and Applications*. He is active in science teacher professional development in the US and West Africa.

David H. Jonassen was Curators' Professor at the University of Missouri where he taught in the areas of Learning Technologies and Educational Psychology. After earning his doctorate in educational media and experimental educational psychology from Temple University, Dr. Jonassen taught at the University of Missouri, Pennsylvania State University, University of Colorado, the University of Twente in the Netherlands, the University of North Carolina at Greensboro, and Syracuse University. He published 35 books and hundreds of articles, papers, and reports on text design, task analysis, instructional design, computer-based learning, hypermedia, constructivism, cognitive tools, and problem solving. His research focused on the cognitive processes engaged by problem solving and models and methods for supporting those processes during learning, culminating in the book, *Learning to Solve Problems: A Handbook for Designing Problem-Solving Learning Environments*.

Robert M. Klein is an associate professor of mathematics education at Ohio University where he teaches mathematics content and pedagogy courses for inservice and preservice teachers at undergraduate and graduate levels. His professional interests include socio-cultural aspects of mathematics education and focuses on rural issues in mathematics teaching and learning. In addition, Dr. Klein studies the use of various technologies and the role of context in mathematics education.

Jerrid W. Kruse is an assistant professor of teaching and learning at Drake University where he teaches Elementary Science Methods, Secondary Science Methods, Scientific Inquiry and the Nature of Science, Educational Technology, Educational Psychology, and various undergraduate physical science courses. His scholarship focuses on the history and nature of science and technology in K-12, college, and teacher education. He is currently the editor of the *Iowa Science Teachers Journal*.

Dale S. Niederhauser is an associate professor of technology and teacher education at Iowa State University where he teaches courses in technology integration, diffusion of change, and learning with technology. He is the recipient of several awards for his research from organizations like The Society for Information Technology and Teacher Education, The National Association for Research in Science Teaching and The American Society of Engineering Education. He has served as President of the International Society for Technology in Education Special Interest Group for Teacher Educators and chair of the American Educational Research Association Technology as an Agent for Change American Special Interest Group. His scholarship is focused on technology-using teacher development, learning from hypertext, and problem-solving.

Joanne K. Olson is an associate professor in science education at Iowa State University. Her research interests include novice science teachers' decision-making and the impact of teacher education programs on its development, the nature of science and technology and their influences on science teaching and learning, and the role of liberal education in STEM education reform efforts. Her work has won awards from the National Association for Research in Science Teaching and the Association for Science Teacher Education. A former elementary and middle school science teacher, she currently co-directs the secondary Master of Arts in Teaching program at Iowa State and teaches courses in advanced pedagogy in science education, curriculum theory, and science methods. She is the president-elect of the Association for Science Teacher Education

Amy Noelle Parks is an associate professor of early childhood education at the University of Georgia where she teaches courses on early childhood pedagogies, ethnographic research with young children, and research perspectives in teacher education. Her research examines equity issues affecting young children, particularly within the context of mathematics. Recent work has been published in the *Journal of Research in Mathematics Education*, *Teachers College Record*, and the *Journal of Curriculum Studies*. Her current research project involves following the mathematical learning of a cohort of children at a rural school as they go from preschool to Grade 1.

Kimberly A. Penning has taught at the university level since 1997, and is currently teaching Microbiology at Minnesota State University. Kimberly has taught science methods and biology for prospective elementary teachers. She received her M.S. in Science Education from Minnesota State University. Kimberly has received academic scholarships that include the Houdek scholarship and graduate research scholarship. She is currently working on a second graduate degree in Microbiology, and is conducting research on age effects on fertility of the earthworm, *C. elegans.*

Neil Postman held the position of University Professor in the School of Education at New York University and chaired the Department of Culture and Communication until 2002, one year before his death. Professor Postman was a prolific writer on education, media studies, and the impact of technology on culture, human thinking and behavior. Of the many books he published, the best known and most influential include *Teaching as a Subversive Activity* (with Charles Weingartner) (1969), *Teaching as a Conserving Activity* (1979), *The Disappearance of Childhood* (1982), *Amusing Ourselves to Death* (1985), *Technopoly* (1992), and *The End of Education* (1995). He also published more than 200 articles in magazines and newspapers. Postman was internationally known for his keen insights regarding technology and education, lectured around the world, and received numerous recognitions for his teaching and scholarship.

Teresa J. Shume has been a faculty member at Minnesota State University Moorhead since 1998 where she teaches courses in science teaching methods and foundations of education, supervises student teachers, and has served as the Secondary and K-12 Education Coordinator. With a background as an award winning high school science teacher, she has offered an array of professional development workshops for science educators and has worked as an educational consultant for organizations such as the American Association for the Advancement of Science Project 2061 and Utah's State Office of Education. Her research interests include conceptions of environmental literacy, children's relationships with nature, the nature of technology, and retention of new science teachers.

John T. Spencer is a middle school teacher in an urban Phoenix school district, where he is piloting a one-to-one English Language Learning initiative. He has experience teaching all subjects from sixth through eighth grade and has worked as an instructional technology specialist. He is the recipient of the Educational Technology Master Student of the Year Award from Northern Arizona University for his capstone project involving a technology-integrated personalized professional development program and for developing a school-wide technology-integrated social studies curriculum using refurbished computers on Linux. He blogs regularly at Education Rethink and is a regular columnist for *Kappan Magazine.*

Johannes Strobel is Director of INSPIRE, Institute for P-12 Engineering Research and Learning and Assistant Professor of Engineering Education & Learning Design and Technology at Purdue University. He received his M.Ed. and Ph.D. in Information Science and Learning Technologies from the University of Missouri-Columbia, USA. NSF and several private foundations fund his research. His research and teaching focuses on policy of P-12 engineering, how to support teachers and students' engineering learning, the measurement and support of the change of 'habits of mind,' particularly in regards to sustainability and empathy and the use of cyber-infrastructure to sensitively and resourcefully provide access to and support learning.

Suvi Tala is a researcher and teacher educator in the Department of Physics at the University of Helsinki. She has a M.Sc. degree in physics and expertise in teaching physics, chemistry, mathematics, philosophy and ethics in secondary schools. Her postgraduate research focuses on the role of technology in science and science education, including the nature of science and technology. She also teaches physics teacher education courses that connect physics, science education and philosophy, history and sociology of science. In addition, she is part of a national project aiming to improve the public's awareness about science and technology.

Heather Tillberg-Webb is Interim Associate Provost of Academic Technology and e-Learning at Lesley University, overseeing instructional design and technology integration in online and blended learning. She teaches graduate-level courses in Online Course Design for Lesley University and Instructional Design for George Mason University. She received her Ph.D. in Instructional Technology from the University of Virginia's Curry School of Education. Her research interests include learning analytics as a mechanism for improving instructional practice and learning outcomes, and the role of ethics in educational technology practice.

Jocelyn Wishart is Senior Lecturer in Education (Science) at the University of Bristol, UK where she teaches on both preservice science teacher education programs and continuing professional development programs such as the MSc in Science & Education. She originally studied psychology and has taught science, psychology and ICT in English secondary schools. She now specializes in the use of new technologies to support teaching and learning and is currently researching teaching science through making stop-frame animations. Her recently published work has focused on the potential of handheld, mobile devices in teaching for enhancing learning in different She currently serves as Membership Secretary of the International Association for Mobile Learning.

RECOMMENDED READING

Apple, M.W. & Jungck, S (1990). You Don't Have to be a Teacher to Teach this Unit: Teaching, technology, and gender in the classroom. *American Educational Research Journal, 27*(2), 227–251.

Arthur, W. B. (2009). *The Nature of Technology: What it is and how it evolves.* Free Press, New York.

Bauerlein, M. (2008). *The Dumbest Generation: How the digital age stupefies young Americans and jeopardizes our future.* Penguin Group, New York.

Bonk, C. J. (2011). *The World Is Open: How web technology is revolutionizing education.* Jossey-Bass, San Francisco.

Bowers, C.A. (2000). *Let Them Eat Data: How computers affect education, cultural diversity, and the prospects of ecological sustainability.* The University of Georgia Press, Athens.

Boyle, J. (2010). *The Public Domain: Enclosing the commons of the mind.* Yale University Press, New Haven, CT.

Brende, E. (2004). *Better off: Flipping the switch on technology.* HarperCollins, New York.

Bugeja, M. (2005). *Interpersonal Divide: The search for community in a technological age.* Oxford University Press, New York.

Carr, N. (2010). *The Shallows: What the internet is doing to our brains.* W.W. Norton & Company, New York.

Christians, C. G. (2011). Universalism versus Communitarianism in Media Ethics. In R. S. Fortner & P. M. Fackler (Eds.). *The Handbook of Global Communication and Media Ethics.* Wiley-Blackwell, Malden, MA.

Collins, A. & Halverson, R. (2009). *Rethinking Education in the Age of Technology: The digital revolution and schooling in America.* Teachers College Press, New York.

Cuban, L. (2001). *Oversold and Underused: Computers in the classroom.* Harvard University Press, Cambridge, MA.

Cuban, L. (1986). *Teachers and Machines: The classroom use of technology since 1920.* Teachers College Press, New York.

Damarin, S. K. (1998). Technology and the Culture of Classrooms. *Theory into Practice, 37(1),* 11–19.

Darling-Hammond, L. & Bransford, J. (Eds.) (2005). *Preparing Teachers for a Changing World: What teachers should learn and be able to do.* Jossey-Bass, San Francisco.

De Vries, M. J. (2005). *Teaching About Technology: An introduction to the philosophy of technology for non-philosophers.* Springer, The Netherlands.

De Zengotita, T. (2005). *Mediated: How the media shapes your world and the way you live in it.* Bloomsbury, New York.

Freitas, E., Nolan, K., & Appelbaum, P. (2008). Afterword: Bootleg mathematics. In E. de Freitas & K. Nolan (Eds.) *Opening the Research Text: Critical insights and in(ter)ventions into mathematics education*, pp. 221–247. Springer, New York.

Garofalo, J., Drier, H.S., Harper, S.R., Timmerman, M.A., & Shockey, T.L. (2000). Promoting Appropriate Uses of Technology in Mathematics Teacher Preparation. *Contemporary Issues in Technology and Teacher Education, 1*(1), 66–88. Available: http://www.citejournal.org/vol1/iss1/currentissues/mathematics/article1.htm

Gilbert, J. and Boulter, C. (Eds.) (2000). *Developing Models in Science Education.* Springer, The Netherlands.

Green, L. R. (2002). *Communication, Technology and Society.* Sage, Thousand Oaks, CA.

Huesemann, M., & Hueseman, J. (2011). *Techno-Fix: Why technology won't save us or the environment.* New Society Publishers, Gabriola Island, BC, Canada.

Hughes, T. (2004). *Human-built World: How to think about technology and culture.* University of Chicago Press, Chicago.

Jonas, H. (1985). *The Imperative of Responsibility: In search of ethics for the technological age.* University of Chicago Press, Chicago.

Keen, A. (2008). *The Cult of the Amateur.* Doubleday, New York.
Klein, R. (2007). Educating in Place: Mathematics and technology. *Philosophical Studies in Education, 38,* 119–130.
Lanier, J. (2011). *You are Not a Gadget.* Vintage, New York.
Lessig, L. (2006). *Code and other Laws of Cyberspace, Version 2.0.* Basic Books, New York.
Laszlo, A., & Castro, K. (1995). Technology and Values: Interactive learning environments for the future generation. *Educational Technology, 35*(2), 7–13
MacKenzie, D. (2001). *Mechanizing Proof: Computing, risk, and trust.* MIT Press, Cambridge, MA.
McLuhan, M. (1964). *Understanding Media: The extensions of man.* MIT Press, Cambridge, MA.
Mitcham, C. (1984). *Thinking through Technology: The path between engineering and philosophy.* The University of Chicago Press, Chicago.
Moor, J. H. (2008). Why We Need Better Ethics for Emerging Technologies. In J. van den Hoven & J. Weckert (Eds.). *Information Technology and Moral Philosophy,* pp. 26-29. Cambridge University Press, Cambridge.
Oppenheimer, T. (2003). *The Flickering Mind: The false promise of technology in the classroom and how learning can be saved.* Random House, New York.
O'Riley, P. (2003). *Technology, Culture, and Socioeconomics: A rhizoanalysis of educational discourses.* Peter Lang Publishers.
Peggy A., E., Anne T., O., Olgun, S., Emine, S., & Polat, S. (2012). Teacher Beliefs and Technology Integration Practices: A critical relationship. *Computers & Education, 59,* 423–435.
Petrina, S. (2000). The Politics of Technological Literacy. *International Journal of Technology and Design Education, 10,* 181–206.
Postman, N. (1985). *Amusing Ourselves to Death: Public discourse in the age of show business.* Penguin, New York.
Postman, N. (1992). *Technopoly: The surrender of culture to technology.* Vintage, New York.
Postman, N. (1995). *The End of Education: Redefining the value of school.* Vintage, New York.
Powers, W. (2010). Hamlet's BlackBerry: A Practical Philosophy for Building a Good Life in the Digital Age. Harper, New York.
Schön, D., Sanyal, B. & Mitchell, W. (Eds.) (1997). High technology and low-income communities: Prospects for the positive use of advanced information technology. The MIT Press, Cambridge, MA.
Selber, S.A. (2004). *Multiliteracies for a Digital Age.* Southern Illinois University Press, Carbondale.
Sutherland, R., Robertson, S. & John, P. (Eds.) (2009). *Improving Classroom Learning with ICT.* Routledge, London.
Turkle, S. (2011). *Alone Together: Why we expect more from technology and less from each other.* Basic Books, New York.
Vander Schee, C.J. & Boyles, D. (2010). "Exergaming," Corporate Interests and the Crisis Discourse of Childhood Obesity. *Sport, Education & Society, 15*(2), 169–185.
Vermaas, P.E., Kroes, P.A., Light, A. & Moore, S. (Eds.) (2009). *Philosophy and Design: From engineering to architecture.* Springer, The Netherlands.
Vincenti, W. G. (1990). *What Engineers Know and How They Know it: Analytical studies from aeronautical history.* Johns Hopkins University Press, Baltimore.
Winner, L. (1995). Citizen Virtues in a Technological Order. In Feenberg, A. & Hannay, A. (Eds.). *Technology and the Politics of Knowledge,* pp. 65–84. Indiana Series in the Philosophy of Technology, Indiana University Press, Indianapolis.
Winner, L. (1986). *The Whale and the Reactor: A search for limits in an age of high technology.* University of Chicago Press, Chicago.
Winner, L. (1980). Do Artifacts Have Politics? *Daedalus* 109, 121–136. Reprinted in Winner, L. (1986) *The Whale and the Reactor: A search for limits in age of high technology,* pp. 19–39. University of Chicago Press, Chicago.
Wishart, J. (2010) The Role of Information and Communications Technology. In R. Toplis (Ed.). *How Science Works: Effective pedagogies and practice for secondary school teachers.* Routledge, London.
Yasukawa, K., Skovsmose, O., & Ravn, O. (2012). Shaping and being Shaped by Mathematics: Examining a Technology of Rationality. In O. Skovsmose & B. Greer (Eds.). *Opening the Cage: Critique and politics of mathematics education,* pp. 265–286. Sense Publishers, Rotterdam, The Netherlands.

INDEX

Anonymity, 275, 276

Character education, 132–133, 154
Citizenship, 88, 94, 132, 348, 435
Computers, 45, 87–90, 96–97, 102, 106, 197, 231, 252, 256, 260–262, 276, 354–356
Conceptions of technology, 89–90, 91, 93
Criticism, 427–436
Culture, 26–29, 109
Curriculum, 3, 86, 96, 169, 195, 218, 223, 244, 254, 255, 286, 358–359, 408
Cyber bullying, 130, 269, 435
Cyber psychology, 275–284

Democracy, 15, 27, 93–94, 224
Design cycle, 64
Devices, 36, 43, 46, 63, 71–72, 75–76, 89, 98, 129, 191, 229, 231, 250, 252, 256, 265, 306, 354, 388, 392, 396, 404, 421, 428–429, 431
Dystopian, 37, 47, 329–332, 335, 340

Eisner, Elliot, 220–222
Elementary education, 136, 141–142, 147–148
Ellul, Jacques, 31, 229, 230, 238, 243, 245, 332
Engineering education, 425
Engineering, 53, 54, 56, 57–58, 61–66, 69–71, 232, 235–236, 240–241, 421–422

Humanistic education, 53, 224, 240, 245, 246, 330, 331, 340–341, 388, 432–433
Humanizing, 330, 331, 340–341, 388, 432–433

Ideologies of educational technology, 329–341
Information and communication technology, 87–89, 93, 96, 98, 113–115, 330
Information glut, 12, 15, 30, 176, 185, 408
Information overload, 165, 176, 181, 183, 230, 329, 387
Intellectual partners, 106, 255
Interactive white boards, 201, 204

Learning, 101–109, 117–118, 124–126, 189–190, 249–264, 299, 349–360
Liberal education, 223–225, 242–246
Literacy, 85–98, 154, 309–310, 314, 405

Mathematics, 164–165
Media literacy, 309–310, 314, 325
Middle school, 391–407
Minimal group paradigm, 280–281
Murrow, Edward R., 40–41

Negroponte, Nicholas, 43

Online disinhibition effect, 275–277

Pedagogy, 209–213, 218, 335
Philosophy, 14–17, 57, 63, 239, 244, 245, 261, 295, 297, 330, 334, 345, 428, 430
Postman, Neil, 2–4, 7–14, 35, 37–40, 163, 189–191, 218–219, 242, 293–325, 373, 388

Post-modernism, 276
Preservice, 130, 135, 149, 154, 202, 207–208, 294, 360–365

School change, 221
Schooling, 217–246
Science education, 225–226
Science teacher education
Science teaching, 115, 126, 190, 300, 375, 420
Scientific knowledge building, 53
Scientific models, 411–412, 418, 420, 422–424
Scientific practice, 425
Simulation, 117–122, 357
Social co-construction, 106, 107–109
Social dominance theory, 280
Social media, 35, 46, 92, 108, 130–154, 434–435

Teacher education, 131, 135–137, 153, 207, 345–365
Teacher questioning, 337–338
Teaching, 349–355, 373–390, 437–439
Teaching practice, 202, 210, 213, 388
Techno science, 51–77
Techno-globalism, 104–105
Technological knowledge, 64–67
Technology criticism, 427–436
Technology diffusion, 42
Technology education, 54–57, 59, 64, 72, 88–89, 154, 198, 359, 379
Technology enhanced learning, 54–77, 249–264
Technology integration, 265, 335, 341, 357
Technology literacy, 85–98, 405

Utopian, 14, 18–19

Vygotsky, 293–325